The Ecology of Plant Secondary Metabolites
From Genes to Global Processes

Plant secondary metabolites (PSMs) such as terpenes and phenolic compounds are known to have numerous ecological roles, notably in defence against herbivores, pathogens and abiotic stresses, and in interactions with competitors and mutualists.

This book reviews recent developments in the field to provide a synthesis of the function, ecology and evolution of PSMs, revealing our increased awareness of their integrative role in connecting natural systems. It emphasises the multiple roles of secondary metabolites in mediating the interactions between organisms and their environment at a range of scales of ecological organisation, demonstrating how genes encoding for PSM biosynthetic enzymes can have effects from the cellular scale within individual plants all the way up to global environmental processes.

A range of recent methodological advances, including molecular, transgenic and metabolomic techniques, are illustrated and promising directions for future studies are identified, making this a valuable reference for researchers and graduate students in the field.

GLENN R. IASON is Head of Chemical and Molecular Ecology at the James Hutton Institute (Aberdeen, UK) with interests in the role of plant secondary metabolites in the nutritional ecology of herbivores, and their wider effects in communities and ecosystems.

MARCEL DICKE is Professor of Entomology at Wageningen University in the Netherlands. His ecological research focuses on the interaction between plants and insects, and he has completed pioneering studies in the area of multi-trophic interactions and community ecology.

SUSAN E. HARTLEY is Professor of Ecology at the University of York and Director of the York Environmental Sustainability Institute. She specialises in the study of plant–animal interactions, particularly the mechanisms by which plant defences affect herbivore performance.

The Ecology of Plant Secondary Metabolites

From Genes to Global Processes

Edited by

GLENN R. IASON
The James Hutton Institute, Aberdeen, UK

MARCEL DICKE
Wageningen University, The Netherlands

SUSAN E. HARTLEY
University of York, UK

CAMBRIDGE UNIVERSITY PRESS
Cambridge, New York, Melbourne, Madrid, Cape Town,
Singapore, São Paulo, Delhi, Mexico City

Cambridge University Press
The Edinburgh Building, Cambridge CB2 8RU, UK

Published in the United States of America by Cambridge University Press, New York

www.cambridge.org
Information on this title: www.cambridge.org/9780521193269

First published 2012

Printed in the United Kingdom at the University Press, Cambridge

A catalogue record for this publication is available from the British Library

ISBN 978-0-521-19326-9 Hardback
ISBN 978-0-521-15712-4 Paperback

Contents

The colour plate section is placed between pages 304 and 305.

Contributors

FRANCESCA BAGNOLI
Consiglio Nazionale delle
Ricerche, Istituto per la Protezione
delle Piante (CNR-IPP),
50019 Sesto Fiorentino (Firenze), Italy
francesca.bagnoli@ipp.cnr.it

JOSEPH K. BAILEY
Department of Ecology
and Evolutionary Biology,
University of Tennessee, Knoxville,
TN 37996, USA
Joe.Bailey@utk.edu

ELISABETH S. BAKKER
Department of Aquatic Ecology,
Netherlands Institute of Ecology
(NIOO-KNAW), Wageningen,
The Netherlands
l.bakker@nioo.knaw.nl

IAN T. BALDWIN
Max Planck Institute for Chemical
Ecology, D-07745 Jena, Germany
baldwin@ice.mpg.de

KASEY E. BARTON
Botany Department
University of Hawaii at Mãnoa,
Honolulu, HI 96822, USA
kbarton@hawaii.edu

JOAN R. BEATON
Ecological Sciences Group, The James
Hutton Institute,
Craigiebuckler, Aberdeen,
AB15 8QH, UK
Joan.Beaton@hutton.ac.uk

MEIKE BUROW
Department of Plant Biology
and Biotechnology,
Faculty of Life Sciences,
Copenhagen University,
DK-1871 Frederiksberg C,
Denmark
mbu@life.ku.dk

COLIN D. CAMPBELL
Ecological Sciences Group,
The James Hutton Institute,
Craigiebuckler, Aberdeen,
AB15 8QH, UK
Colin.Campbell@hutton.ac.uk

NICOLE M. VAN DAM
Radboud University
Nijmegen, Institute for Water
and Wetland Research,
Department of Ecogenomics,
6525 AJ Nijmegen,
The Netherlands
n.vandam@science.ru.nl

JANE L. DEGABRIEL
School of Biological Sciences,
University of Aberdeen,
Aberdeen, AB24 2TZ, UK
janedegabriel@gmail.com

JOERG DEGENHARDT
Institute of Pharmacy,
Martin Luther University
of Halle-Wittenberg,
D-06120 Halle (Saale),
Germany
joerg.degenhardt@pharmazie.
uni-halle.de

MARCEL DICKE
Laboratory of Entomology,
Wageningen University, 6700 EH
Wageningen, The Netherlands
marcel.dicke@wur.nl

RENÉ ESCHEN
CABI, Rue des Grillons 1, CH-2800
Delémont, Switzerland
r.eschen@cabi.org

SILVIA FINESCHI
Consiglio Nazionale delle
Ricerche, Istituto per la
Protezione delle Piante
(CNR-IPP), 50019 Sesto Fiorentino
(Firenze), Italy
silvia.fineschi@ipp.cnr.it

DYLAN G. FISCHER
Environmental Studies
Program, The Evergreen
State College, Olympia,
WA 98505, USA
fischerd@evergreen.edu

BENJAMIN M. FITZPATRICK
Department of Ecology and
Evolutionary Biology, University of
Tennessee, Knoxville, TN 37996, USA
benfitz@utk.edu

ANNA FONTANA
Department of Biochemistry, Max
Planck Institute for Chemical Ecology,
D-07745 Jena, Germany
afontana@ice.mpg.de

JENNIFER SORENSEN FORBEY
Boise State University, Boise, ID, USA
jenniferforbey@boisestate.edu

ALAN C. GANGE
School of Biological Sciences, Royal
Holloway, University of London,
Egham, TW20 0EX, UK
A.Gange@rhul.ac.uk

MARK A. GENUNG
Department of Ecology and
Evolutionary Biology, University of
Tennessee, Knoxville, TN 37996, USA
mgenung@utk.edu

JONATHAN GERSHENZON
Department of Biochemistry,
Max Planck Institute for
Chemical Ecology, D-07745 Jena,
Germany
gershenzon@ice.mpg.de

RIETA GOLS
Laboratory of Entomology,
Wageningen University, 6700
EH Wageningen, The Netherlands
rieta.gols@wur.nl

ELISABETH M. GROSS
Laboratoire Interactions
Ecotoxicologie Biodiversité
Ecosystémes (LIEBE), CNRS UMR 7146,
Université de Lorraine Campus
Bridoux, 57050 Metz, France
egross@univ-metz.fr

SUSAN E. HARTLEY
Department of Biology, University
of York, York, YO10 5DD, UK
sue.hartley@york.ac.uk

ELIZABETH M. HILL
School of Life Sciences, University
of Sussex, Falmer, Brighton,
BN1 9QG, UK
E.M.Hill@sussex.ac.uk

JULIA M. HORWOOD
School of Life Sciences, University
of Sussex, Falmer, Brighton,
BN1 9QG, UK
J.Horwood@sussex.ac.uk

MARK D. HUNTER
University of Michigan, Ann Arbor,
MI, USA
mdhunter@umich.edu

GLENN R. IASON
Ecological Sciences Group, The James
Hutton Institute, Craigiebuckler,
Aberdeen, AB15 8QH, UK
Glenn.Iason@hutton.ac.uk

ARTHUR KEITH
Department of Biological Sciences,
Northern Arizona University,
Flagstaff, AZ 86011, USA
Arthur.Keith@NAU.EDU

JULIA KORICHEVA
School of Biological Sciences,
Royal Holloway, University
of London, Egham,
TW20 0EX, UK
Julia.Koricheva@rhul.ac.uk

JACK J. LENNON
Ecological Sciences Group, The James
Hutton Institute, Craigiebuckler,
Aberdeen, AB15 8QH, UK
Jack.Lennon@hutton.ac.uk

CARRI J. LEROY
Environmental Studies
Program, The Evergreen
State College, Olympia,
WA 98505, USA
LeRoyC@evergreen.edu

RICHARD L. LINDROTH
Department of Entomology,
University of Wisconsin-
Madison, Madison,
WI 53706, USA
lindroth@wisc.edu

FRANCESCO LORETO
Consiglio Nazionale delle
Ricerche, Istituto per la
Protezione delle Piante (CNR-
IPP), 50019 Sesto Fiorentino
(Firenze), Italy
francesco.loreto@ipp.cnr.it

CLARE McARTHUR
School of Biological
Sciences, University
of Sydney, NSW 2006,
Australia
c.mcarthur@usyd.edu.au

BEN D. MOORE
Ecological Sciences Group, The James
Hutton Institute, Craigiebuckler,
Aberdeen, AB15 8QH, UK
B.Moore@uws.edu.au

JULIANNE M. O'REILLY-WAPSTRA
School of Plant Science, University
of Tasmania and CRC for Forestry,
Hobart, Tasmania 7001, Australia
Julianne.OReilly@utas.edu.au

GRAHAM H. R. OSLER
Ecological Sciences Group, The James
Hutton Institute, Craigiebuckler,
Aberdeen, AB15 8QH, UK
g.osler@macaulay.ac.uk

ERIK H. POELMAN
Laboratory of Entomology,
Wageningen University, 6700 EH
Wageningen, The Netherlands
erik.poelman@wur.nl

BRADLEY M. POTTS
School of Plant Science, University of
Tasmania and CRC for Forestry,
Hobart, Tasmania 7001, Australia
B.M.Potts@utas.edu.au

CLARA C. PREGITZER
Department of Ecology and
Evolutionary Biology, University of
Tennessee, Knoxville, TN 37996, USA
cpregitz@utk.edu

BRIAN J. REHILL
Department of Chemistry, US Naval
Academy, Annapolis, MD 21402, USA
rehill@usna.edu

LYNNE ROBINSON
School of Life Sciences, University
of Sussex, Falmer, Brighton,
BN1 9QG, UK
lr84@sussex.ac.uk

JOANNE R. RUSSELL
The James Hutton Institute,
Invergowrie, Dundee, DD2 5DA, UK
Joanne.Russell@hutton.ac.uk

VIVIANE SCHROEDER
School of Biological Sciences,
Royal Holloway, University
of London, Egham,
TW20 0EX, UK
Viviane.Schroeder.2009@live.rhul.ac.uk

MEREDITH C. SCHUMAN
Max Planck Institute for Chemical
Ecology, D-07745 Jena, Germany
mschuman@ice.mpg.de

JENNIFER A. SCHWEITZER
Department of Ecology and
Evolutionary Biology, University of
Tennessee, Knoxville, TN 37996, USA
Jen.Schweitzer@utk.edu

DAVID A. SIM
Ecological Sciences Group, The James
Hutton Institute, Craigiebuckler,
Aberdeen, AB15 8QH, UK
Dave.Sim@hutton.ac.uk

JENNI A. STOCKAN
Ecological Sciences Group, The James
Hutton Institute, Craigiebuckler,
Aberdeen, AB15 8QH, UK
Jenni.Stockan@hutton.ac.uk

FRANCISCO ÚBEDA
Department of Ecology
and Evolutionary Biology,
University of Tennessee,
Knoxville,
TN 37996, USA
fubeda@utk.edu

THOMAS G. WHITHAM
Department of Biological
Sciences, Northern
Arizona University, Flagstaff,
AZ 86011, USA
Thomas.Whitham@nau.edu

UTE WITTSTOCK
Institute of Pharmaceutical Biology,
Technical University of Braunschweig,
D-38106 Braunschweig, Germany
u.wittstock@tu-bs.de

MATTHEW ZINKGRAF
Department of Biological Sciences,
Northern Arizona University, Flagstaff,
AZ 86011, USA, and Merriam-Powell
Center for Environmental Research,
Northern Arizona University, Flagstaff,
AZ 86011, USA
msz2@nau.edu

Preface

Although chemical interactions between plants and other organisms had been documented many years previously (e.g. Stahl, 1888; Verschaffelt, 1910), it was the seminal work of Fraenkel published in 1959 that placed the ecological function of plant secondary metabolites (PSMs) onto the agenda of modern ecology. He recognised that PSMs are not merely a repository for plants' waste products, but rather they had a primary function: they could act as defences against enemies such as phytophagous insects. Crucially, he also realised that these enemies exert a selection pressure on the plants to defend themselves. Since then, numerous ecological roles of PSMs have been elucidated, notably as defences against a broad range of herbivores and pathogens, as mediators of interactions with competitors and mutualists, and as defence against abiotic stress. Recently, emerging developments have taken us well beyond consideration of PSMs in the context of simple interactions between pairs of species. Our view of plant secondary metabolites has shifted significantly in the past 50 years and we now understand the subtlety and scale of their effects, which cross trophic levels, spread throughout ecosystems, and even affect global processes. At the same time, methodological developments, particularly in the '-omics' technologies, have led to a greater understanding of the synthesis and regulation of PSMs. These methodological developments now also facilitate unique tools for the targeted manipulation of both the synthesis of PSMs and their ecological function independently of other phenotypic effects. Consequently, we are now in a position to assess the extent to which PSMs and their effects traverse natural systems from genes upwards, and, in the spirit of Fraenkel, the reciprocal effects of the biotic and abiotic environment on those genes. This is, therefore, the ideal time to take stock of our current understanding of the function, ecology and evolution of PSMs, in order to focus our future efforts to use this knowledge to best effect in science and its application.

This volume arises from a British Ecological Society Symposium entitled *The integrative roles of plant secondary metabolites in natural systems* which was held in 2010 at the University of Sussex, UK. It reviews recent scientific developments and provides a new synthesis of the function, ecology and evolution of PSMs that takes account of recent advances in our awareness of their integrative roles in connecting natural systems. The roles of PSMs span from the molecular to the global: genes control their synthesis, PSMs communicate responses within plants to external conditions and damage, PSMs define interactions with other individuals of their own and other species, and their effects cascade through communities and ecosystems to global processes. Uncovering the evolution of these complex multi-functional, multi-trophic, multi-scale systems goes hand in hand with the potential to manage and exploit them. This is the new challenge for the study of the ecology of plant secondary metabolites in an era that has been facilitated, and will be further developed, by application of exciting recent methodological advances including genomics and metabolomics.

This volume, consisting primarily of invited contributions, does not try to incorporate examples of all types of PSM, cover all biomes or revisit the well-trodden path of interactions between pairs of species. Instead it focuses on areas of where new ideas are currently developing, including our greater understanding of allocation and distribution of PSMs within plants and their effects under changing environmental conditions. Some of these are drawn together in a synthesis chapter, which also identifies some cross-cutting themes and future research directions, including the costs of PSMs, evolutionary and phylogenetic considerations, methodological advances and effects on multiple components within communities or ecosystems. The authors of the individual chapters explicitly point the way for development of research required within their own domain. For the purposes of organisation and logical flow, the contributions to the symposium were grouped into themes, but, given our emerging view of PSMs as integrators, we felt it was inappropriate to include such boundaries within the published volume. Interestingly, the distribution of symposium contributions revealed a relative lack of studies spanning the full range of scales from 'genes to communities or ecosystems'. This suggests that few research groups are able to attempt to integrate across levels of ecological organisation, and those that do attempt this difficult task have no option but to focus on one of a small set of study systems for which we have sufficient knowledge to permit this type of work. This seems one area where more ambitious approaches would be both welcome and fruitful. Increased collaboration between ecologists and researchers from other biological disciplines, such as plant physiology, cell biology, pathology and developmental biology, would also generate novel insights into the multiple roles of PSMs in all facets of plant and ecosystem biology.

We are particularly indebted to the authors of the chapters, and to all those who contributed to the symposium and to this volume. We thank the numerous anonymous reviewers of the chapters (they know who they are!), the Editorial Board of Ecological Reviews, and Cambridge University Press. We hope with their assistance to have moved towards a greater understanding of the multiple roles of PSMs, from genes to global processes.

References

Fraenkel, G. S. (1959) The *raison d'être* of secondary plant substances. *Science*, **129**, 1466–1470.

Stahl, E. (1888) Eine biologische Studie über die Schutzmittel der Pflanzen gegen Schneckenfrass. *Jenaer Zeitung fuer Naturwissenschaften*, **15**, 557–684.

Verschaffelt, E. (1910) The cause determining the selection of food in some herbivorous insects. *Proceedings of the Academy Sciences Amsterdam*, **13**, 536–542.

Glenn Iason,
Marcel Dicke
and Sue Hartley

The integrative roles of plant secondary metabolites in natural systems: a synthesis

GLENN R. IASON

Ecological Sciences Group, The James Hutton Institute

MARCEL DICKE

Laboratory of Entomology, Wageningen University

SUSAN E. HARTLEY

Department of Biology, University of York

1.1 Introduction

Since Fraenkel (1959) recognised that plant secondary metabolites (PSMs) were not simply plant waste products but served to defend them against insect herbivores, numerous ecological roles for these intriguing compounds have been established, notably as defences against a broad range of herbivores and pathogens, but also as mediators of interactions with competitors and mutualists, and as a defence against abiotic stress. A single compound can influence multiple components within an ecological system, and can have effects that act across many different scales. Add to this the huge diversity of PSMs that have now been characterised, and the possible interactive effects among them, and it is clear that PSMs either individually or as groups can no longer be considered only in the context of interactions between the plant and a single other species. They are now recognised as major contributors to the bridge between genes and ecosystems, because (context-dependent) gene expression patterns influence the phenotype of a plant. The effects of PSMs are now known to affect community dynamics and to cascade through ecosystems, driving their composition and function and acting as agents of their evolution (e.g. Whitham *et al.*, 2006). Here, we summarise the key points and emergent themes from the chapters in this book and provide a synthesis of the recent developments in the ecology and evolution of PSMs, illustrating how a range of approaches, including molecular, transgenic and metabolomic techniques, have brought us to the cusp of a new understanding of their integrative roles in ecosystems.

1.2 Distribution, allocation and evolutionary selection for PSMs

The chemical diversity of PSMs, combined with the number and complexity of potential biotic and abiotic interactions in which they are involved, has

The Ecology of Plant Secondary Metabolites: From Genes to Global Processes, eds. Glenn R. Iason, Marcel Dicke and Susan E. Hartley. Published by Cambridge University Press. © British Ecological Society 2012.

hitherto prevented these systems being predictable beyond the outcome of the strongest, pairwise and most well defined of these interactions. However, several recent developments are moving us towards a better understanding and predictability of the roles of PSMs in more complex systems.

The concept of costs has featured highly in our attempts to understand the temporal patterns, distribution and occurrence, and physiological limitations to the synthesis of PSMs (Rhoades, 1979, 1985; Herms & Mattson, 1992; Gershenzon, 1994; Jones & Hartley, 1999). We are now aware of the circumstances under which trade-offs among investment in PSMs and evolutionary fitness-determining characteristics apply (e.g. Siemens *et al.*, 2002; Dicke & Baldwin, 2010). The strong genetic framework for consideration of the effects of PSMs, supported by the use of natural or genetically selected or manipulated variation, paves the way for the study of their mediating roles in biotic and abiotic interactions via the fitness of plants, a currency in which both costs and benefits of PSMs can ultimately be measured (this volume: O'Reilly-Wapstra *et al.*, Chapter 2; Gershenzon *et al.*, Chapter 4; Iason *et al.*, Chapter 13; Schuman & Baldwin, Chapter 15). We are beginning to understand how PSMs and their effectiveness as defences vary across different ontogenetic, phenological and life cycle stages of plants (Iason *et al.*, 2011; O'Reilly-Wapstra *et al.*, Chapter 2; Koricheva & Barton, Chapter 3). Clear patterns emerge, with PSM concentrations increasing during the seedling stage followed by declines during leaf phenological stages. Although the underlying selective forces are often unclear (Koricheva & Barton, Chapter 3), the question is raised as to whether and to what extent selection for PSMs can act by transfer of evolutionary pressures across different life stages of plants, or between different plant tissues. We know that the complement and concentration of PSM in a particular tissue can be generated elsewhere in the plant and altered by signalling pathways (Baldwin & Ohnmeiss, 1994; van Dam, Chapter 10) and can vary markedly between different plant tissues according to their potential 'value' to the plant (Hartley *et al.*, Chapter 11). We currently require a synthesis as to how these temporal and spatial pressures interact to influence variation in PSMs over time and space within plants.

Although single secondary metabolites may act in multiple ways, the diversity of compounds present in any plant provides numerous possible permutations for them to act in combination. Cates (1996) drew attention to the potential for interactive effects of PSMs, but bio-prospecting studies that seek bioactive PSMs for human use have mainly focused on the search for 'the single active compound' (Firn & Jones, 2000). We are now increasingly considering the effects of combinations of PSMs (Gershenzon *et al.*, Chapter 4), and combinations with other traits in natural environments (Agrawal, 2011). It is clear that the use of several types of defence simultaneously by plants (Zarate *et al.*, 2007; Wei *et al.*, 2011) suggests that our older ideas of simple trade-offs between these different types need revision as we gain better insights into which defensive traits may evolve together.

In fact, PSMs can no longer be considered to function simply as plant defences against their enemies. Many interactions between PSMs and organisms exploiting plants are now known to be positive. For example, some soil microbes can use carbon-rich PSMs exuded from plant roots as a carbon source (Smolander *et al.*, 2006). Obviously the net effects of PSMs on some non-adapted target herbivores are negative, but for specialist herbivores, including vertebrates that have co-evolved counter-adaptations (Schoonhoven *et al.*, 2005; Forbey & Hunter, Chapter 5), they can be positive. Associated with this is the important confirmation that evolutionary selective pressure on PSMs can be exerted in opposite directions by herbivores according to their degree of specialism, specifically, positively by generalist and negatively by specialist herbivores (Lankau, 2007). As well as contrasting impacts, we now recognise the contrasting roles of many PSMs. For example, the effects of tannins and other phenolic PSMs may arise not only from their protein precipitating properties, but also from their status as pro- or antioxidants (Salminen & Karonen, 2011). A fuller understanding of the conditions under which these contrasting effects prevail is required.

This volume does not try to incorporate examples of PSM from all biomes, but focuses on areas of recent development of ideas. As our understanding of the distribution, allocation and evolutionary selection for PSMs has advanced, however, it has become clear that we have some major gaps. A review of PSMs in aquatic systems uncovered a stunning lack of knowledge of their prevalence and ecological roles (Gross & Bakker, Chapter 8). Despite seminal studies of the importance of PSMs in macrophytic plants being undertaken in the marine environment, e.g. in phytochemical induction (Cronin & Hay, 1996), there is a similar shortfall in our knowledge of the roles of PSMs, but a strong role of allelopathic interactions in shaping the planktonic community is likely (Czaran *et al.*, 2002; Suikkanen *et al.*, 2004).

1.3   Integrative roles of PSMs in connecting multiple components of ecological systems

PSMs are now recognised as having a major role in mediating interactions between different components of ecological communities, often widely separated in space and time (e.g. van der Putten *et al.*, 2001; Poelman *et al.*, 2008). They also impact on ecosystem processes and so ultimately on the structure and function of ecosystems. One such process that PSMs have long been known to influence is decomposition and hence the interactions between above- and belowground processes: the chemical composition of litter influences its decomposition rate and the recycling of nutrients through the soil–plant system (Bardgett *et al.*, 1998; Iason *et al.*, Chapter 13). But more subtly, PSMs are intimately involved in the integration of above- and belowground processes in plants via internal signalling processes, for example between roots and shoots following action on either, by herbivores, pathogens or mutualists (Kaplan *et al.*, 2008; Gange *et al.*, Chapter 9; van Dam, Chapter 10).

The recent advances beyond PSM-mediated interactions between pairs of species usher us into a new era of multi-trophic and interactive effects dissectible by experimentation involving plants with known genetic back-ground and traits. These methods have already begun to elucidate the role of PSMs (e.g. Kessler *et al.*, 2004; Linhart *et al.*, 2005; Lankau & Strauss, 2008) and other plant traits (Johnson *et al.*, 2009) as determinants of complex interac-tions involving multiple species and genotypes within species. The availabil-ity of transgenic plants modified with respect to their production of PSMs has rapidly facilitated the identification of the effects of particular genes at higher ecological levels, including their mediation of multi-trophic interactions (Kessler *et al.*, 2004; Zhang *et al.*, 2009; Schuman & Baldwin, Chapter 15; Dicke *et al.*, Chapter 16). Better knowledge of genetic variation in PSMs is providing evidence of their impact at higher levels of biological organisation up to the ecosystem level (Bailey *et al.*, Chapter 14).

1.4   Ramifications of PSMs across scales from genes to global effects

Whilst the PSMs vary across global, landscape, geographical and smaller spatial scales (Moles *et al.*, 2011; Moore & DeGabriel, Chapter 12; Iason *et al.*, Chapter 13), there are relatively few studies investigating the nature and significance of this variation for the structure and function of ecosystems. Spatial variation can dominate other sources of variation, for example in structuring invertebrate communities (Tack *et al.*, 2010), and it therefore seems important to explore the role of spatial variation in PSMs further. The availability of cheaper and rapid genomic, metabolomic (Hartley *et al.*, Chapter 11) and other biochemical techniques (such as near-infrared spectro-scopy, e.g. Moore *et al.*, 2010) or remote sensing techniques that can be used to process large numbers of samples or survey large areas (Moore *et al.*, 2010), is likely to facilitate a better understanding of the causes and consequences of large-scale spatial variation in PSMs for other organisms.

Large-scale impacts of PSMs are becoming clear: they both cause and respond to global-scale environmental change (Penuelas & Staudt, 2010). The release of volatile isoprenoids by trees in response to abiotic stress has a global-scale impact in that they are precursors to ozone (O_3) in the tropo-sphere in the presence of anthropogenic nitrogen oxides (Bagnoli *et al.*, Chapter 6). Global environmental changes such as increased atmospheric O_3 and carbon dioxide are hypothesised to feed back to plants and influence their PSM concentrations (e.g. Jones & Hartley, 1999). Although some strong effects and general patterns are detectable (Bezemer & Jones, 1998), there is a surprisingly high degree of inconsistency in the responses of PSMs, and the ecology of the systems they influence, to elevated atmospheric CO_2 and atmospheric ozone (Hartley et al., 2000; Lindroth, Chapter 7). The apparently idiosyncratic effects may be due to relatively few plant species having been studied at a community level, and the fact that most studies have addressed only the phenolic and terpenoid PSMs (Lindroth, Chapter 7).

1.5 Methodological developments

The developments in molecular ecology are progressing rapidly. With more and more species being fully sequenced and with development of novel techniques for global transcriptomic analyses in non-model species, new opportunities to link the PSM phenotype to molecular genetic characteristics will rapidly expand. Molecular techniques that provide new tools and specific genotypes in which to investigate the ecological aspects of PSMs can create molecular model plants to aid understanding of the biochemical and metabolic consequences of genes, such as over-expressed genes or mutant plant material in which specific genes are knocked out (Schuman & Baldwin, Chapter 15; Dicke et al., Chapter 16). These techniques, along with gene silencing, provide exciting possibilities for investigating the role of PSMs in the ecology of various plants such as wild tobacco, aspen or brassicas (e.g. Kessler et al., 2004, 2008; Poelman et al., 2008; Broekgaarden et al., 2010). The exploitation of transgenic approaches to enhance knowledge of the chemical ecology of PSMs and apply this knowledge in crop and forest protection (Degenhardt et al., 2003, 2009; Aharoni et al., 2005; Kos et al., 2009) may offer significant research opportunities in the future.

Despite the current emphasis on the importance of ecosystem approaches, in order to solve real-world environmental problems (Millennium Ecosystem Assessment, www.maweb.org), we still have no examples of large-scale whole-ecosystem studies that allow us to integrate and quantify the multiple roles of PSMs across scales. The ecological ramifications and influences of foundation species on a community are by definition profound. Although the community genetics approach applies well to studies of the extended phenotype, including that mediated by the PSMs of foundation species (Whitham et al., 2003), this approach has, as yet, not found widespread application in practical conservation management for maximising biodiversity.

The increasing availability of molecular phylogenies will facilitate hypothesis testing in terms of the evolutionary drivers of PSM diversity. For example, a phylogenetic approach to track PSM diversity and composition throughout the evolutionary history of the Burseraceae recently illustrated an increasing chemical diversity within the family across evolutionary time (Becerra et al., 2009). Other possibilities include the identification of crucial stages in the evolution of PSMs in freshwater environments, in relation to their evolutionary history of transfer to or from aquatic to terrestrial habitats (Gross & Bakker, Chapter 8) and taxonomic associations of plants that produce tropospheric ozone-producing volatile isoprenoids (Bagnoli *et al.*, Chapter 6).

1.6   Towards further integration

Relatively few studies consider multiple groups of PSMs within a single system. We suggest that this is due to the propensity of scientists to focus on familiar groups of compounds, or taxonomic groups of organisms participating in the interactions, or indeed to study a particular type of interaction, such as plant–herbivore interactions. But it is now clear that the ecology of whole systems may be influenced by PSMs, owing to their impacts on complex multi-trophic interactions (Iason *et al.*, Chapter 13; Bailey *et al.*, Chapter 14). Even fewer studies examine the role of PSMs across the range of scales from genes to landscapes. Future truly integrative, cross-disciplinary case studies are required to fully elucidate their roles in ecological communities. We need to understand the genetic, physiological and evolutionary controls on PSM production as well as the way that impacts of PSMs feed through to the ecosystem at the landscape scale. The most useful of these case studies will involve ecological model species and the most enlightening will not be constrained by taxonomic boundaries, phytochemical groups or confined to adjacent trophic levels.

Ecological effects mediated at least in part by PSMs and that span trophic levels have indeed been identified. Examples include the correlation between herbivory and decomposition (Grime *et al.*, 1996) and the relationship between arbuscular mycorrhizal fungi on the roots and the greater preponderance of specialist herbivores on the shoots of plants (Gange *et al.*, 2002). The association between soil microbial communities and aboveground processes such as leaf herbivory invokes signalling cascades that are analogous to those involved in mediating other above/belowground interactions of plants. This analogy spans plant responses to root-herbivory, plant microbial and plant pathogenic interactions as well as responses to abiotic stressors such as drought (Pineda *et al.*, 2010; Johnson *et al.*, 2011; Gange *et al.*, Chapter 9; van Dam, Chapter 10). Generalisation of the occurrence and relative importance of these processes remains to be established.

The studies presented in this volume emphasise the multiple integrative roles of PSMs in natural systems. They demonstrate the potential we have for understanding the far-reaching ecological consequences of the genes underlying the production of PSMs, and should ultimately guide us in our choices of transgenic and other manipulations that will permit their practical application in environmental management for improved food security or other ecosystem services. The recently emerging ecological and evolutionary roles of PSMs highlighted in this volume point to the trajectories along which future studies may be launched.

References

Agrawal, A. A. (2011) Current trends in the evolutionary ecology of plant defence. *Functional Ecology*, **25**, 420–432.

Aharoni, A., Jongsma, M. A. and Bouwmeester, H. J. (2005) Volatile science? Metabolic engineering of terpenoids in plants. *Trends in Plant Science*, **10**, 594–602.

Baldwin, I. T. and Ohnmeiss, T. E. (1994) Wound-induced changes in root and shoot jasmonic acid pools correlate with induced nicotine synthesis in *Nicotiana sylvestris* Spegazzini and Comes. *Journal of Chemical Ecology*, **20**, 2139–2157.

Bardgett, R. D., Wardle, D. A. and Teates G. W. (1998) Linking above-ground and below-ground interactions: how plant responses to foliar herbivory influences soil organisms. *Soil Biology and Biochemistry*, **30**, 1867–1878.

Becerra, J. X., Noge, K. and Venable, D. L. (2009) Macroevolutionary chemical escalation in an ancient plant–herbivore arms race. *Proceedings of the National Academy of Sciences USA*, **106**, 18062–18066.

Bezemer, M. and Jones, T. H. (1998) Plant–insect herbivore interactions under elevated atmospheric CO_2: quantitative analyses and guild effects. *Oikos*, **82**, 212–222.

Broekgaarden, C., Poelman, E. H., Voorrips, R. E., Dicke, M. and Vosman, B. (2010) Intraspecific variation in herbivore community composition and transcriptional profiles in field-grown *Brassica oleracea* cultivars. *Journal of Experimental Botany*, **61**, 807–819.

Cates, R. G. (1996) The role of mixtures and variation in the production of terpenoids in conifer–insect–pathogen interactions. *Diversity and Redundancy in Ecological Interactions: Recent Advances in Phytochemistry*, **30**, 179–216.

Cronin, G. and Hay, M. E. (1996) Induction of seaweed chemical defenses by amphipod grazing. *Ecology*, **77**, 2287–2301.

Czaran, T. L., Hoekstra, R. F. and Pagie, L. (2002) Chemical warfare between microbes promotes biodiversity. *Proceedings of the National Academy of Sciences USA*, **99**, 786–790.

Degenhardt, J., Gershenzon, J., Baldwin, I. T. and Kessler, A. (2003) Attracting friends to feast on foes: engineering terpene emission to make crop plants more attractive to herbivore enemies. *Current Opinion in Biotechnology*, **14**, 169–176.

Degenhardt, J., Hiltpold, I., Kollner, T. G. *et al.* (2009) Restoring a maize root signal that attracts insect-killing nematodes to control a major pest. *Proceedings of the National Academy of Sciences USA*, **106**, 13213–13218.

Dicke, M. and Baldwin, I. T. (2010) The evolutionary context for herbivore-induced plant volatiles: beyond the 'cry for help'. *Trends in Plant Science*, **15**, 167–175.

Firn, R. D. and Jones, C. G. (2000) The evolution of secondary metabolism – a unifying model. *Molecular Microbiology*, **37**, 989–994.

Fraenkel, G. S. (1959) The *raison d'être* of secondary plant substances. *Science*, **129**, 1466–1470.

Gange, A. C., Stagg, P. G. and Ward, L. K. (2002) Arbuscular mycorrhizal fungi affect phytophagous insect specialism. *Ecology Letters*, **5**, 11–15.

Gershenzon, J. (1994) The metabolic costs of terpenoid accumulation in higher plants. *Journal of Chemical Ecology*, **20**, 1281–1328.

Grime, J. P., Cornelissen, J. H. C., Thompson, K. and Hodgson, J. G. (1996) Evidence of a causal connection between anti-herbivore defence and the decomposition rate of leaves. *Oikos*, **77**, 489–494.

Hartley, S. E., Jones, C. G., Couper, G. C. and Jones, T. H. (2000) Phenolic biosynthesis under elevated CO_2. *Global Change Biology*, **6**, 497–506.

Herms, D. A. and Mattson, W. J. (1992) The dilemma of plants: to grow or to defend. *Quarterly Review of Biology*, **67**, 283–335.

Iason, G. R., O'Reilly-Wapstra, J. M., Brewer, M. J., Summers, R. W. and Moore, B. D. (2011) Do multiple herbivores maintain chemical diversity of Scots pine monoterpenes? *Philosophical Transactions of the Royal Society B*, **366**, 1337–1345.

Johnson, M. T. J., Agrawal, A. A., Maron, J. L and
   Salminen, J. P. (2009) Heritability,
   covariation and natural selection on 24
   traits of common evening primrose
   (*Oenothera biennis*) from a field experiment.
   Journal of Evolutionary Biology, **22**, 1295–1307.
Johnson, S. N., Staley, J. T., McLeod, F. A. L. and
   Hartley, S. E. (2011) Plant-mediated effects of
   soil invertebrates and summer drought on
   above-ground multi-trophic interactions.
   Journal of Ecology, **99**, 57–65.
Jones, C. G. and Hartley, S. E. (1999) A precursor
   competition model for phenolic allocation
   in plants. *Oikos*, **86**, 27–44.
Kaplan, I., Halitschke, R., Kessler, A. *et al.* (2008)
   Physiological integration of roots and
   shoots in plant defense strategies links
   above- and belowground herbivory. *Ecology
   Letters*, **11**, 841–851.
Kessler, A., Halitschke, R. and Baldwin, I. T. (2004)
   Silencing the jasmonate cascade: induced
   plant defenses and insect populations.
   Science, **305**, 665–668.
Kessler, D., Gase, K. and Baldwin, I. T. (2008) Field
   experiments with transformed plants reveal
   the sense of floral scents. *Science*, **321**,
   1200–1202.
Kos, M., Van Loon, J. J. A., Dicke, M. and
   Vet, L. E. M. (2009) Transgenic plants as vital
   components of integrated pest
   management. *Trends in Biotechnology*, **27**,
   621–627.
Lankau, R. A. (2007) Specialist and generalist
   herbivores exert opposing selection on a
   chemical defense. *New Phytologist*, **175**,
   176–184.
Lankau, R. A. and Strauss, S. Y. (2008) Community
   complexity drives patterns of natural
   selection on a chemical defense of
   Brassica nigra. *American Naturalist*,
   171, 150–161.
Linhart, Y. B., Keefover-Ring, K., Mooney, K. A.,
   Breland, B. and Thompson, J. D. (2005)
   A chemical polymorphism in a multitrophic
   setting: thyme monoterpene composition
   and food web structure. *American Naturalist*,
   166, 517–529.

Moles, A. T., Bonser, S. P., Poore, A. G. B.,
   Wallis, I. R. and Foley, W. J. (2011) Assessing
   the evidence for latitudinal gradients in
   plant defence and herbivory. *Functional
   Ecology*, **25**, 380–388.
Moore, B. D., Lawler, I. R., Wallis, I. R.,
   Beale, C. M. and Foley, W. J. (2010)
   Palatability mapping: a koala's eye view of
   spatial variation in habitat quality. *Ecology*,
   91, 3165–3176.
Penuelas, J. and Staudt, M. (2010) BVOCs and
   global change. *Trends in Plant Science*, **15**,
   133–144.
Pineda, A., Zheng, S. J., van Loon, J. J. A.,
   Pieterse, C. M. J. and Dicke, M. (2010) Helping
   plants to deal with insects: the role of
   beneficial soil-borne microbes. *Trends in Plant
   Science*, **15**, 507–514.
Poelman, E. H., Broekgaarden, C., van Loon, J. J. A.
   and Dicke, M. (2008) Early season herbivore
   differentially affects plant defence
   responses to subsequently colonizing
   herbivores and their abundance in the field.
   Molecular Ecology, **17**, 3352–3365.
Rhoades, D. F. (1979) Evolution of plant chemical
   defense against herbivores. In
   G. A. Rosenthal and D. G. Janzen (eds.)
   *Herbivores: Their Interaction with Secondary
   Plant Metabolites*. New York: Academic
   Press, 3–54.
Rhoades, D. F. (1985) Offensive–defensive
   interactions between herbivores and
   plants – their relevance in herbivore
   population-dynamics and ecological theory.
   American Naturalist, **125**, 205–238.
Salminen, J. P. and Karonen, M. (2011) Chemical
   ecology of tannins and other phenolics: we
   need a change in approach. *Functional
   Ecology*, **25**, 325–338.
Schoonhoven, L. M., van Loon, J. J. A. and
   Dicke, M. (2005) *Insect–Plant Biology*. Oxford:
   Oxford University Press.
Siemens, D. H., Garner, S. H., Mitchell-Olds, T.
   and Callaway R. M. (2002) Cost of defense in
   the context of plant competition: *Brassica
   rapa* may grow and defend. *Ecology*, **83**,
   505–517.

Smolander, A., Ketola, R. A., Kotiaho, T. *et al.* (2006) Volatile monoterpenes in soil atmosphere under birch and conifers: effects on soil N transformations. *Soil Biology and Biochemistry*, **38**, 3436–3442.

Suikkanen, S., Fistarol, G. O. and Graneli, E. (2004) Allelopathic effects of the Baltic cyanobacteria *Nodularia spumigena*, *Aphanizomenon flos-aquae* and *Anabaena lemmermannii* on algal monocultures. *Journal of Experimental Marine Biology*, **308**, 85–101.

Tack, A. J. M., Ovaskainen, O., Pulkkinen, P. and Roslin, T. (2010) Spatial location dominates over host plant genotype in structuring an herbivore community. *Ecology*, **91**, 2660–2672.

van der Putten, W. H., Vet, L. E. M., Harvey, J. A. and Wackers, F. L. (2001) Linking above- and belowground multitrophic interactions of plants, herbivores, pathogens, and their antagonists. *Trends in Ecology and Evolution*, **16**, 547–554.

Wei, J. N., Wang, L. H., Zhao, J. H. *et al.* (2011) Ecological trade-offs between jasmonic acid-dependent direct and indirect plant defences in tritrophic interactions. *New Phytologist*, **189**, 557–567.

Whitham, T. G., Young, W. P., Martinsen, G. D. *et al.* (2003) Community and ecosystem genetics: a consequence of the extended phenotype. *Ecology*, **84**, 559–573.

Whitham, T. G., Bailey, J. K., Schweitzer, J. A. *et al.* (2006) A framework for community and ecosystem genetics: from genes to ecosystems. *Nature Reviews Genetics*, **7**, 510–523.

Zarate, S. I., Kempema, L. A. and Walling, L. L. (2007) Silverleaf whitefly induces salicylic acid defenses and suppresses effectual jasmonic acid defenses. *Plant Physiology*, **143**, 866–875.

Zhang, P. J., Zheng, S. J., van Loon, J. J. A. *et al.* (2009) Whiteflies interfere with indirect plant defense against spider mites in Lima bean. *Proceedings of the National Academy of Sciences USA* **106**, 21202–21207.

CHAPTER TWO

Natural selection for anti-herbivore plant secondary metabolites: a *Eucalyptus* system

JULIANNE M. O'REILLY-WAPSTRA and BRAD M. POTTS
School of Plant Science, University of Tasmania
CLARE McARTHUR
School of Biological Sciences, University of Sydney

2.1 Introduction

Since the seminal papers of Fraenkel (1959) and Ehrlich and Raven (1964), much research has demonstrated the role of plant secondary metabolites (PSMs) as defence mechanisms against invertebrate and vertebrate herbivory. These metabolites can act directly on the herbivore as toxins (Theis & Lerdau, 2003; Gershenzon & Dudareva, 2007), digestibility reducers (Ayres *et al.*, 1997; De Gabriel *et al.*, 2009) and deterrents (Pass & Foley, 2000), and they can also act indirectly by, for example, attracting natural enemies of the herbivore (Dicke, 2009). The idea that the herbivores themselves are acting as selective agents on these PSMs has existed since it was first noted that these compounds may serve as anti-herbivore traits, and in some systems it is clear that herbivores may act as agents of natural selection on some specific PSMs (Simms & Rauscher, 1989; Mauricio & Rauscher, 1997; Stinchcombe & Rauscher, 2001; Agrawal, 2005). However, in most systems there is still a dearth of evidence addressing this question, particularly in light of the vast number of herbivores that attack a single plant species across its entire life and the array of PSMs that are expressed in a plant species. Are all of these herbivores agents of selection and have all PSMs evolved because of the selective pressures by the herbivores, or are PSMs driven by selection from other pressures such as abiotic factors (Close & McArthur, 2002)? Knowing the answer to these questions is important when attempting to understand what is driving population divergence within species and the evolution and change in PSMs.

For selection to occur there must be additive genetic-based variability in herbivory within plant populations. This herbivory must correlate with additive genetic-based variation in plant defensive traits, and herbivory must affect plant fitness (see Box 2.1). Key papers in the late 1980s through to the late 1990s clearly demonstrated the evolutionary impact that invertebrate herbivores were having on plant chemical defences in some systems (Rauscher & Simms, 1989; Simms & Rauscher, 1992; Rauscher, 1993; Mauricio & Rauscher,

The Ecology of Plant Secondary Metabolites: From Genes to Global Processes, eds. Glenn R. Iason, Marcel Dicke and Susan E. Hartley. Published by Cambridge University Press. © British Ecological Society 2012.

1997; Juenger & Bergelson, 1998). These papers, and those published more recently (O'Reilly-Wapstra *et al.*, 2004; Agrawal, 2005; Stinchcombe, 2005; Bailey *et al.*, 2007; Johnson *et al.*, 2009; Parker *et al.*, 2010), predominantly utilised quantitative genetics, using plants of known genetic stock grown in common gardens coupled with field damage and/or captive feeding trials to help address the question of selection. The strength of these methods lies in the fact that not only can they investigate the 'bare bones' of evidence for detecting selection (genetic-based variability in PSMs and impacts on plant fitness), but they can also be used to estimate other key pieces of information important to our understanding of natural selection, e.g. heritability of traits, genetic correlations in traits and in herbivore preferences, temporal and spatial variation in selection, and the type and direction of selection.

Box 2.1   Conceptual flow chart for the role of herbivores as agents of selection on plant secondary metabolites (PSMs)

If there is intra-specific variation in herbivory then certain evidence is needed to demonstrate the potential role of herbivores as selective agents on plant resistance. In this diagram, we focus on PSMs, but the trait could be any defensive trait such as leaf toughness. The left hand side of this flow chart shows the lines of evidence. Additive genetic-based variation observed in herbivory must correlate with additive genetic-based variation in PSM concentration in the plant, and there must be an effect of browsing on plant fitness (Mauricio & Rausher, 1997; Lankau, 2007). If one of these criteria is not satisfied then there is unlikely to be selection on plant defensive traits by browsing herbivores. If there is no variation in browsing preferences, there may still be significant effects on plant fitness which could lead to population extinction and changes in community composition and structure (right hand side of flow chart). This chart was modified from O'Reilly-Wapstra and Cowan (2010).

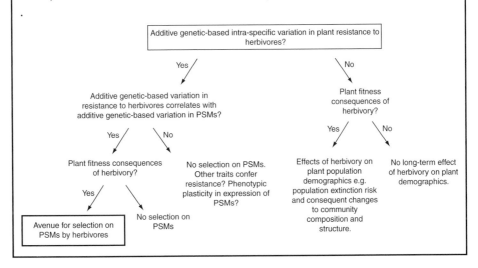

For any one plant species, the strength and direction of selection on traits can vary throughout its evolutionary history and ecological status. Some of the factors that lead to such variation include the presence of multiple selective pressures (Stinchcombe & Rausher, 2001; Parker *et al.*, 2010), the genetic architecture of a trait (Henery *et al.*, 2007; Freeman *et al.*, 2008), heteroblastic foliage change in the species (O'Reilly-Wapstra *et al.*, 2007a; Barton & Koricheva, 2010), genetic correlations between traits and of herbivore preferences (Leimu & Koricheva, 2006) and temporal and geographic variation in selection (Thompson, 2005). Eucalypts offer an ideal model study system to investigate selection by herbivores on PSMs in this complex ecological and evolutionary setting. *Eucalyptus globulus* is one species of eucalypt that has received considerable research attention, particularly in relation to quantitative and molecular genetic variation in key plant traits (Potts *et al.*, 2004) and in relation to biotic interactions with other species (e.g. O'Reilly-Wapstra *et al.*, 2002; Rapley *et al.*, 2004a). Here we use *E. globulus* as a framework to understand the selective impacts of herbivores on PSMs. Recently Rasmann and Agrawal (2009) recommended the use of more than one or two techniques when investigating the physiological role and ecological and evolutionary significance of PSMs in plants. They suggest that a more holistic approach to investigations would be beneficial in elucidating our understanding of plant–herbivore interactions; that is, using combinations of quantitative genetics (Jordan *et al.*, 2002; O'Reilly-Wapstra *et al.*, 2002), molecular genetics (Ralph, 2009), phylogenetic studies (Agrawal, 2007; Futuyma & Agrawal, 2009) and ecological studies (e.g. field and captive bioassays; Lawler *et al.*, 2000; Moore & Foley, 2005). We have applied this philosophy in order to tease apart adaptive traits and the mechanisms and processes driving those traits that have led to the display of plant–herbivore interactions evident in eucalypt systems. With the increasing genomic knowledge in the genus *Eucalyptus* (e.g. Potts *et al.*, 2004; Freeman *et al.*, 2008; Sansaloni *et al.*, 2010; O'Reilly-Wapstra *et al.*, 2011), this genus offers an ideal long-lived woody plant group (like the northern hemisphere *Populus*; Jansson & Douglas, 2007) to complement the available list of model plant organisms used to address fundamental (and applied) questions relating to herbivory and plant defence.

2.2 *Eucalyptus* as a model genus

Eucalypts are part of the Myrtaceae family and are a species-rich genus with over 700 species in the group (Brooker, 2000). They dominate much of the Australian landscape (Williams & Brooker, 1997), and the phenotypic variation in traits (e.g. leaf morphology, physiology, chemistry), habit and reproductive strategies is vast, with trees ranging from dwarf mallee form in semi-desert areas through to the tallest angiosperm in the world in wet forests (Potts & Pederick, 2000). Additionally, many eucalypt species exhibit heteroblastic leaf change from juvenile to adult leaves, again offering distinct variation in foliage characteristics (e.g. PSM concentration, toughness, size, shape) during ontogenetic development (Jordan *et al.*, 2000; O'Reilly-Wapstra *et al.*, 2007a; Jaya *et al.*, 2010). Owing to their dominance in the landscape, *Eucalyptus* trees are involved in many biotic interactions, with a wide range of animal species relying on eucalypts for food and shelter (Majer *et al.*, 1997; Woinarski *et al.*, 1997). The majority of eucalypts are broad-leaved evergreen species, and foliage is infested by a range of invertebrate herbivores such as leaf beetles, sawfly larvae, galling insects, leaf-miners and weevils, and also by fungal pathogens (Elliot & de Little, 1985; Majer *et al.*, 1997). However, only a few arboreal marsupial herbivores utilise eucalypt foliage as a major part of their diet, ranging from two specialist herbivores (the koala, *Phascolarctos cinereus*, and greater glider, *Petaurides volans*) to the more generalist feeders (the common ringtail possum, *Pseudocheirus peregrinus*, the common brushtail possum, *Trichosurus vulpecula*, the short-eared possum, *T. caninus*, and the mountain brushtail possum, *T. cunninghami* (formerly *T. caninus*); Hume, 1999). Eucalypt foliage is a low-quality food as it is characteristically low in nutrients and high in a range of PSMs (Hume, 1999). There are distinct qualitative and quantitative differences in the PSM concentration of foliage both between and within eucalypt species and hence there is a wide variation in suitable forage from which eucalypt-feeding herbivores can choose, and on which selection can potentially act.

The diverse range of PSMs found in eucalypt foliage includes volatile terpenes (e.g. mono and sesquiterpenes), non-volatile terpenes (e.g. triterpenes), phenolic compounds (e.g. hydrolysable and condensed tannins, flavonoids, phloroglucinol adduct compounds), glycosides, sterols and coumarins (Hillis, 1966; Li & Madden, 1995; Brophy & Southwell, 2002; Moore *et al.*, 2004; Gleadow *et al.*, 2008; Koshevoi *et al.*, 2009). Eucalypt foliage is also often very glaucous, being covered in foliar waxes (Jones *et al.*, 2002; Rapley *et al.*, 2004b). Many of these compounds have been shown to be important in reducing browsing by herbivores. For example, studies have demonstrated the importance of formylated phloroglucinol compounds (FPCs) (Wallis *et al.*, 2002; O'Reilly-Wapstra *et al.*, 2004; Loney *et al.*, 2006), terpenes (Wiggins *et al.*, 2003) and tannins (Marsh *et al.*, 2003; O'Reilly-Wapstra *et al.*, 2005) in deterring feeding by the common brushtail possum, *Trichosurus vulpecula*. Formylated phloroglucinol compounds also affect eucalypt foliage choice by other arboreal marsupial folivores (Moore *et al.*, 2004) including the ringtail possum, *Pseudocheirus peregrinus* (Lawler *et al.*, 2000) and koala, *Phascolarctos cinereus* (Moore & Foley, 2005). Two particular groups of FPC compounds, the sideroxylonals and macrocarpals, are important determinants of feeding by brushtail possums, with increasing concentration of these compounds decreasing levels of foliage intake (Wallis *et al.*, 2002; O'Reilly-Wapstra *et al.*,

2010). In native eucalypt stands, spatial heterogeneity in sideroxylonals between individuals (Lawler *et al.*, 2000; Moore & Foley, 2005; Andrew *et al.*, 2007a; Moore *et al.*, 2010) leads to patchiness in preferred feeding habitat of brushtail possums, which in turn influences feeding behaviour and foraging efficiency (Wiggins *et al.*, 2006). More recently, the amount of nitrogen available to the herbivore (as influenced by concentration of condensed tannins in the foliage) has been shown to affect reproductive output of female brushtail possums (De Gabriel *et al.*, 2009).

One particular species of eucalypt, *Eucalyptus globulus*, has been the focus of much research and provides an ideal study organism in which to investigate fundamental questions in plant–herbivore interactions. This species is a dominant tree in native forests in Southeastern Australia and is also the most widely planted commercial hardwood species globally. Fifteen years of research have examined the quantitative and molecular genetics of this species (Dutkowski & Potts, 1999; Steane *et al.*, 2006; Barbour *et al.*, 2009). The extensive genetic diversity detected in *E. globulus* provides a unique genetic framework to pair with ecological studies as the knowledge of its gene pool is amongst the most detailed of any forest tree species (Potts *et al.*, 2004). This quantitative genetic diversity has been summarised by classifying the species into 13 geographic races (see Box 2.2; Dutkowski & Potts, 1999), which molecular studies indicate comprise three major lineages; a mainland Australian lineage, an Eastern Tasmanian lineage and a Western Tasmanian lineage (Steane *et al.*, 2006). Within these races, trees are again classified into different localities (population of trees growing within 10 km of each other) and different families (offspring from a single mother tree) within populations (Potts & Jordan, 1994; Dutkowski & Potts, 1999). The foliage is attacked by a range of vertebrate and invertebrate herbivores and fungal pathogens, with one study estimating up to 65 putative invertebrate and fungal symptoms on the foliage of one tree (Gosney, 2009). *Eucalyptus globulus* foliage contains large quantities of PSMs including terpenoids and phenolic-based compounds (see references below in Section 2.3) and there is considerable intra-specific variation in these compounds, spatially and across life stages (e.g. O'Reilly-Wapstra *et al.*, 2007a, 2010).

In the remainder of this chapter we will address the question of selection of anti-herbivore PSMs by herbivores in this plant species. We will focus on the relationships between *E. globulus* and mammalian herbivore browsing. We synthesise information about specific PSMs that are known to deter herbivory, examine the genetic control and heritability of these PSMs, investigate the genetic correlations in preference of *E. globulus* by multiple herbivores and investigate whether there are fitness consequences associated with browsing. We will examine how these factors change through ontogenetic development and between genetically and geographically distinct populations of *E. globulus*.

Box 2.2   Map showing the location of the 13 genetically distinct native races of *Eucalyptus globulus*

These 13 races were classified based on morphological assessment of 35 different quantitative traits of trees grown in a series of common garden trials (Dutkowski & Potts, 1999). Additionally, these 13 races have been classified into three distinct lineages based on microsatellite analysis (Steane *et al.*, 2006). These lineages comprise the mainland Australian lineage (Western Otways, Eastern Otways, Strzelecki Ranges, Southern Gippsland and Wilsons Promontory Lighthouse), the western-Tasmanian lineage (Western Tasmania and King Island) and the eastern-Tasmanian lineage (Furneaux Group, Northeastern Tasmania, Southeastern Tasmania, Southern Tasmania, Recherche Bay and Mt Dromedary). Solid lines delineate races, dashed lines delineate sub-races. Figure adapted from Dutkowski and Potts (1999) and Barbour *et al.* (2009).

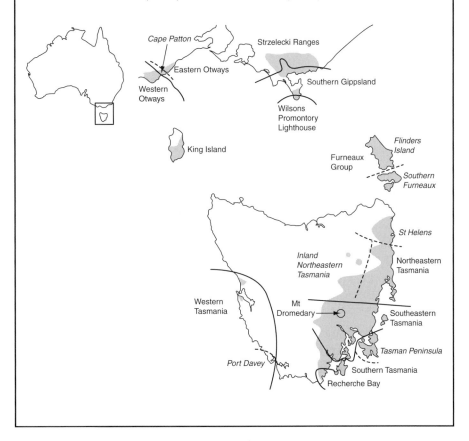

2.3   Genetic variation and control of anti-herbivore PSMs

Herbivores may show variation in feeding behaviour if there is variation in the key plant phenotypic traits influencing their feeding choices. Matched with this, natural selection by herbivores can only elicit an evolutionary change in these plant traits if variation is genetically determined. Genetic variation in traits occurs at multiple scales. Plant secondary metabolites can vary between evolutionary lineages of a species, between species within a lineage and also at multiple scales within species (e.g. between different populations and families within populations). These metabolites can also show variation through ontogenetic development of the plant within a species. There has been a recent resurgence in the literature examining the macroevolutionary patterns of plant defence against herbivores (Agrawal & Fishbein, 2008; Agrawal *et al.*, 2009; Becerra *et al.*, 2009) and some studies have discussed these patterns in light of correlative macroevolutionary patterns in dependent herbivores (Futuyma & Agrawal, 2009) illustrating potential co-evolution (Wheat *et al.*, 2007). For example, macroevolutionary patterns have shown that in the Pierinae–Brassicales system, the evolution of a nitrile-specifier protein (NSP) glucosinolate detoxification gene was an innovation that allowed Pierinae to host-shift to the Brassicales soon after the appearance of the Brassicales plant order (Wheat *et al.*, 2007). While there have been several studies examining the taxonomic presence/absence and patterns of certain PSMs (Li *et al.*, 1995; Eschler *et al.*, 2000; Gleadow *et al.*, 2008) in the genus *Eucalyptus*, no studies to date have examined macroevolutionary patterns, divergence and radiation in defensive PSMs in relation to the divergence and radiation of dependent vertebrate or invertebrate herbivores (but see Steinbauer (2010) for latitudinal trends in eucalypt foliar oils and correlative patterns with leaf beetles). Consequently, we detail below how PSMs in *Eucalyptus* exhibit variation at multiple microevolutionary genetic scales, with a predominant focus on scales of variation in *E. globulus*. Owing to the extensive quantitative (e.g. numerous common garden field trials) and molecular genetic (QTL maps, chloroplast and nuclear markers, microsatellites, DArT markers) resources available for *E. globulus*, there is an excellent genetic framework on which we can address the question of selection by herbivores on *E. globulus* PSMs.

2.3.1   Genetic variation within species; race, locality and family variation

'Quantitative genetics…has largely dominated our thinking on plant defense evolution for the past 20 years' (Agrawal, 2007). As this quote indicates, quantitative genetics has been a leading tool in investigating aspects of the microevolutionary patterns between plants and their herbivores. Most studies utilising a quantitative genetics approach focus on within-species variation in quantitative traits and measure these traits in trees of known pedigree growing in common gardens. In *E. globulus* we have used this approach not only to address questions such as selection on PSMs by herbivores, but also to provide a tool to address other fundamental questions such as evolutionary selection imposed by multiple herbivores (O'Reilly-Wapstra *et al.*, 2010), whether selection is stable over the lifetime of the plant (O'Reilly-Wapstra *et al.*, 2007a) and whether there is variation in plant traits and herbivore responses (hence variation in the selection surface) over the geographic distribution of the plant species (O'Reilly-Wapstra *et al.*, 2010).

Variation in key PSMs in *E. globulus* is under relatively straightforward genetic control. Numerous trials have shown clear variation between genotypes (between different races within a species, different localities within races and/or different families within localities) in certain PSMs. For example, the FPCs, which are known to influence mammalian herbivore preferences, differ significantly between races, localities and families within localities in *E. globulus* (O'Reilly-Wapstra *et al.*, 2004, 2007b; Freeman *et al.*, 2008). Sideroxylonal, one of the key compounds influencing intake of *E. globulus* by possums, is quantitatively lower in the northeastern-Tasmanian populations, compared with southern-Tasmanian and mainland Australian populations (O'Reilly-Wapstra *et al.*, 2004, 2010). Other compounds (e.g. terpenes) linked to browsing preferences by mammals for *E. globulus* also show clear genetic-based intra-specific variation (O'Reilly-Wapstra *et al.*, 2005), as do foliar waxes, which influence insect feeding and oviposition preferences (Jones *et al.*, 2002; Rapley *et al.*, 2004a).

An indication of the strength of genetic control of PSMs can be obtained from heritability estimates. Narrow-sense heritability gives an indication of the proportion of additive genetic variation in the trait on which selection can act and cause evolutionary change in a population (Falconer & Mackay, 1996). Some FPCs exhibit moderate to high narrow-sense heritability in wild populations of *E. melliodora* (sideroxylonal h^2 = 0.89; Andrew *et al.*, 2005) and in common garden trials of *E. tricarpa* (sideroxylonal h^2 = 0.60 and 0.39–0.62; Andrew *et al.*, 2007b, 2010, respectively). Broad-sense heritabilities (includes additive variation and non-additive variation due to dominance and epistatic effects) for these compounds in *E. globulus* are also high (sideroxylonal h^2 = 0.79, macrocarpal A h^2 = 0.51; Freeman *et al.*, 2008). Similarly, broad-sense heritabilities (h^2 = 0.37–0.79) for some mono- and sesquiterpenes in *E. globulus* are moderate to high (O'Reilly-Wapstra *et al.*, 2011) with narrow-sense heritability for 1,8-cineole (a dominant monoterpene in many eucalypt species) also documented as quite high (h^2 = 0.83) in 2-year-old foliage from *E. kochii* (Barton *et al.*, 1991). It is clear from these studies that there is strong enough genetic variation in FPCs and some terpenes on which selection can act. Apart from the cited examples of heritability estimates for some FPCs and terpenes in *E. globulus* and other eucalypt species, very few studies have estimated the heritability of any other PSMs in eucalypts (but see Goodger and Woodrow (2002) for heritability estimates of cyanogenic glycosides). Therefore, the genetic variation on which selection can act in other PSMs in eucalypts is largely unknown.

2.3.2   Quantitative trait loci and candidate genes for the expression of eucalypt PSMs

Combining quantitative and molecular genetic tools can aid our understanding of the evolution of plant traits (Freeman *et al.*, 2008; Rasmann & Agrawal, 2009). For example, identification of quantitative trait loci (QTL) has the potential to further our understanding of the genetic control of variation in quantitative traits by providing information on the number, location and magnitude of effects of influential loci (Sewell & Neale, 2000). In *E. globulus* QTL have been identified for several PSMs, two FPCs and a suite of mono- and sesquiterpenes. A study utilising 112 clonally duplicated progenies from an outcross F_2 of *E. globulus* identified two unlinked QTL for macrocarpal G, and another unlinked QTL for sideroxylonal A (an isomer of sideroxylonal) (Freeman *et al.*, 2008). The sideroxylonal A QTL co-located with one for total sideroxylonal previously reported in adult *E. nitens* foliage (Henery *et al.*, 2007), providing independent validation in a different evolutionary lineage and a different ontogenetic stage. Variation in sesquiterpenes in *E. globulus* appears to be influenced by only a few loci, since QTL for six of the seven different sesquiterpenes assayed in one study co-locate (O'Reilly-Wapstra *et al.*, 2011). This, in combination with highly significant correlations between these compounds, argues that their variation is influenced by a QTL early in the biosynthetic pathway with pleiotropic effect.

More recently, molecular genetic analyses have focused on identifying candidate genes for traits coupled with associational genetic analysis to identify which genes are responsible for the expression of particular resistance and/or PSM traits (Dracatos *et al.*, 2009; Ralph, 2009 for a review in the well-studied *Populus* system). This field of research is in its infancy in eucalypt systems, but this will undoubtedly change once the *E. globulus* genome is sequenced. Single nucleotide polymorphisms (SNPs) were recently characterised in four eucalypt species (*E. globulus*, *E. nitens*, *E. camaldulensis* and *E. loxophleba*) from the same sub-genus, *Symphyomyrtus* (Kulheim *et al.*, 2009). This study looked at 23 genes, from four biosynthetic pathways leading to the formation of secondary metabolites, and identified a high frequency of SNPs (8631) across the four species, providing a good comparative study of polymorphism in these eucalypt species. Many of these SNPs were common between the species, suggesting that selective forces have maintained them over evolutionary time.

2.3.3   Stability of genetic-based PSMs

It is clear that PSMs in *E. globulus* populations show phenotypic variation and that this variation is genetically determined, with some showing high levels of heritability. However, an important question is: how stable is this genetic-based defensive chemistry? Expression of PSMs in plant species may vary with changes in the environment (e.g. season, nutrient levels) (Close *et al.*, 2003; Donaldson & Lindroth, 2007) and with changes in ontogenetic development of the plant, from seedling to adult life stages (Barton & Koricheva, 2010). This variability could result in variation in the strength and direction of selection on traits. Obviously the more heritable PSM traits are more likely to show stability in the rankings of genotypes across different environments. This has been observed in a straightforward trial that examined expression of PSMs and resistance of *E. globulus* seedlings of known genetic stock grown under two different soil nutrient treatments (O'Reilly-Wapstra *et al.*, 2005). This trial showed a clear genetic × environment interaction where more resistant genotypes under the low nutrient treatment 'lost' their comparative resistance when placed under the high nutrient treatment, while the more susceptible genotypes remained unchanged. In this study, nutrient levels altered the nitrogen and condensed tannin content; and it has been well documented that tannins are phenotypically quite plastic (Osier & Lindroth, 2001; Close *et al.*, 2003). However, the FPCs, which are known to have relatively high heritabilities (Andrew *et al.*, 2005; Freeman *et al.*, 2008), remained unchanged between the two treatments. The stability in FPC concentration in *E. globulus* saplings (1 year old; O'Reilly-Wapstra *et al.*, unpublished data) and in other eucalypt species (e.g. *E. tricarpa*; Andrew *et al.*, 2010) across different common environment field trials has also been shown. These observations indicate that if selection by herbivores is occurring on environmentally stable traits such as FPCs, then patterns of selection may be consistent across different environments (all things being equal), although the influence of other variables on the strength and direction of selection by the herbivore (e.g. variable herbivore numbers and the presence/absence of other herbivores) may also play a role.

The expression of PSMs in plants varies considerably over their lifetime (Boege & Marquis, 2005; Barton & Koricheva, 2010), as does the range of herbivores that feed on the plant. Understanding the stability of PSMs across life stages is important in understanding the lifetime evolutionary selective pressures on a plant. Lack of consistency in anti-herbivore PSMs across different life stages may be a response to past selective pressures imposed by multiple herbivores on the different stages. Alternatively, consistent expression of defence across multiple stages may arise if plants suffer attack from a single herbivore across multiple stages or if they utilise a similar suite of defences to protect themselves against multiple herbivores (Leimu &

Koricheva, 2006). One might expect increased variability in the expression of PSMs in plants that undergo marked changes in leaf morphology (hetero-blasty) during ontogenetic development from juvenile to adult leaves. In essence, these leaves may represent completely different habitats for the herbivores. However, in *E. globulus* the heritable FPCs appear to be relatively stable across the heteroblastic life stages of the plant (O'Reilly-Wapstra *et al.*, 2007a). While there are quantitative differences in the amounts of FPCs between juvenile (including coppice, i.e. re-growth foliage after foliage loss) and adult stages – juvenile coppice foliage had higher sideroxylonal A concentrations – the ranking of the genotypes remains stable. In contrast, other groups of PSMs such as terpenes, total phenolics and condensed tannins showed no stability between juvenile and adult leaf phases; in fact tannins, although present in the adult leaf stage, were not present at all in the coppice juvenile leaf phase (O'Reilly-Wapstra *et al.*, 2007a). These observations suggest that if selection by herbivores is occurring on the ecologically important FPCs, and if variation in adult and coppice (juvenile) defensive chemistry is under the control of the same genes, then patterns are consistent with selection by herbivore browsing at either life phase. On the other hand, if there was no common genetic control, this pattern could still emerge through independent selection at each life stage. It is currently unclear which of these processes is occurring in this system.

2.4   Genetic-based PSMs as drivers of herbivore responses

In order for herbivores to be selective agents on anti-herbivore PSMs there needs to be not only genetic-based variability in the PSM, but a causal relationship between PSM concentration and impacts on herbivore feeding. Several groups of compounds in *E. globulus* have been shown to affect herbivory: for vertebrate herbivores (see examples and references above) these include tannins, FPCs and volatile terpenes, and for invertebrate herbivores such as autumn gum moth (*Mnesempala privata*) they include tannins (Rapley *et al.*, 2007) and foliar waxes (Jones *et al.*, 2002; Rapley *et al.*, 2004b). The FPCs, particularly sideroxylonal and macrocarpals, are important in reducing intake of *E. globulus* by brushtail possums and Tasmanian pademelons (*Thylogale billardierii*, a ground-dwelling macropod species) (O'Reilly-Wapstra *et al.*, 2004, 2010). Figure 2.1 shows the relationship between intake of *E. globulus* by each of these browsing herbivores and concentration of sideroxylonal. These patterns also hold up in large-scale field trials where, despite a multitude of other factors (e.g. multiple food sources, weather, animal competition, anti-predatory behaviours) influencing the complex browsing decisions by these herbivores, sideroxylonal still played a role in determining which individual *E. globulus* plants the herbivores would eat (Miller *et al.*, 2009).

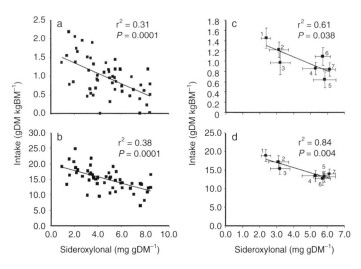

Figure 2.1 Relationship between sideroxylonal A concentration (mg per gram dry matter, DM) and intake of foliage from individual *E. globulus* trees by (a) *Thylogale billardierii* and (b) *Trichosurus vulpecula*, and mean (±SE) intake of foliage from seven distinct *E. globulus* populations by (c) *Thylogale billardierii* and (d) *Trichosurus vulpecula*. Intake of 54 individual coppiced trees from seven populations to feeding by six *T. billardierii* and six *T. vulpecula* was assessed in a no-choice captive feeding trial (O'Reilly-Wapstra *et al.*, 2002). Intake was measured as amount (gDM kgBM^{-1}) of foliage removed from the plant after feeding. Numbers refer to the seven populations sourced from the three races: 1, St Helens; 2, German Town; 3, South Geeveston; 4, Royal George; 5, South Bruny; 6, Blue Gum Hill; 7, Jeeralang North. St Helens, German Town and Royal George are from the Northeastern Tasmanian race; South Geeveston, Blue Gum Hill and South Bruny are from the Southern Tasmanian race; Jeeralang North is from the Strzelecki Ranges race. Panel B is taken from O'Reilly-Wapstra *et al.* (2004).

In systems where key or dominant PSMs can be manipulated, for example by adding purified or synthetic individual compounds to control diets, or manipulating the gene responsible for the expression of a single PSM, then the direct causal link between food intake and a single PSM can be investigated. Increasing concentration of sideroxylonal, added to otherwise benign basal diet, directly decreased intake of the diet by brushtail possums (Wallis *et al.*, 2002). Additionally, laboratory studies looking at the effects of 1,8-cineole (the quantitatively dominant monoterpene in *E. globulus*) on intake of a control diet by the common brushtail possum have shown that in high quantities, this compound directly reduces intake of the diet (Wiggins *et al.*, 2003; Boyle & McLean, 2004). Feeding behaviour (e.g. number of feeding bouts, length of individual feeding bouts) during a night of feeding was also significantly affected by high concentration of this monoterpene. However, there is often a strong correlation between concentration of 1,8-cineole and

FPCs in eucalypts (Lawler *et al.*, 1999; Moore *et al.*, 2004), including in *E. globulus* (O'Reilly-Wapstra *et al.*, 2004), and animals may use volatile terpenes as a cue to the presence of more toxic, non-volatile, compounds such as the FPCs (Lawler *et al.*, 1999; Moore *et al.*, 2004). This raises the possibility that, if PSMs are strongly correlated, then some may be under direct selection and others under indirect selection due to the strong correlation with the trait under direct selection (Falconer & Mackay, 1996). A combination of univariate (slope estimates total selection) and multiple (partial regression slopes estimate direct selection) regression analysis, regressing plant fitness (after feeding by a particular herbivore) against standardised PSM compounds can help address this question (Conner & Hartl, 2004). For example, Bailey *et al.* (2007) demonstrated through multiple regression selection analyses that selective herbivory by elk (*Cervus canadensis*) on aspen (*Populus tremuloides*) was acting directly on one particular compound, tremulacin, and indirectly on another compound, salicortin. To date, the teasing apart of direct and indirect selection on PSMs in eucalypts has received little to no attention despite the large array of PSMs present in the genus. This is probably because few studies have quantified fitness consequences of herbivory (see below), matched with variation in the defensive compounds.

2.5   Correlated multi-species responses to PSMs; diffuse versus pairwise selection

More often than not, plants are eaten and infested by a wide range of herbivores and pathogens throughout their entire life-history, and such interactions occur both above and below ground (Strauss & Irwin, 2004; Wardle, 2006; Wise, 2009; Parker *et al.*, 2010). Hence, research examining the selective pressures of any one herbivore on plant PSMs should consider the possible interactive effects of other herbivores or pathogens on that evolutionary interaction. In cases where the evolutionary selective pressure of a single herbivore on a plant is not affected by other interacting species, the relationship is considered pairwise selection or pairwise co-evolution if reciprocal selective pressures are occurring (Juenger & Bergelson, 1998). However, when other species do influence the relationship between a herbivore and a plant, then this is considered to be diffuse selection or diffuse co-evolution (Stinchcombe & Rausher, 2001; Anderson & Paige, 2003). For example, selection on deer resistance in the ivyleaf morning glory, *Ipomoea hederacea*, depends on the presence or absence of insect herbivores; it is only in the presence of the insects that greater deer resistance is favoured (Stinchcombe & Rausher, 2001). Analysis of the genetic correlations in preferences of herbivores can indicate whether such interactive relationships occur (Leimu & Koricheva, 2006). A lack of correlation suggests that the relationships are independent and pairwise, presumably with multiple herbi-vores responding to different defensive traits. In contrast, a significant corre-lation implies that the relationships are not independent and hence different herbivores or other selective agents may influence the same defensive traits.

To date, no published research has examined the true additive genetic correlations in preferences of multiple organisms for *E. globulus* foliage. Estimating the correlations in preferences of each organism to the same open-pollinated genetic families, both across and within field sites, would provide great insight into diffuse versus pairwise selective patterns in this system. If we synthesise the published studies of preferences of these organ-isms at the broader population and race levels, then there appear to be some similarities and differences in the preferences for foliage by different organ-isms (O'Reilly-Wapstra *et al.*, 2002, 2010; Rapley *et al.*, 2004a; Milgate *et al.*, 2005). For example, O'Reilly-Wapstra *et al.* (2010) illustrate a significant gene-tically based correlation among populations between relative resistance of *E. globulus* to both brushtail possums and Tasmanian pademelons, indicating that the relationship between these herbivores, and any subsequent selection on the plant resistance traits, does not appear to be independent. These two species prefer genetically distinct populations from the northeast of Tasmania (O'Reilly-Wapstra *et al.*, 2002), and this preference appears consis-tent for different Tasmanian populations of herbivores tested in reciprocal trials (N. L. Wiggins *et al.*, unpublished data). In contrast, Northeastern Tasmanian races are less preferred by the autumn gum moth, which prefers (measured as higher rates of oviposition) populations from the Furneaux race (Rapley *et al.*, 2004a). If these organisms exert fitness costs to *E. globulus*, the mammal and the insect species may be having independent effects on this tree species. Correlated responses with other selective pressures, however, cannot be discounted.

2.6   Consequences of browsing for plant fitness

As well as genetic-based variation in key defensive traits, evidence for selec-tion by brushtail possums on eucalypt defences needs to show clear fitness consequences of browsing on plants. Perhaps one of the greatest challenges to addressing fundamental questions relating to plant–herbivore theory in long-lived woody species is quantifying accurate measures of plant fitness. While this is relatively straightforward in annual or short-lived plants, many studies examining similar questions in woody plant systems have to rely on surrogate measures of lifetime fitness such as estimating seed output at one time period and assessing growth rates or survival (Dahlgren *et al.*, 2009; O'Reilly-Wapstra *et al.*, 2010; Sampaio & Scariot, 2010).

The effect of marsupial browsing on eucalypts has been more extensively documented in managed systems than in native forests. As an accurate measurement of lifetime fitness is difficult, surrogates of fitness such as survival, growth rates and capsule and/or flowering output assessed at one time period have been measured. While eucalypts are relatively robust to foliage loss, through coppicing and re-growth (Whittock *et al.*, 2003; Rapley *et al.*, 2009), repeated browsing has clear fitness effects. For example, browsing on seedlings removed up to 80% of foliage after a 16-week period (Miller *et al.*, 2006) and repeated browsing in the field reduced seedling survival and growth rates (Bulinski & McArthur, 1999). Repeated browsing also significantly reduced the reproductive output (number of umbels) of *E. nitens* at 10 years of age compared with trees that received no mammalian herbivore browsing (Figure 2.2). In addition to this, the brushtail possum has also been implicated in the deaths of many adult eucalypt trees in agricultural areas, thus significantly contributing to rural tree decline (Neyland, 1996). Evidence would suggest then that browsing by this mammalian herbivore does incur fitness costs to eucalypt species and this, matched with genetic-based variation in the trait correlating with the genetic-based variation in herbivore preferences, suggests that the brushtail possum may be an agent of selection on PSM in *E. globulus*.

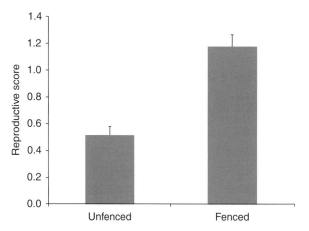

Figure 2.2 Mean (+SE) reproductive output of 10-year-old *E. nitens* trees that were repeatedly browsed by mammalian herbivores (unfenced) or not browsed by mammalian herbivores (fenced) over a 10-year period. *Eucalyptus nitens* seedlings (5 per plot) were planted in either fenced (10 per trial) or unfenced (10 per trial) plots in five different field trials in Tasmania, Australia. At age 10, trees were assessed for reproductive output (estimate of the number of buds and capsules on a 5-point scale; 0 = no buds/capsules, 1 = 0–10 buds/capsules, 2 = 10–100 buds/capsules, 3 = 100–1000 buds/capsules, 4 = >1000 buds/capsules). Analysis shows a significant difference in mean reproductive output of trees between fenced ($n = 38$, mean = 1.13) and unfenced ($n = 35$, mean = 0.53) plots (Kruskal Wallis non-parametric test, $\chi = 6.21$, $P = 0.01$).

If this is the case, it appears that the strength of selection varies across the geographic range of *E. globulus*, as a recent study showed that the strength of the relationship between relative resistance to common brushtail possums and one of the FPC groups, the macrocarpal compounds, in *E. globulus* was not uniform within races (O'Reilly-Wapstra *et al.*, 2010). The concept of a geographic mosaic in selection and in the possible co-evolutionary relationships between plants and animals across the geographic range of the plant species has received much attention in recent years (Benkman, 1999; Thompson, 2005; Gomez *et al.*, 2009; Laine, 2009) but has been little studied in eucalypts. Eucalypts belong to a genetically diverse genus that covers a wide range of habitats and geographic range. Even single species can cover a wide environmental gradient, often exhibiting associated clines in their adaptive traits such as vegetative phase change (Jordan *et al.*, 2000) where marked genetic differentiation between populations within a species is typical (Potts & Pederick, 2000). It is highly likely that similar variation occurs in PSM concentration between genetically distinct populations in eucalypt species other than *E. globulus* (Andrew *et al.*, 2010; O'Reilly-Wapstra *et al.*, 2010) and this variation may lead to a geographic mosaic in selection in other eucalypt species, but this has not yet been investigated.

2.7   Conclusions and future directions

This chapter demonstrates that the studies of the *E. globulus* system have led to a better understanding of the ecological and evolutionary processes underlying plant–herbivore interactions. There is quantitative variation in key PSMs (the FPCs) in this tree species; this variation and the correlated variation in herbivore intake of the foliage is under genetic control and there are fitness consequences of browsing by these herbivores. The satisfaction of these conditions leads us to suggest that variation in some PSMs may have evolved in *E. globulus* in response to selective pressures from mammalian herbivores. The genetic ranking of the expression of FPCs appears stable across environments (different common garden field sites and nursery conditions) and also stable across ontogenetic life stages of the plant, indicating that selection by herbivores in one life stage could elicit evolutionary change in the other life stage. However, the strength of selection by herbivores appears to vary across the geographic range of *E. globulus*, perhaps through the influence of other variables on the strength of selection such as herbivore numbers, the presence or absence of other herbivores or other selective forces. Hence, despite this evidence for the role of mammalian herbivores as selective agents on PSMs in *E. globulus*, specific questions still remain.

For example, as inferred in Section 2.5, *E. globulus* may be under selective pressure from a range of organisms and abiotic factors. The independent, additive or interactive roles of these factors in driving the evolution of PSMs in

E. globulus are unknown and this also remains an area where research is needed in most other plant systems. While previous research has built a picture of the role of herbivores as agents of selection in some capacity (Ehrlich & Raven, 1964; Rausher & Simms, 1989; Rausher, 2001; O'Reilly-Wapstra *et al.*, 2010), there is still little evidence to resolve the fundamental question of whether plant enemies are the main drivers of evolutionary diversification or are simply responding to underlying genetic variation in their host plants (Futuyma & Agrawal, 2009). Another fundamental question relating to plant–herbivore interactions in the literature is, 'Why is there such a great number and vast diversity of PSMs in any one plant species?' (Romeo *et al.*, 1996; Becerra *et al.*, 2009). Having insight into the suite of selective pressures, the strength and direction of selection on individual PSMs and whether selective agents are driving evolutionary diversification in plant systems would help address these questions. For example, the strength of selection by herbivores would vary significantly (temporally and spatially) because of the multitude of other factors (other plant traits, animal behaviours, abiotic factors) that influence feeding. The trade-offs that most likely exist for *E. globulus* in producing PSMs (e.g. under possible diversifying selection from multiple pressures) are also largely unknown. Additionally, a comprehensive understanding of how the strength and direction of selection on PSMs varies across the geographic distribution of *E. globulus* is missing, and a move towards causal rather than correlative studies is required. Evidence for the reciprocal co-evolutionary selective pressure of plant species on behavioural and physiological mechanisms of mammalian herbivores to tolerate PSMs (that is, demonstrating genetic-based variation in behavioural and/or physiological responses to PSMs with resulting fitness consequences for the herbivore) has also been largely neglected in most woody plant systems (De Gabriel *et al.*, 2009). Up-scaling all of this research to investigate community-level selection and possible community co-evolution, particularly in dominant species such as *E. globulus*, are also areas gaining momentum (Whitham *et al.*, 2008).

A large proportion of the *E. globulus* genome will be re-sequenced. This of course will provide a wealth of opportunities to investigate the major genes controlling changes in PSMs due to selection from biotic and abiotic pressures. In the age of advanced gene technology, the main focus in many plant–herbivore systems has shifted to identifying the genes responsible for change in the observed traits. While this enhances our understanding of how such systems function, the forces (i.e. the selective pressures) driving evolutionary change are often not investigated. Advances in one area, such as identifying the genes responsible for trait changes, should not come at the expense of advances in other areas, such as knowing what organisms or factors are driving the change. To reiterate Rasmann and Agrawal (2009), the greatest advances in fully elucidating the complex world of plant–herbivore interactions will occur when we incorporate multiple tools into our research such as field and laboratory ecological studies, molecular genetics, quantitative genetics and phylogenetic analyses.

Acknowledgements
We would like to thank James Bulinski for the establishment of the *E. nitens* trials (Figure 2.2) and Hugh Fitzgerald, Paul Tilyard and Stephen Patterson for assistance with trials (Figure 2.1). We also thank Erik Wapstra for comments on the manuscript. Animals used in trials (Figure 2.1) were caught and maintained under the University of Tasmania Animal Ethics Committee, project no. 99038; and Parks and Wildlife Service permit no. FA 99053. We thank the CRC for Forestry, the Australian Research Council and the UTAS Rising Stars Programme for funding.

References

Agrawal, A. A. (2005) Natural selection on common milkweed (*Asclepias syriaca*) by a community of specialized insect herbivores. *Evolutionary Ecology Research*, **7**, 651–667.

Agrawal, A. A. (2007) Macroevolution of plant defense strategies. *Trends in Ecology and Evolution*, **22**, 103–109.

Agrawal, A. A. and Fishbein, M. (2008) Phylogenetic escalation and decline of plant defense strategies. *Proceedings of the National Academy of Sciences USA*, **105**, 10057–10060.

Agrawal, A. A., Fishbein, M., Halitschke, R. *et al.* (2009) Evidence for adaptive radiation from a phylogenetic study of plant defenses. *Proceedings of the National Academy of Sciences USA*, **106**, 18067–18072.

Anderson, L. L. and Paige, K. N. (2003) Multiple herbivores and coevolutionary interactions in an *Ipomopsis* hybrid swarm. *Evolutionary Ecology*, **17**, 139–156.

Andrew, R. L., Peakall, R., Wallis, I. R. *et al.* (2005) Marker-based quantitative genetics in the wild? The heritability and genetic correlation of chemical defenses in *Eucalyptus*. *Genetics*, **171**, 1989–1998.

Andrew, R. L., Peakall, R., Wallis, I. R. *et al.* (2007a) Spatial distribution of defense chemicals and markers and the maintenance of chemical variation. *Ecology*, **88**, 716–728.

Andrew, R. L., Wallis, I. R., Harwood, C. E. *et al.* (2007b) Heritable variation in the foliar secondary metabolite sideroxylonal in *Eucalyptus* confers cross-resistance to herbivores. *Oecologia*, **153**, 891–901.

Andrew, R. L., Wallis, I. R., Harwood, C. E. *et al.* (2010) Genetic and environmental contributions to variation and population divergence in a broad-spectrum foliar defence of *Eucalyptus tricarpa*. *Annals of Botany*, **105**, 707–717.

Ayres, M. P., Clausen, T. P., MacLean, S. F. *et al.* (1997) Diversity of structure and antiherbivore activity in condensed tannins. *Ecology*, **78**, 1696–1712.

Bailey, J., Schweitzer, J., Rehill, B. *et al.* (2007) Rapid shifts in the chemical composition of aspen forests: an introduced herbivore as an agent of natural selection. *Biological Invasions*, **9**, 715–722.

Barbour, R. C., O'Reilly-Wapstra, J. M., De Little, D. W. *et al.* (2009) A geographic mosaic of genetic variation within a foundation tree species and its community-level consequences. *Ecology*, **90**, 1762–1772.

Barton, A. F. M., Cotterill, P. P. and Brooker, M. I. H. (1991) Short note: heritability of cineole yield in *Eucalyptus kochii*. *Silvae Genetica*, **40**, 37–38.

Barton, K. E. and Koricheva, J. (2010) The ontogeny of plant defense and herbivory: characterizing general patterns using meta-analysis. *American Naturalist*, **175**, 481–493.

Becerra, J. X., Noge, K. and Venable, D. L. (2009) Macroevolutionary chemical escalation in an ancient plant–herbivore arms race. *Proceedings of the National Academy of Sciences USA*, **106**, 18062–18066.

Benkman, C. W. (1999) The selection mosaic and diversifying coevolution between crossbills and Lodgepole pine. *American Naturalist*, **153**, S75–S91.

Boege, K. and Marquis, R. J. (2005) Facing herbivory as you grow up: the ontogeny of resistance in plants. *Trends in Ecology and Evolution*, **20**, 441–448.

Boyle, R. R. and McLean, S. (2004) Constraint of feeding by chronic ingestion of 1,8-cineole in the brushtail possum (*Trichosurus vulpecula*). *Journal of Chemical Ecology*, **30**, 757–775.

Brooker, M. I. H. (2000) A new classification of the genus *Eucalyptus* L'Her. (Myrtaceae). *Australian Systematic Botany*, **13**, 79–148.

Brophy, J. J. and Southwell, I. A. (2002) *Eucalyptus* chemistry. In J. J. W. Coppen (ed.) *Eucalyptus: The Genus Eucalyptus*. London, Taylor & Francis, 102–160.

Bulinski, J. and McArthur, C. (1999) An experimental field study of the effects of mammalian herbivore damage on *Eucalyptus nitens* seedlings. *Forest Ecology and Management*, **113**, 241–249.

Close, D. C. and McArthur, C. (2002) Rethinking the role of many plant phenolics – protection from photodamage not herbivores? *Oikos*, **99**, 166–172.

Close, D. C., McArthur, C., Paterson, S. *et al.* (2003) Photoinhibition: a link between effects of the environment on eucalypt seedling leaf chemistry and herbivory. *Ecology*, **84**, 2952–2966.

Conner, J. K. and Hartl, D. L. (2004) *A Primer of Ecological Genetics*. Sunderland, MA: Sinauer Associates.

Dahlgren, J., Oksanen, L., Olofsson, J. *et al.* (2009) Plant defences at no cost? The recovery of tundra scrubland following heavy grazing by grey-sided voles, *Myodes rufocanus*. *Evolutionary Ecology Research*, **11**, 1205–1216.

De Gabriel, J. L., Moore, B. D., Foley, W. J. *et al.* (2009) The effects of plant defensive chemistry on nutrient availability predict reproductive success in a mammal. *Ecology*, **90**, 711–719.

Dicke, M. (2009) Behavioural and community ecology of plants that cry for help. *Plant, Cell and Environment*, **32**, 654–665.

Donaldson, J. R. and Lindroth, R. L. (2007) Genetics, environment, and their interaction determine efficacy of chemical defense in trembling aspen. *Ecology*, **88**, 729–739.

Dracatos, P. M., Cogan, N. O. I., Sawbridge, T. I. *et al.* (2009) Molecular characterisation and genetic mapping of candidate genes for qualitative disease resistance in perennial ryegrass (*Lolium perenne* L.). *BMC Plant Biology*, **9**, 22.

Dutkowski, G. W. and Potts, B. M. (1999) Geographic patterns of genetic variation in *Eucalyptus globulus* ssp. *globulus* and a revised racial classification. *Australian Journal of Botany*, **47**, 237–263.

Ehrlich, P. R. and Raven, P. H. (1964) Butterflies and plants: a study in coevolution. *Evolution*, **18**, 586–608.

Elliot, H. J. and de Little, D. W. (1985) *Insect Pests of Trees and Timber in Tasmania*. Hobart, Tasmania: Forestry Commission.

Eschler, B. M., Pass, D. M., Willis, R. *et al.* (2000) Distribution of foliar formylated phloroglucinol derivatives amongst *Eucalyptus* species. *Biochemical Systematics and Ecology*, **28**, 813–824.

Falconer, D. S. and Mackay, T. C. (1996) *Introduction to Quantitative Genetics*. Harlow: Pearson Education.

Fraenkel, G. S. (1959) The *raison d'être* of secondary plant substances. *Science*, **129**, 1466–1470.

Freeman, J. S., O'Reilly-Wapstra, J. M.,
Vaillancourt, R. E. *et al.* (2008) Quantitative
trait loci for key defensive compounds
affecting herbivory of eucalypts in Australia.
New Phytologist, **178**, 846–851.

Futuyma, D. J. and Agrawal, A. A. (2009)
Macroevolution and the biological diversity
of plants and herbivores. *Proceedings of the
National Academy of Sciences USA*, **106**,
18054–18061.

Gershenzon, J. and Dudareva, N. (2007) The
function of terpene natural products in the
natural world. *Nature Chemical Biology*, **3**,
408–414.

Gleadow, R. M., Haburjak, J., Dunn, J. E. *et al.*
(2008) Frequency and distribution of
cyanogenic glycosides in *Eucalyptus* L'Herit.
Phytochemistry, **69**, 1870–1874.

Gomez, J. M., Perfectti, F., Bosch, J. *et al.* (2009) A
geographic selection mosaic in a
generalized plant–pollinator–herbivore
system. *Ecological Monographs*, **79**,
245–263.

Goodger, J. Q. D. and Woodrow, I. E. (2002)
Cyanogenic polymorphism as an indicator
of genetic diversity in the rare species
Eucalyptus yarraensis (Myrtaceae). *Functional
Plant Biology*, **29**, 1445–1452.

Gosney, B. (2009) Linking host genetic
variation with dependent communities:
the case of *Eucalyptus globulus*. Unpublished
Masters thesis, Hobart, University of
Tasmania.

Henery, M. L., Moran, G. F., Wallis, I. R. *et al.* (2007)
Identification of quantitative trait loci
influencing foliar concentrations of
terpenes and formylated phloroglucinol
compounds in *Eucalyptus nitens*. *New
Phytologist*, **176**, 82–95.

Hillis, W. E. (1966) Variation in polyphenol
composition within species of *Eucalyptus*
l'Herit. *Phytochemistry*, **5**, 541–556.

Hume, I. D. (1999) *Marsupial Nutrition*.
Cambridge: Cambridge University Press.

Jansson, S. and Douglas, C. J. (2007) *Populus*: A
model system for plant biology. *Annual
Review of Plant Biology*, **58**, 435–458.

Jaya, E., Kubien, D. S., Jameson, P. E. *et al.*
(2010) Vegetative phase change and
photosynthesis in *Eucalyptus occidentalis*:
architectural simplification prolongs
juvenile traits. *Tree Physiology*, **30**,
393–403.

Johnson, M. T. J., Agrawal, A. A., Maron, J. L. *et al.*
(2009) Heritability, covariation and natural
selection on 24 traits of common evening
primrose (*Oenothera biennis*) from a field
experiment. *Journal of Evolutionary Biology*, **22**,
1295–1307.

Jones, T. H., Potts, B. M., Vaillancourt, R. E.
et al. (2002) Genetic resistance of
Eucalyptus globulus to autumn gum moth
defoliation and the role of cuticular
waxes. *Canadian Journal of Forest Research*,
32, 1961–1969.

Jordan, G. J., Potts, B. M., Chalmers, P. *et al.* (2000)
Quantitative genetic evidence that the
timing of vegetative phase change in
Eucalyptus globulus ssp. *globulus* is an
adaptive trait. *Australian Journal of Botany*, **48**,
561–567.

Jordan, G. J., Potts, B. M. and Clarke, A. R. (2002)
Susceptibility of *Eucalyptus globulus* ssp.
globulus to sawfly (*Perga affinis* ssp. *insularis*)
attack and its potential impact on
plantation productivity. *Forest Ecology and
Management*, **160**, 189–199.

Juenger, T. and Bergelson, J. (1998) Pairwise
versus diffuse natural selection and the
multiple herbivores of scarlet gilia, *Ipomopsis
aggregata*. *Evolution*, **52**, 1583–1592.

Koshevoi, O. N., Komissarenko, A. N.,
Kovaleva, A. M. *et al.* (2009) Coumarins from
Eucalyptus viminalis leaves. *Chemistry of Natural
Compounds*, **45**, 532–533.

Kulheim, C., Yeoh, S. H., Maintz, J. *et al.* (2009)
Comparative SNP diversity among four
Eucalyptus species for genes from secondary
metabolite biosynthetic pathways. *BMC
Genomics*, **10**, 11.

Laine, A. L. (2009) Role of coevolution in
generating biological diversity: spatially
divergent selection trajectories. *Journal of
Experimental Botany*, **60**, 2957–2970.

Lankau, R. A. (2007) Specialist and generalist herbivores exert opposing selection on a chemical defense. *New Phytologist*, **175**, 176–184.

Lawler, I. R., Stapley, J., Foley, W. J. *et al.* (1999) Ecological example of conditioned flavor aversion in plant–herbivore interactions: effect of terpenes of *Eucalyptus* leaves on feeding by common ringtail and brushtail possums. *Journal of Chemical Ecology*, **25**, 401–415.

Lawler, I. R., Foley, W. J. and Eschler, B. M. (2000) Foliar concentration of a single toxin creates habitat patchiness for a marsupial folivore. *Ecology*, **81**, 1327–1338.

Leimu, R. and Koricheva, J. (2006) A meta-analysis of genetic correlations between plant resistances to multiple enemies. *American Naturalist*, **168**, E15–E37.

Li, H.-F. and Madden, J. L. (1995) Analysis of leaf oils from a *Eucalyptus* species trial. *Biochemical Systematics and Ecology*, **23**, 167–177.

Li, H.-F., Madden, J. L. and Potts, B. M. (1995) Variation in volatile leaf oils of the Tasmanian *Eucalyptus* species – 1. Subgenus *Monocalyptus*. *Biochemical Systematics and Ecology*, **23**, 299–318.

Loney, P. E., McArthur, C., Sanson, G. *et al.* (2006) How do soil nutrients affect within-plant patterns of herbivory in seedlings of *Eucalyptus nitens*? *Oecologia*, **150**, 409–420.

Majer, J. D., Recher, H. F., Wellington, B. *et al.* (1997) Invertebrates of eucalypt formations. In J. Williams and J. C. Z. Woinarski (eds.) *Eucalypt Ecology. Individuals to Ecosystems*. Cambridge: Cambridge University Press.

Marsh, K. J., Wallis, I. R. and Foley, W. J. (2003) The effect of inactivating tannins on the intake of *Eucalyptus* foliage by a specialist *Eucalyptus* folivore (*Pseudocheirus peregrinus*) and a generalist herbivore (*Trichosurus vulpecula*). *Australian Journal of Zoology*, **51**, 31–42.

Mauricio, R. and Rausher, M. D. (1997) Experimental manipulation of putative selective agents provides evidence for the role of natural enemies in the evolution of plant defence. *Evolution*, **51**, 1435–1444.

Milgate, A. W., Potts, B. M., Joyce, K. *et al.* (2005) Genetic variation in *Eucalyptus globulus* for susceptibility to *Mycosphaerella nubilosa* and its association with tree growth. *Australasian Plant Pathology*, **34**, 11–18.

Miller, A. M., McArthur, C. and Smethurst, P. J. (2006) Characteristics of tree seedlings and neighbouring vegetation have an additive influence on browsing by generalist herbivores. *Forest Ecology and Management*, **228**, 197–205.

Miller, A. M., O'Reilly-Wapstra, J. M., Potts, B. M. *et al.* (2009) Non-lethal strategies to reduce browse damage in eucalypt plantations. *Forest Ecology and Management*, **259**, 45–55.

Moore, B. D. and Foley, W. (2005) Tree use by koalas in a chemically complex landscape. *Nature*, **435**, 488–490.

Moore, B. D. Wallis, I. R., Pala-Paul, J. *et al.* (2004) Antiherbivore chemistry of *Eucalyptus* – cues and deterrents for marsupial folivores. *Journal of Chemical Ecology*, **30**, 1743–1769.

Moore, B. D., Lawler, I. R., Wallis, I. R. *et al.* (2010) Palatability mapping: a koala's eye view of spatial variation in habitat quality. *Ecology*, **91**, 3165–3176.

Neyland, M. (1996) *Tree Decline in Tasmania*. Hobart: Land and Water Management Council, 44.

O'Reilly-Wapstra, J. and Cowan, P. (2010) Native plant/herbivore interactions as determinants of the ecological and evolutionary effects of invasive mammalian herbivores: the case of the common brushtail possum. *Biological Invasions*, **12**, Special Issue, 373–387.

O'Reilly-Wapstra, J. M., McArthur, C. and Potts, B. M. (2002) Genetic variation in resistance of *Eucalyptus globulus* to marsupial browsers. *Oecologia*, **130**, 289–296.

O'Reilly-Wapstra, J. M., McArthur, C. and Potts, B. M. (2004) Linking plant genotype, plant defensive chemistry and mammal browsing in a *Eucalyptus* species. *Functional Ecology*, **18**, 677–684.

O'Reilly-Wapstra, J. M., Potts, B. M., McArthur, C. et al. (2005) Effects of nutrient variability on the genetic-based resistance of *Eucalyptus globulus* to a mammalian herbivore and on plant defensive chemistry. *Oecologia*, **142**, 597–605.

O'Reilly-Wapstra, J., Humphreys, J. and Potts, B. (2007a) Stability of genetic-based defensive chemistry across life stages in a *Eucalyptus* species. *Journal of Chemical Ecology*, **33**, 1876–1884.

O'Reilly-Wapstra, J. M., Iason, G. R. and Thoss, V. (2007b) The role of genetic and chemical variation of *Pinus sylvestris* seedlings in influencing slug herbivory. *Oecologia*, **152**, 82–91.

O'Reilly-Wapstra, J. M., Bailey, J. K., McArthur, C. et al. (2010) Genetic- and chemical-based resistance to two mammalian herbivores varies across the geographic range of *Eucalyptus globulus*. *Evolutionary Ecology Research*, **12**, 1–16.

O'Reilly-Wapstra, J. M., Freeman, J. S., Davies, N. W. et al. (2011) Quantitative trait loci for foliar terpenes in a global eucalypt species. *Tree Genetics and Genomes*, **7**, 485–498.

Osier, T. L. and Lindroth, R. L. (2001) Effects of genotype, nutrient availability, and defoliation on aspen phytochemistry and insect performance. *Journal of Chemical Ecology*, **27**, 1289–1313.

Parker, J. D., Salminen, J. P. and Agrawal, A. A. (2010) Herbivory enhances positive effects of plant genotypic diversity. *Ecology Letters*, **13**, 553–563.

Pass, G. J. and Foley, W. J. (2000) Plant secondary metabolites as mammalian feeding deterrents: separating the effects of the taste of salicin from its post-ingestive consequences in the common brushtail possum (*Trichosurus vulpecula*). *Journal of Comparative Physiology B*, **170**, 185–192.

Potts, B. M. and Jordan, G. J. (1994) The spatial pattern and scale of variation in *Eucalyptus globulus* ssp. *globulus*: variation in seedling abnormalities and early growth. *Australian Journal of Botany*, **42**, 471–492.

Potts, B. M. and Pederick, L. A. (2000) Morphology, phylogeny, origin, distribution and genetic diversity of the eucalypts. In P. J. Keane, G. A. Kile, F. D. Podger and B. N. Brown (eds.) *Diseases and Pathogens of Eucalypts*. Collingwood: CSIRO Publishing, 11–34.

Potts, B. M., Vaillancourt, R. E., Jordan, G. J. et al. (2004) Exploration of the *Eucalyptus globulus* gene pool. In N. M. G. Borralho, J. S. Pereira, C. Marques et al. (eds.) *Eucalyptus in a Changing World*. Proceedings of IUFRO Conference, Aveiro, 11–15 October. Portugal: RAIZ Instituto Investigação da Floresta e Papel, 46–61.

Ralph, S. G. (2009) Studying *Populus* defences against insect herbivores in the post-genomic era. *Critical Reviews in Plant Science*, **28**, 335–345.

Rapley, L., Allen, G. R. and Potts, B. M. (2004a) Genetic variation of *Eucalyptus globulus* in relation to autumn gum moth *Mnesampela privata* (Lepidoptera: Geometridae) oviposition preference. *Forest Ecology and Management*, **194**, 169–175.

Rapley, L., Allen, G. R. and Potts, B. M. (2004b) Susceptibility of *Eucalyptus globulus* to *Mnesampela privata* defoliation in relation to a specific foliar wax compound. *Chemoecology*, **14**, 157–163.

Rapley, L. P., Allen, G. R., Potts, B. M. et al. (2007) Constitutive or induced defences – how does *Eucalyptus globulus* defend itself from larval feeding? *Chemoecology*, **17**, 235–243.

Rapley, L. P., Potts, B. M., Battaglia, M. et al. (2009) Long-term realised and projected growth impacts caused by autumn gum moth defoliation of 2-year-old *Eucalyptus nitens* plantation trees in Tasmania, Australia. *Forest Ecology and Management*, **258**, 1896–1903.

Rasmann, S. and Agrawal, A. A. (2009) Plant defense against herbivory: progress in identifying synergism, redundancy, and antagonism between resistance traits. *Current Opinion in Plant Biology*, **12**, 473–478.

Rausher, M. D. (1993) Patterns of selection on phytophage resistance in *Ipomoea purpurea*. *Evolution*, **47**, 970–976.

Rausher, M. D. (2001) Co-evolution and plant resistance to natural enemies. *Nature*, **411**, 857–864.

Rausher, M. D. and Simms, E. L. (1989) The evolution of resistance to herbivory in *Ipomoea purpurea*. I. Attempts to detect selection. *Evolution*, **43**, 563–572.

Romeo, J. T., Saunders, J. A. and Barbosa, P. (1996) *Phytochemical Diversity and Redundancy in Ecological Interactions*. New York: Plenum Press.

Sampaio, M. B. and Scariot, A. (2010) Effects of stochastic herbivory events on population maintenance of an understorey palm species (*Geonoma schottiana*) in riparian tropical forest. *Journal of Tropical Ecology*, **26**, 151–161.

Sansaloni, C., Petroli, C., Carling, J. *et al.* (2010) A high-density Diversity Arrays Technology (DArT) microarray for genome-wide genotyping in Eucalyptus. *Plant Methods*, **6**, 16.

Sewell, M. M. and Neale, D. B. (2000) Mapping quantitative traits in forest trees. In S. M. Jain and S. C. Minocha (eds.) *Molecular Biology of Woody Plants* Volume 1. The Netherlands: Kluwer Academic Publishers, 407–423.

Simms, E. L. and Rausher, M. D. (1989) The evolution of resistance to herbivory in *Ipomoea purpurea*. II. Natural selection by insects and costs of resistance. *Evolution*, **43**, 573–585.

Simms, E. L. and Rausher, M. D. (1992) Uses of quantitative genetics for studying the evolution of plant resistance. In R. S. Fritz and E. L. Simms (eds.) *Plant Resistance to Herbivores and Pathogens. Ecology, Evolution and Genetics*. Chicago, IL: University of Chicago Press, 42–68.

Steane, D. A., Conod, N., Jones, R. C. *et al.* (2006) A comparative analysis of population structure of a forest tree, *Eucalyptus globulus* (Myrtaceae), using microsatellite markers and quantitative traits. *Tree Genetics and Genomes*, **2**, 30–38.

Steinbauer, M. J. (2010) Latitudinal trends in foliar oils of eucalypts: environmental correlates and diversity of chrysomelid leaf-beetles. *Austral Ecology*, **35**, 205–214.

Stinchcombe, J. R. (2005) Measuring natural selection on proportional traits: comparisons of three types of selection estimates for resistance and susceptibility to herbivore damage. *Evolutionary Ecology*, **19**, 363–373.

Stinchcombe, J. R. and Rausher, M. D. (2001) Diffuse selection on resistance to deer herbivory in the ivy leaf morning glory, *Ipomoea hederacea*. *American Naturalist*, **158**, 376–388.

Strauss, S. Y. and Irwin, R. E. (2004) Ecological and evolutionary consequences of multispecies plant–animal interactions. *Annual Review of Ecology and Systematics*, **35**, 435–466.

Theis, N. and Lerdau, M. (2003) The evolution of function in plant secondary metabolites. *International Journal of Plant Sciences*, **164**, S93–S102.

Thompson, J. N. (2005) *The Geographic Mosaic of Coevolution*. Chicago, IL: University of Chicago Press.

Wallis, I. R., Watson, M. L. and Foley, W. J. (2002) Secondary metabolites in *Eucalyptus melliodora*: field distribution and laboratory feeding choices by a generalist herbivore, the common brushtail possum. *Australian Journal of Zoology*, **50**, 507–519.

Wardle, D. A. (2006) The influence of biotic interactions on soil biodiversity. *Ecology Letters*, **9**, 870–886.

Wheat, C. W., Vogel, H., Wittstock, U. *et al.* (2007) The genetic basis of a plant–insect coevolutionary key innovation. *Proceedings of the National Academy of Sciences USA*, **104**, 20427–20431.

Whitham, T. G., DiFazio, S. P., Schweitzer, J. A. *et al.* (2008) Extending genomics to natural communities and ecosystems. *Science*, **320**, 492–495.

Whittock, S. P., Apiolaza, L. A., Kelly, C. M. *et al.* (2003) Genetic control of coppice and lignotuber development in *Eucalyptus globulus*. *Australian Journal of Botany*, **51**, 57–67.

Wiggins, N. L., McArthur, C., McLean, S. *et al.* (2003) Effects of two plant secondary metabolites, cineole and gallic acid, on nightly feeding patterns of the common brushtail possum. *Journal of Chemical Ecology*, **29**, 1447–1464.

Wiggins, N. L., Marsh, K. J., Wallis, I. R. *et al.* (2006) Sideroxylonal in *Eucalyptus* foliage influences foraging behaviour of an arboreal folivore. *Oecologia*, **147**, 272–279.

Williams, J. E. and Brooker, M. I. H. (1997) Eucalypts: an introduction. In J. E. Williams and J. C. Z. Woinarski (eds.) *Eucalypt Ecology: Individuals to Ecosystems*. Cambridge: Cambridge University Press, 278–302.

Wise, M. J. (2009) Competition among herbivores of *Solanum carolinense* as a constraint on the evolution of host-plant resistance. *Evolutionary Ecology*, **23**, 347–361.

Woinarski, J. C. Z., Recher, H. F. and Majer, J. D. (1997) Vertebrates of eucalypt formations. In J. E. Williams and J. C. Z. Woinarski (eds.) *Eucalypt Ecology. Individuals to Ecosystems*. Cambridge: Cambridge University Press, 303–341.

CHAPTER THREE

Temporal changes in plant secondary metabolite production: patterns, causes and consequences

JULIA KORICHEVA

School of Biological Sciences, Royal Holloway, University of London

KASEY E. BARTON

Botany Department, University of Hawaii at Mānoa

3.1   Introduction

Both overall concentrations and composition of plant secondary metabolites (PSMs) may vary strongly not only among different plant species or different plant individuals within a species, but also within a single individual plant over time. We distinguish two types of temporal changes in plants: *ontogenetic* and *seasonal*. Ontogeny refers to genetically programmed developmental changes that take place during a plant's life; the terms 'maturation' and 'phase change' are also sometimes used to describe the same process (Kozlowski, 1971; Poethig, 1990). Commonly recognised ontogenetic stages in plants include seedlings, juvenile plants (saplings) and mature plants. Each stage is characterised by distinctive anatomical, morphological and biochemical features (e.g. dependence on seed reserves in seedlings, and ability to flower and set seeds in mature plants). Since plants are modular organisms composed of numerous semi-autonomous units (e.g. branches and leaves), in addition to ontogenetic changes at the whole-plant level, each plant module/organ also experiences ontogenetic changes as it develops and matures. For instance, leaves formed during a specific phase of shoot growth retain morphological and physiological features characteristic of that phase, resulting in variation in the character of structures along the axis of the shoot; this phenomenon is known as heteroblasty (Poethig, 1990; O'Reilly-Wapstra *et al.*, Chapter 2). While these within-module ontogenetic changes are genetically programmed just as whole-plant ontogenetic changes are, they are not necessarily synchronised with the latter because new modules are produced continuously through the plant's lifetime. As a result, long-lived plants like trees may represent a complex mosaic or a gradient of plant tissues at different ontogenetic phases of development (sometimes referred to as a 'developmental stream') (Kearsley & Whitham, 1998).

The Ecology of Plant Secondary Metabolites: From Genes to Global Processes, eds. Glenn R. Iason, Marcel Dicke and Susan E. Hartley. Published by Cambridge University Press. © British Ecological Society 2012.

In addition to ontogenetic changes, in seasonal environments, plant modules experience a series of physiological changes over the season caused by changes in temperature and the availability of light (photoperiod), water and nutrients. We refer to these changes as seasonal changes. In annual and short-lived herbaceous plants, seasonal and ontogenetic changes are confounded, but in long-lived woody plants like trees they can be distinguished because both juvenile and mature ontogenetic stages last many seasons, and within-plant ontogenetic changes such as leaf expansion usually occur within a relatively short time interval during the season (e.g. a few weeks in spring in temperate ecosystems). Both ontogenetic and seasonal changes may affect PSM production in plants. In this chapter we review the patterns of ontogenetic and seasonal changes in constitutive and induced PSMs and discuss their proximate (mechanistic) and ultimate (evolutionary) causes as well as their ecological consequences for other organisms and ecosystem processes (Schuman & Baldwin, Chapter 15). The majority of studies on the topic have been conducted on foliar chemistry, and therefore our review will focus largely on temporal changes in PSM production in leaves. Other types of temporal changes in PSM production (e.g. diurnal and nocturnal changes and circadian rhythmicity in floral scent emission) are outside the scope of this chapter.

3.2   Temporal patterns in PSM concentrations

In order to reveal general patterns of temporal changes in PSMs, we have conducted meta-analyses of published literature on whole-plant ontogenetic changes in constitutive and induced PSMs, and on changes in constitutive PSM concentrations during leaf maturation and through the season. Detailed description of the methods is provided in Barton and Koricheva (2010). In brief, relevant studies were identified via keyword searches in the Web of Science citation index, data from individual studies were converted into effect size metrics and meta-analyses were conducted using the MetaWin 2.1 statistical software (Rosenberg *et al.*, 2000). For constitutive PSMs, the size of the effect was calculated as Hedges' d (Gurevitch & Hedges, 2001), the standardised mean difference between mean PSM concentrations in the older and the younger ontogenetic or seasonal stages. Positive effects indicate higher concentrations of PSMs in tissues of the older plants. The analyses of ontogenetic patterns in induced defence were conducted on Hedges' d values calculated from means, standard deviations and sample sizes of four groups (herbivore treatment and control groups for both older and younger plants) as demonstrated in Gurevitch *et al.* (2000). Positive effects indicate greater induced increases in PSMs in older plants. Bias-corrected bootstrap 95% confidence intervals around the effect size were generated from 4999 iterations (Adams *et al.*, 1997), and effects were considered to be significant when 95%

confidence intervals did not include 0. For those mean-effect sizes that were significantly different from 0, we calculated a fail-safe number (n_{fs}) using the weighted method of Rosenberg (2005). This number represents how many additional studies of null effect and mean weight would need to be added to reduce the significance level of the observed mean effect to 0.05. In general, a fail-safe number greater than $5n + 10$, where n is the original number of studies in the analysis, is considered robust against publication bias (Rosenthal, 1991).

The first meta-analysis reviewed whole-plant ontogenetic changes in constitutive and induced PSMs, and included data from 67 studies published in 1969–2009 and conducted on 63 plant species from 19 plant families ($n = 156$) (Barton & Koricheva, 2010). Four types of comparisons of PSM concentrations were included: in young versus old seedlings, in seedlings versus juveniles or young juvenile versus older juvenile, in juvenile versus mature plants and in early versus late mature plants. The results of this analysis revealed that constitutive concentrations of all types of PSM increase significantly through the seedling stage in both herbaceous and woody plants (Figure 3.1). Subsequent whole-plant ontogenetic changes in herbaceous plants depended on the type of plant and type of PSM. In herbs, constitutive concentrations of alkaloids and phenolics did not change significantly through juvenile and mature stages whereas concentrations of terpenoids increased through the juvenile stage and concentrations of defensive proteins increased through juvenile–mature transition, but decreased in mature plants (Figure 3.1a). In woody plants, no significant changes in constitutive concentrations of any type of PSM occurred through juvenile stage and juvenile–mature transition whereas phenolic concentrations declined during the mature stage (Figure 3.1b). We also examined ontogenetic changes in the induction of PSMs by herbivores. Most of these studies have been conducted on herbaceous plants, and juvenile herbs showed significantly higher induction of PSMs than mature plants (Figure 3.2). In contrast, the only study that investigated ontogenetic patterns in PSM induction in woody plants (Boege, 2005) found that mature plants show higher induction than juveniles. Low sample sizes prevented us from investigating whether ontogenetic patterns in induction vary among classes of PSMs.

We have also conducted two new meta-analyses on changes in constitutive PSM concentrations during leaf maturation and through the season. Because it is difficult to separate these developmental processes from whole-plant ontogeny in short-lived herbaceous species, these analyses were restricted to studies on long-lived perennials (woody plants) only. The leaf development meta-analysis included data from 30 studies published in 1980–2009 and conducted on 94 woody plant species from 38 plant families ($n = 130$). In this analysis, PSM concentrations in young expanding leaves were compared with

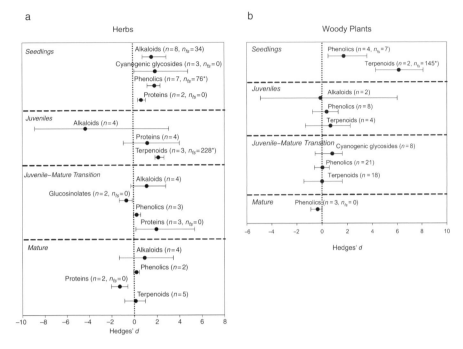

Figure 3.1 Whole-plant ontogenetic changes in constitutive PSM concentrations in herbaceous species (a) and woody plant species (b) (modified from Barton & Koricheva, 2010). Hedges' d is the standardised mean difference between PSM concentrations in ontogenetically older and younger plants; positive effects indicate higher concentrations of PSMs in older plants. Error bars are 95% bias-corrected confidence intervals; significant patterns occur when the confidence intervals fail to include zero, which is marked with a dotted line. n, sample sizes (number of comparisons); n_{fs}, fail-safe numbers. Asterisks indicate that the fail-safe number is robust to publication bias ($>5n+10$).

those in mature fully expanded leaves. For most classes of PSM, including total phenolics, flavonoids, hydrolysable tannins, phenolic glycosides, defensive proteins and monoterpenes, concentrations were higher in young leaves than in fully expanded mature leaves (Figure 3.3). Concentrations of cyanogenic glycosides, condensed tannins and lignin did not differ between the young and mature leaves. The only PSMs which were found to increase significantly during leaf development were sesquiterpenes.

The meta-analysis of seasonal changes in constitutive concentrations of PSMs in leaves included data from 42 studies published in 1970–2009 and conducted on 44 plant species from 20 plant families ($n=119$). Studies often measured PSM concentrations more than twice during the growing season, and we calculated the effect size to be the greatest magnitude of change across a season. In species with synchronous leaf development, such

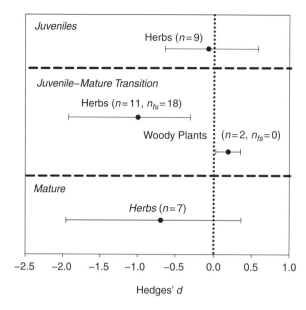

Figure 3.2 Whole-plant ontogenetic patterns in the induction of PSMs in herbs and woody plant species. Positive effects indicate greater induced increases in PSMs in older plants. Data are expressed as in Figure 3.1.

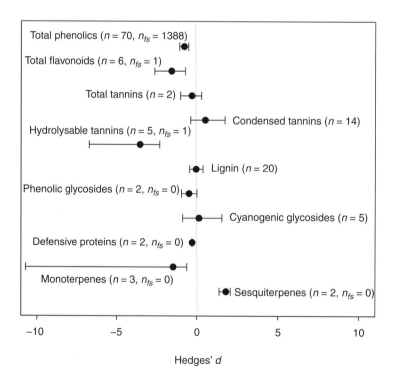

Figure 3.3 Changes in constitutive PSM concentrations during leaf development in woody plants. Data are expressed as in Figure 3.1

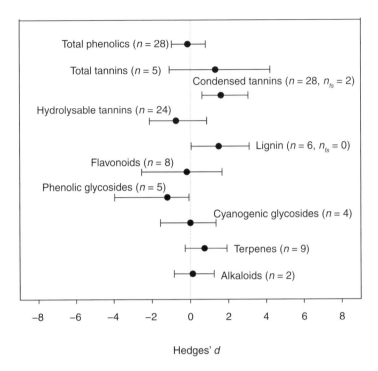

Figure 3.4 Changes in constitutive PSM concentrations during the growing season in woody plants. Data are expressed as in Figure 3.1.

as spring-flushing temperate trees, seasonality and leaf ontogeny are confounded, so whenever possible, comparisons of mature leaves at different times of the season were analysed. This meta-analysis revealed quite different patterns from the one on leaf maturation (Figure 3.4). Concentrations of most classes of PSM showed no significant changes over time. Exceptions included significant increases in condensed tannins and lignin, and significant decreases in phenolic glycosides. Interestingly, temporal patterns differed between different groups of phenolics. Total phenolics and hydrolysable tannins generally decrease during leaf development (Figure 3.3) but show no significant trend across the growing season (Figure 3.4). In contrast, condensed tannins had no overall pattern during leaf development (Figure 3.3) but significantly increased across the growing season (Figure 3.4).

In general, it appears that for most types of PSM, concentrations decrease rapidly during leaf maturation, but then remain relatively stable in mature leaves through the season, although seasonal changes in composition of various types of PSM may occur (e.g. Salminen *et al.*, 2001, 2004).

3.3   Proximate causes of temporal patterns in PSM concentrations

The accumulation of any PSM in plant tissues is controlled by the balance between the rates of its gain and loss. For instance, in *Mentha* × *piperita* (Lamiaceae), a model system for the study of monoterpene metabolism, the rate of synthesis was shown to be the principal factor explaining high mono-terpene concentrations in young leaves, whereas rates of monoterpene losses through volatilisation and catabolism were negligible (Gershenzon *et al.*, 2000). By using $^{14}CO_2$ labelling, the above study showed that monoterpene biosynthesis in *M. piperita* is restricted to a brief period early in leaf develop-ment, and synthesised monoterpenes are stored in leaves with monoterpene content remaining constant for 40 days. Activities of the enzymes responsible for biosynthesis of *M. piperita*'s monoterpenes and the expression of corre-sponding structural genes had very similar developmental profiles (McConkey *et al.*, 2000), suggesting that the developmental biosynthesis of monoterpenes is regulated largely at the level of gene expression.

If biosynthesis of PSMs occurs early in leaf development and stops before leaf expansion is completed, passive dilution of concentrations of PSMs by accumulation of other plant metabolites such as carbohydrates and proteins may occur (Koricheva, 1999). For instance, in leaves of a tropical tree *Inga umbellifera* (Fabaceae) flavonoids are synthesised only until an intermediate expansion stage and, as leaf biomass continues to increase until leaf matura-tion, significant dilution of flavonoid concentrations occurs while total con-tent of flavonoids per leaf remains the same (Brenes-Arguedas *et al.*, 2006). In another *Inga* species, *I. goldmanii*, synthesis of flavonoids continues through-out leaf development as indicated by a continuous increase in flavonoid total content, and although growth dilution of flavonoid concentrations still occurs, it is much less pronounced than in *I. umbellifera* (Brenes-Arguedas *et al.*, 2006).

Sometimes temporal changes in PSM concentrations may be due to PSM transport from one plant tissue or organ to another. For instance, a decrease in the concentrations of pyrrolizidine alkaloids in older leaves of *Cynoglossum officinale* (Boraginaceae) was shown to be due not to catabolism, but to trans-location of alkaloids from ageing leaves to young newly formed ones (van Dam *et al.*, 1995). In *Senecio* species (Asteraceae), pyrrolizidine alkaloids are synthesised in roots in the form of senecionine N-oxide; during plant ontog-eny this compound is exported into aboveground organs, structurally diversi-fied and slowly allocated between tissues yielding the species-specific alkaloid bouquets (Hartmann & Dierich, 1998; Hartmann, 2007).

Temporal changes in concentrations of PSMs which are synthesised and stored in specialised plant structures such as glandular trichomes or resin ducts may be directly related to the ontogeny of those organs. For instance, cells of leaf glandular trichomes in *Betula* species (Betulaceae) differentiate before the leaf epidermis and produce PSMs (e.g. flavonoid aglycones) only during the relatively short period of leaf unfolding and expansion (Valkama *et al.*, 2004). In fully expanded leaves, glandular trichomes are in the post-secretory phase and function mainly as storage organs. Similarly, mono-terpene synthesis in *Mentha × piperita* is localised to the secretory cells of leaf glandular trichomes. However, because gland development takes about 60 hours in total (Turner *et al.*, 2000), monoterpene synthesis in *M. piperita* begins only after leaves are already a few days old (Gershenzon *et al.*, 2000). PSM types which do not require special synthesis or storage organs can be synthesised without a time lag. For instance, pyrrolizidine alkaloids in *Senecio* roots are synthesised from the onset of seedling growth (Schaffner *et al.*, 2003).

PSMs may be lost from plant tissues through volatilisation or leaching. Emission rates of volatile PSMs such as monoterpenes may exhibit pro-nounced seasonal changes, which are largely driven by changes in temper-ature (Lerdau *et al.*, 1995; Peñuelas & Llusia, 1997). Volatilisation may also contribute to changes in the relative proportions of individual compounds within plant tissues, as the more volatile PSMs are more abundant in emis-sions than in the total plant content (Peñuelas & Llusia, 1997). Leaching rates of PSM may also change through the season. For instance, the dehydrostilbene batatasin-III, the major allelopathic compound of the dwarf shrub *Empetrum hermaphroditum* (Ericaceae), is continuously leached by rain and dew from foliage into the soil, and its concentrations in the through-fall are highest in August, indicating increased leaching from foliage at the end of the growing season (Wallstedt *et al.*, 2000).

Finally, concentrations of PSMs may decrease with time if compounds undergo catabolism, polymerisation or insolubilisation. In general, rapid turn-over and degradation are not common for plant PSMs (Mihaliak *et al.*, 1991; Gershenzon *et al.*, 1993; Baldwin *et al.*, 1994; Hartmann & Dierich, 1998; Ruuhola & Julkunen-Tiitto, 2000). However, rapid turnover and polymerisation of phlorotannins have been observed in marine brown algae (Arnold & Targett, 1998, 2000), and in birch an increase in the formation of insoluble ellagitannins has been reported through the growing season (Salminen, 2002).

Not just the total content of PSMs, but also their composition may change through leaf development (Gershenzon *et al.*, 2000; Brenes-Arguedas *et al.*, 2006), shoot development (Brilli *et al.*, 2009) or through the season (Salminen *et al.* 2001, 2004), and sometimes these changes are also due to subsequent transformation of PSMs. If each of the synthesised compounds in the metabolic pathway serves as a precursor for the following one, then each individual compound occurs only during the specific developmental stage and is replaced later by the next compound in the pathway, as has been observed for gallotannins in *Betula* leaves through the season (Salminen *et al.*, 2001). On the other hand, if each of the synthesised compounds is the end-product which does not undergo further transformation or rapid turnover, then the resulting temporal pattern is diversification of PSMs, as has been observed for pyrrolizidine alkaloids in *Senecio* (Hartmann & Dierich, 1998).

3.4   Ultimate causes of temporal changes in PSM production

While the proximate causes of temporal changes in PSMs are relatively well understood, ultimate evolutionary causes behind those changes are still heavily debated. Two types of hypotheses have been put forward to explain the evolution of patterns of seasonal and ontogenetic variation in PSM production. The first type of hypothesis considers the above patterns to be by-products of resource allocation constraints. These resource-based theories emphasise that it might be metabolically more costly to defend young plant tissues. In developing tissues, PSM production can be constrained by the lack of enzymatic machinery for their synthesis and specialised structures required for compartmentalisation of PSMs. The growth–differentiation balance hypothesis (Herms & Mattson, 1992) predicts therefore that there is a physiological trade-off between growth and differentiation processes which limit production of PSMs in meristematic cells. However, Herms and Mattson (1992) point out that this ontogenetic constraint on PSM production can be circumvented by secretion of stored PSMs over newly expanding foliage or precocious maturation of cells specialised in synthesis and accumulation of PSMs, as has been observed in leaf glandular trichomes in *Betula* (Valkama *et al.*, 2004). Another resource-based hypothesis, the protein competition model of phenolic allocation (Jones & Hartley, 1999), is based on the assumption that protein and phenolic syntheses compete for the common precursor phenylalanine. The 'developmental variant' of the above model predicts that through leaf development, phenolic concentrations should be low during the early phases of primordial cell division and leaf expansion because total protein demands are highest during these periods. The results of our meta-analyses provide little support for the growth–differentiation balance hypothesis and the protein competition model because the majority of PSM types were at highest concentrations in young expanding leaves (Figure 3.3). In cases where PSM concentrations decreased during leaf expansion, the most likely cause was growth dilution rather than allocation constraints (Koricheva, 1999; Brenes-Arguedas *et al.*, 2006).

With respect to the whole-plant ontogenetic patterns of PSM production, Boege and Marquis (2005) proposed a model of ontogenetic changes in defence based on the assumption that expression of defence is mainly constrained by resources. They predicted that at the cotyledon stage, young seedlings are relatively well defended against herbivores because they have stored resources in seeds and cotyledons. But once these reserves are depleted and plants enter the true-leaf seedling stage, they should have low defence concentrations because large root:shoot ratios do not enable them to accumulate resources for functions other than growth. Results of our meta-analyses (Barton & Koricheva, 2010) do not support this prediction because a rapid increase in all types of PSMs in seedlings was one of the most pronounced patterns (Figure 3.1).

The second type of hypothesis explaining evolution of temporal changes in PSMs regards them as adaptive responses to selection pressure imposed by herbivores. For instance, Feeny (1976) hypothesised that the type and amount of PSMs produced by plants depends on the degree of plant apparency – that is, the likelihood of its discovery by natural enemies such as herbivores. According to the apparency hypothesis, unapparent plants (e.g. short-lived herbaceous species) are less likely to be discovered by herbivores and therefore allocate fewer resources to chemical defences. The most effective defensive strategy in this case is to possess PSMs which are active in small concentrations as toxins against non-adapted enemies. In contrast, apparent plants like trees are 'bound to be found' by herbivores and allocate larger proportions of their resources to defence. Those plants are likely to have PSMs that are effective against both generalist and specialist herbivores and are not readily susceptible to counter-adaptation. Rhoades and Cates (1976) extended Feeny's apparency hypothesis to seasonal distribution of PSMs and predicted that quantitative chemical defences (digestibility-reducing compounds such as condensed tannins) will occur in higher concentrations in apparent/predictable plant tissues such as mature leaves whereas qualitative defences effective at low concentrations (e.g. toxins like alkaloids) will prevail in less apparent and more ephemeral tissues such as young leaves. The results of our meta-analysis partly support these predictions. For instance, we found that while the majority of PSMs occurred at higher concentrations in young developing leaves (Figure 3.3), concentrations of condensed tannins (quantitative defences *sensu* Feeny) were higher in mature leaves late in the season (Figure 3.4). The apparency hypothesis can also be extended to changes in PSM concentrations through plant ontogeny. Seedlings can be considered as a less apparent ontogenetic stage as compared to saplings and mature plants. Therefore, one would predict an increase in plant allocation to defence through ontogeny, which was indeed the pattern that we have observed (Figures 3.1 and 3.5).

The optimal defence theory (Rhoades, 1979), which predicts that plants allocate defences in direct proportion to the risk from herbivores and the value of the particular tissue in terms of plant fitness, and in inverse proportion to the cost of defence, can also be extended to predict temporal changes in PSM production. For instance, the value of a leaf for a plant is highest when it is young because of greater future return on photo-assimilates (Harper,

1989), and thus it can be predicted that plants should preferentially allocate PSMs into younger leaves rather than older leaves (Iwasa *et al.*, 1996). The results of our meta-analysis on changes in PSMs with leaf age support this prediction because the majority of PSMs occurred at higher concentrations in younger leaves (Figure 3.3).

Bryant *et al.* (1992) suggested that higher concentrations of PSMs in juvenile plants in boreal forests have evolved in response to browsing by mammals. While mammalian browsing is an important source of mortality for tree seedlings and saplings, most mammalian herbivores in boreal forests (moose, deer, hare, voles) cannot climb the trees and hence mature trees escape from their herbivory. Therefore, boreal trees may relax their defences against mammalian browsing through ontogeny. On the other hand, many mammalian herbivores of temperate and tropical ecosystems are either good climbers (e.g. arboreal primates and marsupials) or have other adaptations to reach branches of tall trees (elephant's trunk, giraffe's neck). Therefore, one would not expect a reduction in defences to mammalian herbivores in tropical trees through ontogeny. The results of our meta-analysis support this prediction. We have found that PSM concentrations were higher in juvenile boreal trees than in mature boreal trees, but there were no significant changes in PSM concentrations in temperate and tropical trees through the juvenile–mature transition (Barton & Koricheva, 2010).

Another adaptive hypothesis, the host heterogeneity hypothesis (Whitham, 1983; Whitham *et al.*, 1984), considers temporal and spatial variation in foliage quality within plants as an important defence mechanism in itself, which makes it more difficult for herbivores to adapt to resistance mechanisms of plants. This hypothesis may explain why short-lived and rapidly evolving pests and pathogens are unable to breach the resistance of long-lived hosts such as trees (Whitham & Schweitzer, 2002; Ruusila *et al.*, 2005). However, in the case of host heterogeneity, it is important to consider that not just PSMs, but also nutrients (water, nitrogen, sugars) vary within a host, contributing to patterns in herbivore preference and performance (Haukioja, 2003). Moreover, this defence strategy may not work when herbivores can increase performance through dietary mixing of different-age foliage, presumably by diluting harmful PSMs and/or increasing nutrient uptake, as has been shown for a generalist herbivore *Orygia leucostigma* (Lepidoptera: Lymantriidae) feeding on *Abies balsamea* (Johns *et al.*, 2009).

3.5 Ecological consequences of temporal and ontogenetic changes in PSM production
3.5.1 Effects on herbivores
Whole-plant ontogenetic changes in PSMs affect the amount of herbivory received by plants. Our meta-analysis (Barton & Koricheva, 2010) has shown

good correspondence between plant defence and herbivory patterns during the seedling stage: an increase in plant defences was accompanied by a decrease in herbivory, which was particularly pronounced in herbs. During the juvenile stage, there was also a good correspondence between the increase in chemical defences and the significant decrease in herbivory in herbs. However, no clear patterns in herbivory during the juvenile stage were found in woody plants or grasses. During the juvenile–mature plant transition, an increase in most types of PSMs was observed in herbs (Figure 3.1), and herbivores (largely molluscs) showed significant preference for juveniles over mature plants. In contrast, mammalian herbivores feeding on boreal trees showed significant preference for mature over juvenile plant tissues (Swihart & Bryant, 2001), presumably because defences in boreal trees have decreased during the juvenile–mature plant transition (Barton & Koricheva, 2010). Temporal changes in PSM concentrations through leaf development and through season also affect herbivore performance. For instance, Ruusila *et al.* (2005) have shown that temporal changes in the quality of growing *Betula* leaves, including marked changes in PSM levels, explained a higher proportion of variation in the consumption and growth of *Epirrita autumnata* (Lepidoptera: Geometridae) larvae than did the identity of the host tree.

Temporal changes in PSM production may also affect the timing of herbivore feeding. In his seminal paper, Feeny (1970) suggested that an increase in the concentrations of tannins in *Quercus robur* (Fagaceae) leaves through the season, together with an increase in leaf toughness and a decrease in water and nitrogen content, cause feeding by Lepidoptera on oaks to be concentrated in the spring when oak leaves are still young and of higher nutritional quality. A similar phenomenon has been described for herbivory on shade-tolerant tree species in humid tropics: although leaves of tropical trees live for several years, most of the herbivore damage occurs during the short period of leaf expansion, and rapid leaf toughening due to the accumulation of cellulose or lignin and other PSMs has been identified as the main cause of the decrease in herbivore damage on mature leaves (Coley, 1983; Coley & Kursar, 1996; Kursar & Coley, 2003). Therefore, changes in PSM production during leaf development may determine timing of herbivore feeding in both temperate and tropical ecosystems. The difference is that in temperate and boreal ecosystems and in tree species with one flush of leaves per year, rapid changes in PSM concentrations during leaf development narrow the window of opportunity for most herbivores to only a few weeks. This in turn can constrain seasonal phenology of herbivores. For instance, the late seasonal appearance of leaf-miners on oak has been attributed to a high abundance of externally feeding caterpillars in spring and the resulting asymmetrical competition between caterpillars and leaf-miners (West, 1985). In the tropics, where new leaves are typically produced year-round, herbivory on young leaves may not be constrained to a particular time of year.

It is difficult, however, to attribute the timing of herbivore feeding to seasonal changes in PSMs alone because many other plant parameters are also changing through the season and with leaf development (Haukioja, 2003). To address this problem, Roslin and Salminen (2009) used an artificial diet with added leaf powder to control for changes in physical factors such as leaf toughness and water content as well as leaf size. They showed that the variation in chemical traits alone suffices to create differences in growth rates between larvae fed on young and mature oak foliage. Still, it remains to be shown whether it is the variation in PSMs or in nutrients like amino acids and sugars (or both) that drives differences in herbivore performance. Several lines of evidence suggest that timing of herbivore feeding is associated more with changes in nutrients than in changes in PSM levels. For instance, the results of our meta-analysis showed that young leaves contain higher levels of most types of PSM than older leaves (Figure 3.3), which suggests that the high prevalence of herbivore feeding on young leaves cannot be explained by PSM levels. Furthermore, a more recent study of *Q. robur* foliar chemistry by Salminen *et al.* (2004) has shown that, contrary to Feeny's original results, hydrolysable rather than condensed tannins represent the main fraction of oak tannins and their concentrations are higher in early summer. Another recent study by the same group has shown that vescalagin (a quantitatively dominant hydrolysable tannin of *Q. robur*) had a much stronger negative effect on generalist oak herbivores than condensed tannins (Roslin & Salminen, 2008). If hydrolysable tannins are the dominant anti-herbivore PSMs in oak, and their concentrations are highest in spring, then widespread spring feeding among oak herbivores cannot be attributed to seasonal changes in tannins and is more likely to be due to changes in nutrients or leaf toughness. Finally, seasonal patterns in hydrolysable tannin concentrations differ among tree species, with some species showing seasonal increases (Schroeder, 1986; Yarnes *et al.*, 2008), and others showing seasonal decreases (Faeth, 1986; Baldwin *et al.*, 1987; Rossiter *et al.*, 1988). It remains to be shown whether seasonal patterns of herbivory on these tree species correspond to the seasonal patterns in PSMs.

3.5.2   Effects on natural enemies of herbivores

Volatile PSMs emitted from vegetative tissues act as part of a plant defence system and may directly repel herbivores or act as indirect defences by attracting natural enemies of herbivores (Kant *et al.*, 2009; Dicke *et al.*, Chapter 16). Both leaf age and plant ontogenetic stage have been shown to affect the systemic release of volatiles by herbivore-infested plants. For instance, *Glycine max* (Fabaceae) plants in the vegetative stage emitted 10-

fold more volatiles per biomass than reproductive plants and young leaves emitted over 2.6 times more volatiles than old leaves (Rostas & Eggert, 2008). Such ontogenetic differences in the quantity and composition of volatiles emitted by herbivore-infested plants have been shown to affect the attractiveness of plants and leaves of different ages to parasitic wasps (Mattiacci *et al.*, 2001; Hou *et al.*, 2005) and predatory mites (Takabayashi *et al.*, 1994). As a rule, ontogenetically young tissues usually produce more volatiles on a mass basis than older tissues (Takabayashi *et al.*, 1994; Rostas & Eggert, 2008), which is consistent with the optimal defence theory predictions and with the higher monoterpene concentrations generally found in younger leaves (Figure 3.3). However, in young *Brassica oleracea* var. *gemmifera* (Brassicaceae) plants, older leaves produced more systemically induced volatiles and were more attractive to parasitic wasps (Mattiacci *et al.*, 2001). It has been suggested that young leaves of fast-growing species have a lower risk of exposure to herbivory than those of slow-growing species and hence may allocate less to defence (Mattiacci *et al.*, 2001). In addition, the composition of plant volatiles in a given plant species may affect the direction of ontogenetic changes in volatile production. For instance, our meta-analysis has shown that while monoterpene concentrations are higher in young leaves, sesquiterpene concentrations increase with leaf age (Figure 3.3).

3.5.3   Effects on pathogens

Changes in PSM production through plant development may also affect plant resistance to pathogens. Age-related resistance (ARR) to pathogens occurs in many plant species and usually results in increased disease resistance as plants mature (Whalen, 2005; Develey-Riviere & Galiana, 2007). Reuveni *et al.* (1986) have proposed that the activation of ARR may be associated with the accumulation of antimicrobial substances inhibiting the pathogen development, and several recent studies have demonstrated the link between ARR and developmental changes in PSM production. For instance, ARR in *Arabidopsis* to *Pseudomonas syringae* is associated with the accumulation of intercellular salicylic acid during the transition to flowering (Kus *et al.*, 2002; Carviel *et al.*, 2009). Kus *et al.* (2002) suggested that salicylic acid may act as a signal molecule, stimulating the production and secretion of antibacterial compounds into the intercellular space, and/or that its accumulation contributes to the antibacterial activity observed in plants displaying ARR. The above study also indicated that ARR in *Arabidopsis* is a whole-plant phenomenon, and both young and older leaves on mature plants exhibit ARR (Kus *et al.*, 2002). Similarly, leaves of tobacco acquire resistance to *Phytophthora parasitica* when the plant becomes committed to flowering; the mechanism limiting the spread of fungal hyphae appears to be similar to systemic acquired resistance and involves the accumulation of salicylic acid and PR1 proteins (Hugot *et al.*,

1999). So far, ARR has been studied primarily in herbaceous plants (especially crops), and the common pattern of increased disease resistance through ontogeny agrees well with the ontogenetic increase in constitutive concentrations of most PSM types in herbs (Figure 3.1a).

Sometimes, however, ARR results in a reduction in resistance with plant or leaf age. For instance, Zhang and Chen (2009) have shown that volatiles released by young *Solanum lycopersicum* (Solanaceae) leaves have higher antifungal activity against spore germination and hyphal growth by *Botrytis cinerea* and *Fusarium oxysporum* than volatiles emitted by mature leaves, and suggested that these volatiles may play a role in ARR by tomato. Again, this pattern agrees well with the results of our meta-analysis that most PSMs decrease in concentration during leaf maturation (Figure 3.3).

3.5.4   Effects on pollinators and seed dispersers

Volatile PSMs released by flowers and fruits attract pollinators and seed disseminators, thereby ensuring plant reproductive and evolutionary success (Pichersky & Gershenzon, 2002). Volatile emission rates are usually low in newly open flowers, but peak when flowers are ready for pollination (Dudareva *et al.*, 2000); number and duration of visits by pollinators often follow closely the patterns of volatile emissions (Jones *et al.*, 1998). Composition of volatile PSMs may also change through flower and fruit development. For instance, in *Ficus* (Moraceae), flowers are borne inside the syconium, which will ripen to form a fig. This system is unique because the syconium is the unit of attraction for both pollinators (fig wasps) and seed dispersers (birds and mammals). However, pollinators and seed dispersers should be attracted to the fig during different time windows (receptive and dispersal phases, respectively). Borges *et al.* (2008) have shown that, as predicted, volatile profile changes greatly during the fig developmental cycle, and volatiles produced by figs in receptive and dispersal phases are very different. Given the high specificity of developmental changes in floral and fruit volatiles, it is tempting to believe that these changes are adaptive rather than simply consequences of resource constraints.

3.5.5   Effects on plant–plant interactions

Many PSMs have phytotoxic properties, and temporal changes in their production may thus affect the allelopathic potential of plants. For instance, in *Pteridium aquilinum*, inhibitory effects of water extracts of fronds and of volatile compounds on seed germination and seedling growth of *Populus tremula* and *Pinus sylvestris* are strongest in May and June, which coincides with the beginning of the growing season when bracken fronds are still immature and vulnerable to interference from other species (Dolling *et al.*, 1994). Young foliar tissues of *Artemisia vulgaris* (Asteraceae), an invasive perennial weed in

North America, contain up to 28-fold higher concentrations of monoterpenes than older tissues and have stronger phytotoxic effects on other plants (Barney *et al.*, 2005), suggesting a potential role for terpenoids in *Artemisia* establishment and proliferation in introduced habitats. Similarly, concentrations of major allelopathically active polyphenol compounds in a perennial submerged macrophyte *Myriophyllum verticillatum* (Haloragaceae) and their inhibitory effects on cyanobacteria are highest in spring and early summer, when the macrophytes have not reached the water surface yet and the competition for light is strongest (Bauer *et al.*, 2009). Indeed, results of our meta-analysis showing that PSM concentrations increase rapidly during the early ontogenetic stages suggest that the allelopathic potential of most plant species may be highest during the crucial early stages of plant stand development. However, Nilsson *et al.* (1998) have shown that the concentrations of the dehydrostilbene batatasin-III, the major allelopathic compound of the dwarf shrub *Empetrum hermaphroditum* (Ericaceae), increase rather than decrease through the season within the first year shoots, similar to the pattern observed for condensed tannins (Figure 3.4). Therefore, depending on the type of the allelopathic compound, phytotoxic effects may be stronger earlier or later in the season.

Volatile PSMs emitted by herbivore-infested plants may also mediate the expression of resistance to herbivores and pathogens in undamaged neighbouring plants (Heil & Karban, 2010). Although airborne communication between plants is a rapidly growing field of plant chemical ecology, to our knowledge the role of ontogenetic and seasonal changes in PSM emissions on the ability of plants to emit and receive volatiles has been examined only by Shiojiri and Karban (2006), who have shown that in *Artemisia tridentata* (Asteraceae), young plants are more effective emitters of volatiles and also more responsive receivers of volatiles from other plants.

3.6 Conclusions and future directions

The concentrations and composition of PSMs change greatly over time as plants develop and grow. Despite variability in these developmental trajectories among plant species and PSM types, there are some clear and robust general patterns that our meta-analyses have uncovered (Figure 3.5). For example, PSM concentrations generally increase during seedling growth but decrease during leaf development. As we have discussed, the proximate causes for these temporal patterns in PSM concentrations have been extensively investigated and well documented. However, it remains unclear why these patterns have evolved. The results from our meta-analyses provide little support for hypotheses based on resource allocation constraints, but there have been relatively few studies rigorously testing for age-dependent trade-offs between growth and PSM production (Barton, 2007). Our analyses provide

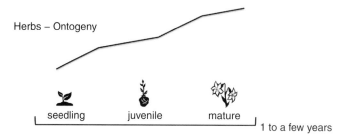

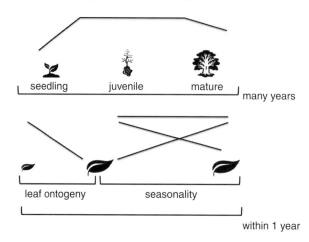

Figure 3.5 Overall patterns of temporal changes in PSM concentrations. Lines represent significant general patterns in PSM concentrations, but do not reflect specific PSM groups.

more support for the idea that herbivory drives the ontogeny of plant defence, but despite the apparent consistency between studies examining ontogenetic patterns in PSM concentrations and studies measuring ontogenetic patterns in herbivore damage or preference (Barton & Koricheva, 2010), there is still a need for experiments specifically designed to measure selection pressure imposed by herbivores at different ontogenetic stages. Moreover, our review shows that the ecological consequences of temporal changes in PSM concentrations are not restricted to herbivores. This suggests that herbivores may not be the only evolutionary agents driving temporal changes in PSM concentrations; pollinators, pathogens and competing plants are the other likely candidates. Experiments that simultaneously examine selection pressure by competing plants, pathogens, multiple kinds of herbivores (including mammalian browsers and insect folivores, for example) and pollinators will provide the most rigorous tests of the agents driving temporal patterns in PSMs. Combining these selection experiments with allocation costs and trade-off analyses would provide the first conclusive test of the competing hypotheses about the ultimate causes of the temporal patterns in plant PSM concentrations. Finally, we have very little information about the population genetics of the ontogeny of PSMs and whether variation exists upon which selection may act (Schappert & Shore, 2000; Donaldson *et al.*, 2006; Rehill *et al.*, 2006; Barton, 2007). Future research exploring the evolutionary dynamics of the ontogeny of PSMs and the ecological consequences and drivers will provide key new insights into this interesting and universal phenomenon.

Acknowledgements

We are grateful to John Maurer for help with producing Figure 3.5. Comments by the editors and anonymous referees have greatly improved the chapter. Financial support for this study was provided by a Natural Environment Research Council Postdoctoral Fellowship to K. E. B. (NERC NE/E012418/1).

References

Adams, D. C., Gurevitch, J. and Rosenberg, M. S. (1997) Resampling tests for meta-analysis of ecological data. *Ecology*, **78**, 1277–1283.

Arnold, T. M. and Targett, N. M. (1998) Quantifying in situ rates of phlorotannin synthesis and polymerization in marine brown algae. *Journal of Chemical Ecology*, **24**, 577–595.

Arnold, T. M. and Targett, N. M. (2000) Evidence for metabolic turnover of polyphenolics in tropical brown algae. *Journal of Chemical Ecology*, **26**, 1393–1410.

Baldwin, I. T., Schultz, J. C. and Ward, D. (1987) Patterns and sources of leaf tannin variation in yellow birch (*Betula allegheniensis*) and sugar maple (*Acer saccharum*). *Journal of Chemical Ecology*, **13**, 1069–1078.

Baldwin, I. T., Karb, M. J. and Ohnmess, T. E. (1994) Allocation of ^{15}N from nitrate to nicotine: production and turnover of a damage-induced mobile defense. *Ecology*, **75**, 1703–1713.

Barney, J. N., Hay, A. G. and Weston, L. A. (2005) Isolation and characterization of allelopathic volatiles from mugwort (*Artemisia vulgaris*). *Journal of Chemical Ecology*, **31**, 247–265.

Barton, K. E. (2007) Early ontogenetic patterns in chemical defense in *Plantago* (Plantaginaceae): genetic variation and trade-offs. *American Journal of Botany*, **94**, 56–66.

Barton, K. E. and Koricheva, J. (2010) The ontogeny of plant defense and herbivory: characterizing general patterns using meta-analysis. *American Naturalist*, **175**, 481–493.

Bauer, N., Blaschke, U., Beutler, E. *et al.* (2009) Seasonal and interannual dynamics of polyphenols in *Myriophyllum verticillatum* and their allelopathic activity on *Anabaena variabilis*. *Aquatic Botany*, **91**, 110–116.

Boege, K. (2005) Influence of plant ontogeny on compensation to leaf damage. *American Journal of Botany*, **92**, 1632–1640.

Boege, K. and Marquis, R. J. (2005) Facing herbivory as you grow up: the ontogeny of resistance in plants. *Trends in Ecology and Evolution*, **20**, 441–448.

Borges, R. M., Bessiere, J. M. and Hossaert-McKey, M. (2008) The chemical ecology of seed dispersal in monoecious and dioecious figs. *Functional Ecology*, **22**, 484–493.

Brenes-Arguedas, T., Horton, M. W., Coley, P. D. *et al.* (2006) Contrasting mechanisms of secondary metabolite accumulation during leaf development in two tropical tree species with different leaf expansion strategies. *Oecologia*, **149**, 91–100.

Brilli, F., Ciccioli, P., Frattoni, M. *et al.* (2009) Constitutive and herbivore-induced monoterpenes emitted by *Populus x euroamericana* leaves are key volatiles that orient *Chrysomela populi* beetles. *Plant, Cell and Environment*, **32**, 542–552.

Bryant, J. P., Reichardt, P. B., Clausen, T. P., Provenza, F. D. and Kuropat, P. J. (1992) Woody plant–mammal interactions. In G. A. Rosenthal and M. R. Berenbaum (eds.) *Herbivores: Their Interactions with Secondary Plant Metabolites*. San Diego, CA: Academic Press, 343–370.

Carviel, J. L., Al-Daoud, F., Neumann, M. et al. (2009) Forward and reverse genetics to identify genes involved in the age-related resistance response in *Arabidopsis thaliana*. *Molecular Plant Pathology*, **10**, 621–634.

Coley, P. D. (1983) Herbivory and defensive characteristics of tree species in a lowland tropical forest. *Ecological Monographs*, **53**, 209–233.

Coley, P. D. and Kursar, T. A. (1996) Anti-herbivore defenses of young tropical leaves: physiological constraints and ecological tradeoffs. In A. P. Smith, S. S. Mulkey and R. L. Chazdon (eds.) *Tropical Forest Plant Ecophysiology*. New York: Chapman & Hall, 305–336.

Develey-Riviere, M. P. and Galiana, E. (2007) Resistance to pathogens and host developmental stage: a multifaceted relationship within the plant kingdom. *New Phytologist*, **175**, 405–416.

Dolling, A., Zackrisson, O. and Nilsson, M. C. (1994) Seasonal variation in phytotoxicity of bracken (*Pteridium aquilinum* L. Kuhn). *Journal of Chemical Ecology*, **20**, 3163–3172.

Donaldson, J. R., Stevens, M. T., Barnhill, H. R. and Lindroth, R. L. (2006) Age-related shifts in leaf chemistry of clonal aspen (*Populus tremuloides*). *Journal of Chemical Ecology*, **32**, 1415–1429.

Dudareva, N., Murfitt, L. M., Mann, C. J. *et al.* (2000) Developmental regulation of methyl benzoate biosynthesis and emission in snapdragon flowers. *Plant Cell*, **12**, 949–961.

Faeth, S. H. (1986) Indirect interactions between temporally separated herbivores mediated by the host plant. *Ecology*, **67**, 479–494.

Feeny, P. (1976) Plant apparency and chemical defense. *Recent Advances in Phytochemistry*, **10**, 1–40.

Feeny, P. P. (1970) Seasonal changes in oak leaf tannins and nutrients as a cause of spring feeding by winter moth caterpillars. *Ecology*, **51**, 565–581.

Gershenzon, J., Murtagh, G. J. and Croteau, R. (1993) Absence of rapid terpene turnover in several diverse species of terpene-accumulating plants. *Oecologia*, **96**, 583–592.

Gershenzon, J., McConkey, M. E. and Croteau, R. B. (2000) Regulation of monoterpene accumulation in leaves of peppermint. *Plant Physiology*, **122**, 205–213.

Gurevitch, J. and Hedges, L. V. (2001) Meta-analysis: combining results of independent experiments. In S. M. Schneider and J. Gurevitch (eds.) *Design and Analysis of Ecological Experiments*. Oxford: Oxford University Press, 347–369.

Gurevitch, J., Morrison, J. A. and Hedges, L. V. (2000) The interaction between competition and predation: a meta-analysis of field experiments. *American Naturalist*, **155**, 435–453.

Harper, J. L. (1989) The value of a leaf. *Oecologia*, **80**, 53–58.

Hartmann, T. (2007) From waste products to ecochemicals: fifty years research of plant secondary metabolism. *Phytochemistry*, **68**, 2831–2846.

Hartmann, T. and Dierich, B. (1998) Chemical diversity and variation of pyrrolizidine alkaloids of the senecionine type: biological need or coincidence? *Planta*, **206**, 443–451.

Haukioja, E. (2003) Putting the insect into the birch–insect interaction. *Oecologia*, **136**, 161–168.

Heil, M. and Karban, R. (2010) Explaining evolution of plant communication by airborne signals. *Trends in Ecology and Evolution*, **25**, 137–144.

Herms, D. A. and Mattson, W. J. (1992) The dilemma of plants: to grow or defend. *Quarterly Review of Biology*, **67**, 283–335.

Hou, M., Takabayashi, J. and Kainoh, Y. (2005) Effect of leaf age on flight response of a parasitic wasp *Cotesia kariyai* (Hymenoptera: Braconidae) to a plant–herbivore complex. *Applied Entomology and Zoology*, **40**, 113–117.

Hugot, K., Aime, S., Conrod, S., Poupet, A. and Galiana, E. (1999) Developmental regulated mechanisms affect the ability of a fungal pathogen to infect and colonize tobacco leaves. *Plant Journal*, **20**, 163–170.

Iwasa, Y., Kubo, T., van Dam, N. and de Jong, T. J. (1996) Optimal level of chemical defense decreasing with leaf age. *Theoretical Population Biology*, **50**, 124–148.

Johns, R., Quiring, D. T., Lapointe, R. and Lucarotti, C. J. (2009) Foliage-age mixing within balsam fir increases the fitness of a generalist caterpillar. *Ecological Entomology*, **34**, 624–631.

Jones, C. G. and Hartley, S. E. (1999) A protein competition model of phenolic allocation. *Oikos*, **86**, 27–44.

Jones, K. N., Reithel, J. S. and Irwin, R. E. (1998) A trade-off between the frequency and duration of bumblebee visits to flowers. *Oecologia*, **117**, 161–168.

Kant, M. R., Bleeker, P. M., Van Wijk, M., Schuurink, R. C. and Haring, M. A. (2009) Plant volatiles in defence. In L. C. Van Loon (ed.) *Plant Innate Immunity*. Burlington, VT: Academic Press, 613–666.

Kearsley, M. J. C. and Whitham, T. G. (1998) The developmental stream of cottonwoods affects ramet growth and resistance to galling aphids. *Ecology*, **79**, 178–191.

Koricheva, J. (1999) Interpreting phenotypic variation in plant allelochemistry: problems with the use of concentrations. *Oecologia*, **119**, 467–473.

Kozlowski, T. T. (1971) *Growth and Development of Trees*. New York: Academic Press.

Kursar, T. A. and Coley, P. D. (2003) Convergence in defense syndromes of young leaves in tropical rainforests. *Biochemical Systematics and Ecology*, **31**, 929–949.

Kus, J. V., Zaton, K., Sarkar, R. and Cameron, R. K. (2002) Age-related resistance in *Arabidopsis* is a developmentally regulated defense response to *Pseudomonas syringae*. *Plant Cell*, **14**, 479–490.

Lerdau, M., Matson, P., Fall, R. and Monson, R. (1995) Ecological controls over monoterpene emissions from Douglas-fir (*Pseudotsuga menziesii*). *Ecology*, **76**, 2640–2647.

Mattiacci, L., Rudelli, S., Rocca, B. A., Genini, S. and Dorn, S. (2001) Systematically-induced response of cabbage plants against a spacialist herbivore, *Pieris brassicae*. *Chemoecology*, **11**, 167–173.

McConkey, M. E., Gershenzon, J. and Croteau, R. B. (2000) Developmental regulation of monoterpene biosynthesis in the glandular trichomes of peppermint. *Plant Physiology*, **122**, 215–223.

Mihaliak, C. A., Gershenzon, J. and Croteau, R. (1991) Lack of rapid monoterpene turnover in rooted plants: implications for theories of plant chemical defense. *Oecologia*, **87**, 373–376.

Nilsson, M. C., Gallet, C. and Wallstedt, A. (1998) Temporal variability of phenolics and batatasin-III in *Empetrum hermaphroditum* leaves over an eight-year period: interpretations of ecological function. *Oikos*, **81**, 6–16.

Peñuelas, J. and Llusia, J. (1997) Effects of carbon dioxide, water supply, and seasonality on terpene content and emission by *Rosmarinus officinalis*. *Journal of Chemical Ecology*, **23**, 979–993.

Pichersky, E. and Gershenzon, J. (2002) The formation and function of plant volatiles: perfumes for pollinator attraction and defense. *Current Opinion in Plant Biology*, **5**, 237–243.

Poethig, R. S. (1990) Phase change and the regulation of shoot morphogenesis in plants. *Science*, **250**, 923–930.

Rehill, B. J., Whitham, T. G., Martinsen, G. D. *et al.* (2006) Developmental trajectories in cottonwood phytochemistry. *Journal of Chemical Ecology*, **32**, 2269–2285.

Reuveni, M., Tuzun, S., Cole, J. S., Siegel, M. R. and Kuc, J. (1986) The effects of plant age and leaf position on the susceptibility of tobacco to blue mold caused by *Peronosprora tabacina*. *Phytopathology*, **76**, 455–458.

Rhoades, D. F. (1979) Evolution of plant chemical defense against herbivores. In G. A. Rosenthal and D. N. Janzen (eds.) *Herbivores: Their Interactions with Secondary Plant Metabolites*. New York: Academic Press, 3–54.

Rhoades, D. F. and Cates, R. G. (1976) Toward a general theory of plant antiherbivore chemistry. *Recent Advances in Phytochemistry*, **10**, 168–213.

Rosenberg, M. S. (2005) The file-drawer problem revisited: a general weighted method for calculating fail-safe numbers in meta-analysis. *Evolution*, **59**, 464–468.

Rosenberg, M. S., Adams, D. C. and Gurevitch, J. (2000) MetaWin: statistical software for meta-analysis. Sunderland, MA: Sinauer Associates.

Rosenthal, R. (1991) *Meta-analytic Procedures for Social Research*. Newbury Park, CA: Sage.

Roslin, T. and Salminen, J. P. (2008) Specialization pays off: contrasting effects of two types of tannins on oak specialist and generalist moth species. *Oikos*, **117**, 1560–1568.

Roslin, T. and Salminen, J. P. (2009) A tree in the jaws of a moth – temporal variation in oak leaf quality and leaf-chewer performance. *Oikos*, **118**, 1212–1218.

Rossiter, M., Schultz, J. C. and Baldwin, I. T. (1988) Relationships among defoliation, red oak phenolics, and gypsy moth growth and reproduction. *Ecology*, **69**, 267–277.

Rostas, M. and Eggert, K. (2008) Ontogenetic and spatio-temporal patterns of induced volatiles in *Glycine max* in the light of the optimal defence hypothesis. *Chemoecology*, **18**, 29–38.

Ruuhola, T. M. and Julkunen-Tiitto, R. (2000) Salicylates of intact *Salix myrsinifolia* plantlets do not undergo rapid metabolic turnover. *Plant Physiology*, **122**, 895–905.

Ruusila, V., Morin J. P., van Ooik, T. *et al.* (2005) A short-lived herbivore on a long-lived host: tree resistance to herbivory depends on leaf age. *Oikos*, **108**, 99–104.

Salminen, J. P. (2002) *Birch Leaf Hydrolysable Tannins: Chemical, Biochemical and Ecological Aspects*. Turku, Finland: University of Turku, 163.

Salminen, J. P., Ossipov, V., Haukioja, E. and Pihlaja, K. (2001) Seasonal variation in the content of hydrolyzable tannins in leaves of *Betula pubescens*. *Phytochemistry*, **57**, 15–22.

Salminen, J. P., Roslin, T., Karonen, M. *et al.* (2004) Seasonal variation in the content of hydrolyzable tannins, flavonoid glycosides, and proanthocyanidins in oak leaves. *Journal of Chemical Ecology*, **30**, 1693–1711.

Schaffner, U., Vrieling, K. and van der Meijden, E. (2003) Pyrrolizidine alkaloid content in *Senecio*: ontogeny and developmental constraints. *Chemoecology*, **13**, 39–46.

Schappert, P. J. and Shore, J. S. (2000) Cyanogenesis in *Turnera ulmifolia* L. (Turneraceae): II. Developmental expression, heritability and cost of cyanogenesis. *Evolutionary Ecology Research*, **2**, 337–352.

Schroeder, L. A. (1986) Changes in tree leaf quality and growth-performance of Lepidopteran larvae. *Ecology*, **67**, 1628–1636.

Shiojiri, K. and Karban, R. (2006) Plant age, communication, and resistance to herbivores: young sagebrush plants are better emitters and receivers. *Oecologia*, **149**, 214–220.

Swihart, R. K. and Bryant, J. P. (2001) Importance of biogeography and onthogeny of woody plants in winter herbivory by mammals. *Journal of Mammalogy*, **82**, 1–21.

Takabayashi, J., Dicke, M., Takahashi, S., Posthumus, M. A. and Vanbeek, T. A. (1994) Leaf age affects composition of herbviore-induced synomones and attraction of predatory mites. *Journal of Chemical Ecology*, **20**, 373–386.

Turner, G. W., Gershenzon, J. and Croteau, R. B. (2000) Development of peltate glandular trichomes of peppermint. *Plant Physiology*, **124**, 665–679.

Valkama, E., Salminen, J. P., Koricheva, J. and Pihlaja, K. (2004) Changes in leaf trichomes and epicuticular flavonoids during leaf development in three birch taxa. *Annals of Botany*, **94**, 233–242.

van Dam, N. M., Witte, L., Theuring, C. and Hartmann, T. (1995) Distribution, biosynthesis and turnover of pyrrolizidine alkaloids in *Cynoglossum officinale*. *Phytochemistry*, **39**, 287–292.

Wallstedt, A., Nilsson, M. C., Zackrisson, O. and Odham, G. (2000) A link in the study of chemical interference exerted by *Empetrum hermaphroditum*: quantification of batatasin-III in soil solution. *Journal of Chemical Ecology*, **26**, 1311–1323.

West, C. (1985) Factors underlying the seasonal appearance of the lepidopterous leaf-mining guild on oak. *Ecological Entomology*, **10**, 111–120.

Whalen, M. C. (2005) Host defence in a developmental context. *Molecular Plant Pathology*, **6**, 347–360.

Whitham, T. G. (1983) Host manipulation of parasites: within-plant variation as a defense against rapidly evolving pests. In R. F. Denno and M. S. McClure (eds.) *Variable Plants and Herbivores in Natural and Managed Systems*. New York: Academic Press, 15–41.

Whitham, T. G. and Schweitzer, J. A. (2002) Leaves as islands of spatial and temporal variation: consequences for plant herbivores, pathogens, communities and ecosystems. In S. E. Lindow, E. I. Hecht-Poinar and V. J. Elliott (eds.) *Phyllosphere Microbiology*. St Paul, MN: APS Press, 279–298.

Whitham, T. G., Williams, A. G. and Robinson, A. M. (1984) The variation principle: individual plants as temporal and spatial mosaics of resistance to rapidly evolving pests. In P. W. Price, C. N. Slobodchikoff and W. S. Gaud (eds.) *A New Ecology: Novel Approaches to Interactive Systems*. New York: John Wiley & Sons, 15–51.

Yarnes, C. T., Boecklen, W. J. and Salminen, J. P. (2008) No simple sum: seasonal variation in tannin phenotypes and leaf-miners in hybrid oaks. *Chemoecology*, **18**, 39–51.

Zhang, P. and Chen, K. (2009) Age-dependent variations of volatile emissions and inhibitory activity toward *Botrytis cinerea* and *Fusarium oxysporum* in tomato leaves treated with chitosan oligosaccharide. *Journal of Plant Biology*, **52**, 332–339.

CHAPTER FOUR

Mixtures of plant secondary metabolites: metabolic origins and ecological benefits

JONATHAN GERSHENZON and ANNA FONTANA
Department of Biochemistry, Max Planck Institute for Chemical Ecology
MEIKE BUROW
Department of Plant Biology and Biotechnology, Copenhagen University
UTE WITTSTOCK
Institute of Pharmaceutical Biology, Technical University of Braunschweig
JOERG DEGENHARDT
Institute of Pharmacy, Martin Luther University of Halle-Wittenberg

4.1 Introduction

Plants produce a large variety of secondary metabolites which are usually considered to function as defences against herbivores and pathogens, as many other contributions to this volume attest. Among the most characteristic features of these compounds are their vast number and enormous chemical diversity. Reports on plant secondary metabolites (PSMs) are replete with phrases describing their 'tremendous array' (Morrissey, 2009), 'bewildering proliferation' (Schoonhoven *et al.*, 2005) or 'extraordinary diversity' (Howe & Jander, 2008). The diversity of secondary metabolites is apparent not only in their chemical structures, but also in their distribution in plants. The composition of secondary metabolites in plants varies at many levels of organisation, such as among different plant taxa (Wink, 2003), among different populations of the same taxon (Kliebenstein *et al.*, 2001a) and between individuals of the same species (Pakeman *et al.*, 2006). Within a plant, there is also variation among different organs (Brown *et al.*, 2003), developmental stages (Lambdon *et al.*, 2003) and environmental conditions (Engelen-Eigles *et al.*, 2006), as well as the frequent presence of complex mixtures of secondary metabolites in individual organs. Most secondary metabolites, including alkaloids (Waffo *et al.*, 2007), phenolics (Ashihara *et al.*, 2010) and terpenes (Köllner *et al.*, 2004), invariably occur in mixtures rather than as individual, isolated substances. The chemistry and distribution of secondary metabolites is so diverse that being able to explain the patterns of diversity seems an essential requirement for understanding their roles in plants.

The Ecology of Plant Secondary Metabolites: From Genes to Global Processes, eds. Glenn R. Iason, Marcel Dicke and Susan E. Hartley. Published by Cambridge University Press. © British Ecological Society 2012.

This review will consider both the generation of secondary metabolite chemical diversity by the plant's biosynthetic machinery and the functional importance of such diversity. We focus on the function of secondary metabolites in defence against herbivores, because this has received the most attention from researchers. Since there are so many levels of diversity in secondary metabolites, we will limit ourselves to just one: the occurrence of mixtures of a single class of compounds, such as alkaloids or terpenes, in individual organs.

4.2   Metabolic origins of secondary metabolite mixtures

Theoretically, all it would take to form a mixture of secondary metabolites of a single class is a simple linear pathway catalysed by slow, inefficient enzymes that allow the build-up of intermediates. However, the mixtures that have been biosynthetically investigated to date all arise from highly branched pathways with multiple endpoints. In fact, the occurrence of multiple branching is a chief characteristic of secondary metabolism. Biochemical mechanisms that promote branching and mixture formation have been previously mentioned (Jones & Firn, 1991), but this topic is worth reconsidering in the light of recent advances in biosynthetic knowledge. Here, we explore some of the chief biosynthetic mechanisms for generating mixtures.

First, it is worth noting that mixtures do not develop simply because secondary pathways are supplied with many different precursors from other parts of plant metabolism (Figure 4.1). The pathways of secondary metabolism arise from only a few sources in primary plant metabolism. For example, all of the alkaloids, glucosinolates and other nitrogen-containing products arise from just a small number of amino acids, while the phenolics arise almost exclusively from the aromatic amino acid phenylalanine or acetyl-CoA via the polyketide pathway. Acetyl-CoA is also the single carbon source for one of the two terpene pathways; the other begins with pyruvate and glyceraldehyde-3-phosphate. Finally, fatty acid derivatives are produced from selected fatty acids (also originally derived from acetyl-CoA). Thus, the diversity of secondary metabolites is not due to the supply of many different precursors from other parts of plant metabolism, but to diversity created within secondary metabolism.

4.2.1   Repeated addition of core units

One mechanism for generating product diversity is that secondary metabolic pathways often proceed by repeated addition of the same basic building blocks to make a wealth of intermediates of different sizes. For example, in terpene formation sequential addition of a branched C_5 'isoprenoid' unit produces intermediates of 10, 15, 20, 25, 30 or 40 carbon atoms that serve as substrates in further biosynthetic steps (Gershenzon & Kreis, 1999). Similarly,

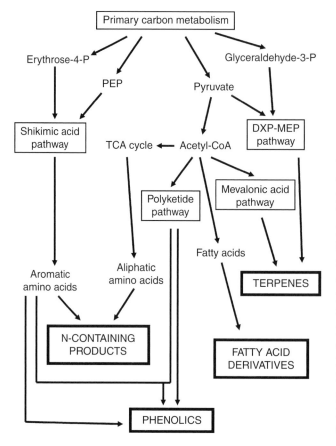

Figure 4.1 Simplified scheme of the origin of pathways of secondary metabolism from primary metabolism in plants. The four large boxes in the lower part of the figure depict the four major classes of plant secondary metabolites. Despite the great diversity of secondary metabolic products, their pathways arise from only a few sources in primary metabolism (Buchanan *et al.*, 2000; Taiz & Zeiger, 2002). PEP, phosphoenol pyruvate; DXP-MEP, deoxyxylulose phosphate-methylerythritol phosphate; TCA, tricarboxylic acid.

in the polyketide pathway sequential rounds of the addition of a C_2 acetate unit (-CH_2CO-) create longer and longer poly-β-keto intermediates for subsequent cyclisation, aromatisation and other reactions that form a variety of phenolic compounds in plants (Austin & Noel, 2003).

Repeated addition of basic building blocks is also a characteristic of many other pathways of secondary metabolism. For example, many glucosinolates are derived from amino acids that have been elongated by repeated insertion of methylene groups into their side chains (Halkier & Gershenzon, 2006) (Figure 4.2). For the methionine-derived glucosinolates of *Arabidopsis thaliana*, this has been shown to occur by a three-step cycle involving condensation of acetyl-CoA to a 2-oxo acid followed by an isomerisation and a decarboxylation step to give a net gain of one carbon atom per turn of the cycle (Graser *et al.*, 2000). The result is a series of precursors leading to glucosinolates with different side-chain lengths ranging from 3 to 10 carbon atoms (Fahey *et al.*, 2001).

To generate an additional level of structural diversity, the basic building blocks of more than one pathway can be combined. Well-known examples

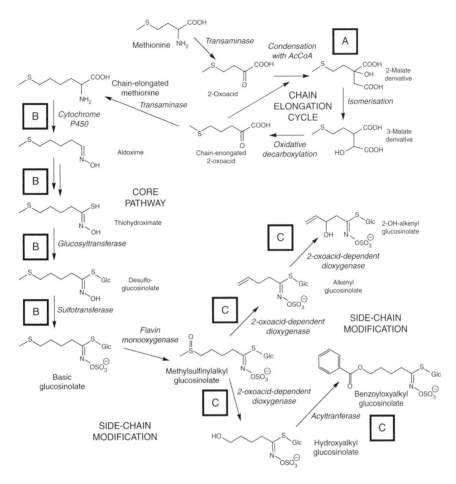

Figure 4.2 Outline of the pathway of aliphatic glucosinolate biosynthesis in *Arabidopsis thaliana*, showing some of the characteristic features responsible for generating mixtures. Abbreviations: A, repeated addition of core units; B, pathway steps with low substrate specificity; C, steps for skeletal modification catalysed by genes from large families of oxidising and group transfer enzymes.

include the addition of a C_{10} terpene unit (the iridoid intermediate secologanin) to intermediates in alkaloid biosynthesis, such as dopamine or tryptamine (Nomura *et al.*, 2008), and addition of polyketide pathway intermediates to phenylalanine-derived phenylpropanoids or isoprenoid units of C_5 or C_{10} (Springob & Kutchan, 2009).

4.2.2 Formation of diverse carbon skeletons

After assembly of the building blocks that determine the general size of the final product, some secondary pathways create a profusion of different carbon skeletons. This is especially true for the terpenes, for which the large enzyme family of terpene synthases converts acyclic prenyl diphosphates of different sizes (C_5, C_{10}, C_{15} and C_{20}) and squalene (C_{30}) into a multitude of cyclic and acyclic carbon skeletons (Tholl, 2006). The diversity of products arises not only from the vast number of terpene synthases, but also from the ability of these enzymes to form multiple products from a single substrate (Degenhardt *et al.*, 2009). Approximately half of all characterised C_{10} and C_{15} terpene synthases catalyse the formation of more than one product in significant amounts. This ability allows multiple carbon skeleton formation without multiple enzymes and, thus, could lower the metabolic costs of making mixtures.

The diversity of carbon skeletons of plant phenolics formed via the polyketide pathway can be attributed to another large enzyme family of secondary metabolism, the polyketide synthases, which reach their greatest diversity in microbes (Hertweck, 2009). Plant polyketide synthases not only catalyse the formation of a number of phenolic ring systems, but also mediate the prior assembly of the required two-carbon acetate core units. These core units are frequently attached to starter molecules, such as benzyl, cinnamoyl or fatty acid moieties. Skeletal variation is generated by assembling varying numbers of acetate units, use of different starter molecules, and different types of cyclisation and aromaticisation (Austin & Noel, 2003).

4.2.3   Multiple paths for skeletal modification

Once the basic carbon skeleton of a compound class is established, variation is often generated by two large families of position-specific, oxidative enzymes, cytochrome P450 monooxygenases (Bolwell *et al.*, 1994) and 2-oxoacid-dependent dioxygenases (Prescott & Lloyd, 2000). Oxidation is then frequently followed by methyl or acyl group transfer mediated by two other families of position-specific catalysts, the O-methyltransferases (Ibrahim *et al.*, 1998) and the acyltransferases (D'Auria, 2006). Every plant species studied has proved to contain large gene families for each of these enzyme types, and many of the individual enzymes are thought to participate in biosynthesis of secondary metabolites.

Formation of mixtures is promoted when a single intermediate serves as a substrate for more than one of these enzymes. For example, in glucosinolate metabolism the side chain of aliphatic glucosinolates can be modified by two different 2-oxoacid-dependent dioxygenases, one giving rise to alkenyl glucosinolates and the other to hydroxyalkyl side chains (Kliebenstein *et al.*, 2001b) (Figure 4.2) and mixtures of these different compounds arise in plants (Kliebenstein *et al.*, 2001a). In indole glucosinolate metabolism, indol-3-ylmethyl glucosinolate is apparently subject to oxidation by at least two different P450 monooxygenases, one specific to the carbon atom at position 4

(Bednarek *et al.*, 2009; Pfalz *et al.*, 2009) and another probably specific to the nitrogen atom at position 1, followed by the position-specific action of methyltransferases. As a result, glucosinolate-producing plants usually contain mixtures of three indolic glucosinolates, the parent indol-3-ylmethyl compound and derivatives methylated at positions 1 and 4 (Kliebenstein *et al.*, 2001a).

4.2.4 Lack of specificity in the late steps of biosynthesis

The production of mixtures of secondary metabolites would also be favoured if a large number of diverse intermediates created in early steps of a pathway were all converted to final products by a few later enzymes with broad substrate specificity. For glucosinolate biosynthesis, this would be the case if intermediates of different chain lengths were all suitable substrates for later enzymes of the pathway (Figure 4.2). In fact, the glucosyltransferases of the glucosinolate pathway glucosylate a wide range of substrates regardless of side-chain length or functional groups (Grubb *et al.*, 2004). Each of the sulfotransferases characterised from this pathway also displays a broad range of specificity (Klein & Papenbrock, 2009). In some glucosinolate-forming plant species, all of the late steps seem to have broad specificity. When these species were fed a variety of natural and synthetic aldoximes (early intermediates of the pathway), they were able to catalyse the following four to five steps for all of these compounds to transform them into glucosinolate end-products (Grootwassink *et al.*, 1990). The presence of enzymes of such broad substrate specificity late in the pathway can be viewed as a way to make mixtures at low metabolic cost by generating more products with fewer enzymes. Given that many classes of enzymes in plant secondary metabolism display such broad specificity (Schwab, 2003), this mechanism may be widespread.

Glucosinolate metabolism also has another level of biochemical machinery designed to diversify end-product composition, as the glucosinolates themselves are not the final biologically active defence compounds (Halkier & Gershenzon, 2006). Upon plant damage, glucosinolates are hydrolysed to create a series of products responsible for virtually all of the biological activities of this compound class. Hydrolysis of the thioglucoside linkage by an enzyme known as myrosinase leads to the formation of free glucose and an unstable aglycone that frequently rearranges to an isothiocyanate (Bones & Rossiter, 2006). However, depending on the structure of the side chain and the presence of additional proteins, the aglycone can rearrange instead to form nitriles, epithionitriles, thiocyanates or oxazolidine-2-thiones (Lambrix *et al.*, 2001; Burow *et al.*, 2007). Thus, the diversity of the parent glucosinolate structures in a plant mixture is further increased by conversion of each into different hydrolysis products. Other plant defence compounds stored as glycosides may also be converted into a variety of different products upon hydrolysis (Morant *et al.*, 2008).

As we have seen in this section, the formation of mixtures can be attributed to several characteristic, diversity-generating features of the biosynthetic machinery for plant defence compounds that operate at different phases of the pathways. Although no single defence pathway possesses all of these features, each has at least some of them, suggesting that the ability to fashion mixtures of related compounds is an important trait of secondary metabolism. Why such mixtures are fashioned is the subject of the next section.

4.3   Ecological benefits of mixtures of secondary plant metabolites

The prevalence of mixtures of secondary metabolites in plants has prompted considerable interest in whether these confer any true defensive benefits on the organism that produces them. Mixtures could simply result from enzymes that are inefficient or have broad substrate ranges, and whose products are functionally redundant or even antagonistic to each other. However, there are a number of ways in which mixtures might be more effective in defence than the equivalent amount of a single compound.

4.3.1   Simultaneous protection against different enemies

Plants in natural environments are subject to attack by numerous herbivores and pathogens. Hence mixtures could prove advantageous if the different components target different enemies. We recently had the opportunity to test this proposition using transgenic plants altered in their spectrum of glucosinolate hydrolysis products. As mentioned earlier (Section 4.2.4), glucosinolate diversity arises both from the production of many parent glucosinolates and the conversion of these compounds into an assortment of hydrolysis products upon plant damage. The hydrolysis products are generally responsible for the biological activity of glucosinolates. We have been studying the defensive roles of glucosinolate hydrolysis products in *A. thaliana* which produces two major classes of these substances, isothiocyanates and nitriles (Lambrix *et al.*, 2001; Burow *et al.*, 2009). The Columbia-0 ecotype yields predominantly isothiocyanates, but we created transgenic lines over-expressing a protein involved in nitrile formation to afford plants making predominantly nitriles. This allowed direct testing of the effects of isothiocyanates versus nitriles *in planta* in a common genetic background (Burow *et al.*, 2006).

Bioassays performed with these transgenic lines and the generalist lepidopteran herbivore, *Spodoptera littoralis*, demonstrated that larvae that fed on nitrile-producing plants grew faster and reached the pupal stage sooner than larvae that fed on isothiocyanate-producing plants (Burow *et al.*, 2006). Isothiocyanate toxicity and deterrence to herbivores are frequently reported in the literature (Seo & Tang, 1982; Agrawal & Kurashige, 2003), and these substances can be considered more effective defences against generalist herbivores than nitriles (Wittstock *et al.*, 2003).

Next, we conducted bioassays with larvae of *Pieris rapae*, a lepidopteran that specialises on glucosinolate-containing plants. In contrast to *S. littoralis*, the growth rate and development time of *P. rapae* did not differ between isothiocyanate-producing and nitrile-producing plant lines (Mumm *et al.*, 2008). Moreover, *P. rapae* larvae did not discriminate between isothiocyanates and nitriles in feeding choice assays. Thus, isothiocyanates generated upon glucosinolate hydrolysis, though effective against generalist herbivores, had no impact on specialists. If isothiocyanates are active only against generalists, but nitriles are neither active against generalists nor against specialists, this raises the question of why plants bother making nitriles at all.

To explore this problem further, we considered the role of glucosinolate hydrolysis products in indirect defence, the attraction of herbivore enemies to the plant. The larval parasitoid, *Cotesia rubecula*, a specialist on *P. rapae* larvae, was more attracted to *P. rapae* feeding on nitrile-releasing plants than larvae feeding on isothiocyanate-releasing plants, which could provide an advantage for nitrile-producers when *P. rapae* herbivore pressure is high. Nitriles probably serve as a cue for these parasitoids because they are the major volatiles emitted from *P. rapae* faeces, which are known to attract *C. rubecula* (Geervliet *et al.*, 1994). The distinct odour of *P. rapae* faeces is due to the detoxification system of this specialist caterpillar, which diverts all glucosinolate hydrolysis taking place in the gut to nitriles instead of isothiocyanates (Wittstock *et al.*, 2004). Plants may exploit the innate attraction of the parasitoid to nitriles by releasing some nitriles of their own as a way of enhancing the signal given off by feeding and faeces-producing *P. rapae* larvae.

Another way that nitriles could serve in anti-herbivore defence is by deterring oviposition. Tests with gravid *P. rapae* females revealed a strong oviposition preference for isothiocyanate-producing versus nitrile-producing plant lines when plants had been damaged previously (Mumm *et al.*, 2008), a result also obtained in another study (de Vos *et al.*, 2008). The lack of preference for nitriles by *P. rapae* females probably arises because nitriles are good indicators that the plant is already infested with other *P. rapae* caterpillars, since nitriles are emitted from larval faeces. Nitriles may also signal an increased risk of parasitism because *C. rubecula*, a specialist parasitoid of *P. rapae*, is attracted to these compounds, as described above. To recapitulate, of the glucosinolate hydrolysis products released by *A. thaliana*, isothiocyanates are toxic and feeding deterrents to generalist, but not specialist herbivores. However, nitriles can reduce specialist damage by deterring oviposition and attracting parasitoid enemies. Thus, the two types of hydrolysis products are not redundant or antagonistic defences. Producing them as a mixture allows the plant to target two different types of herbivores simultaneously.

Interestingly, after the *A. thaliana* Columbia-0 ecotype is fed upon by the specialist *P. rapae* in the laboratory, it actually increases its proportion of nitriles to isothiocyanates formed upon tissue damage (Burow *et al.*, 2009). This response may be adaptive since nitriles are more effective defences against *P. rapae* and perhaps have the same impact on other specialist feeders that *A. thaliana* encounters under natural conditions. *Pieris rapae* may not be a natural enemy of *A. thaliana* (Harvey *et al.*, 2007). Since mixtures of isothiocyanates and nitriles are found as glucosinolate hydrolysis products in many other Brassicaceae species besides *A. thaliana* (Cole, 1976; Falk & Gershenzon, 2007; Williams *et al.*, 2009), the strategy of using such mixtures to defend against multiple herbivores may be widespread.

Generalist and specialist herbivores are commonly sensitive to different types of chemical defences (Brattsten, 1992), so mixtures produced to target both types of herbivores may occur frequently in plant species subject to regular attack by both types. Another example involving glucosinolates concerns the typical mixture of aliphatic and indole glucosinolates found in *A. thaliana* and many other species of the Brassicaceae (Fahey *et al.*, 2001; Reichelt *et al.*, 2002). Both aliphatic and indole glucosinolates appear to be necessary for defence against a full spectrum of herbivores. Among generalist lepidopteran herbivores tested, *Spodoptera exigua* is sensitive to both aliphatic and indole glucosinolates, while *Manduca sexta* and *Trichoplusia ni* are only sensitive to aliphatic glucosinolates (Müller *et al.*, 2010). However, indole glucosinolates are necessary too because only these compounds and not aliphatic glucosinolates are effective defences against aphids (Kim *et al.*, 2008).

4.3.2   More specific communication with mutualists

In many plants, volatile compounds released in response to herbivore damage are attractive to herbivore enemies. These volatiles are typically present as mixtures of terpenes, green leaf volatiles and benzenoids. The complexity of such mixtures may be important for communication with herbivore enemies by allowing plants to send more specific messages.

To test the function of individual herbivore-induced volatiles and mixtures, we have been investigating the terpene blend released by *Zea mays* (maize) after herbivore damage by *S. littoralis* and its attractiveness to the hymenopteran parasitoid, *Cotesia marginiventris*, of which the females lay eggs in lepidopteran larvae. For testing various portions of the maize terpene blend, we have carried out olfactometer bioassays on *C. marginiventris* using *A. thaliana* lines transformed to produce subsets of the total volatile blend of more than 30 components. Transformation was carried out by over-expressing genes encoding maize terpene synthase enzymes resulting in the production of mixtures of terpene volatiles. Since most of these volatile compounds are not conveniently available commercially or via laboratory synthesis, terpene synthase transformants are a very valuable vehicle for manufacturing volatiles or volatile mixtures for bioassays. In our initial work, we demonstrated that *A. thaliana* transformed to release volatiles of maize terpene synthase 10 (TPS10) was more attractive to *C. marginiventris* than untransformed *A. thaliana* controls after the females had an oviposition experience and had learned to associate these volatiles with host presence (Schnee *et al.*, 2006). TPS10 produces a blend of six compounds containing most of the herbivore-induced sesquiterpenes of maize leaves.

In our latest series of olfactometer bioassays (A. Fontana, M. Held, C. Assefa Fantaye, T. Turlings, J. Degenhardt and J. Gershenzon, unpublished results), we first tested whether the constitutively emitted sesquiterpenes of maize, produced by TPS8, were as attractive to *C. marginiventris* as the herbivore-induced sesquiterpenes. Transgenic *A. thaliana* transformed with TPS8, just like plants transformed with TPS10, were more attractive to this parasitoid than untransformed control plants after association with an oviposition experience. Next, we compared the attractiveness of sesquiterpenes from TPS8 transformants with those from TPS10 transformants and with those emanating from both TPS8 and TPS10 transformants, while ensuring that the total amounts of sesquiterpenes remained constant. In olfactory bioassays, *C. marginiventris* with prior oviposition experience were most attracted to the blend of TPS8 + TPS10 volatiles (Generalized linear model (GLM), $P = 0.009$) (Figure 4.3). Naïve wasps did not have a significant preference for any of the odours offered ($F_{2,21} = 0.85$, $P = 0.44$). The more attractive TPS8 + TPS10 mixture represents the full natural blend of constitutive and herbivore-induced sesquiterpenes released by the plant, and thus it is perhaps not surprising that it was found to be more attractive than the partial blends. The value of such a mixture to both the plant (signal emitter) and the parasitoid (signal receiver) may arise from its greater information content in comparison with individual compounds or simpler blends, or greater chance that the full blend contains compounds that happen to have innate attractiveness. In this case, the volatiles emanating constitutively from maize may serve as a general signal to indicate the presence of a suitable habitat for the larval hosts of the parasitoid when signals from the host are not perceptible. In addition, the volatiles released by *S. littoralis*-damaged maize signify the presence of an actual feeding larva, the host of the parasitoid. Together the two blends of volatiles convey more information than one alone. In a similar way, the egg parasitoid, *Chrysonotomyia ruforum*, which is attracted to twigs of *Pinus sylvestris* upon which the sawfly *Diprion pini* has oviposited, responds to a combination of the induced sesquiterpene volatile, (*E*)-β-farnesene, and the constitutive

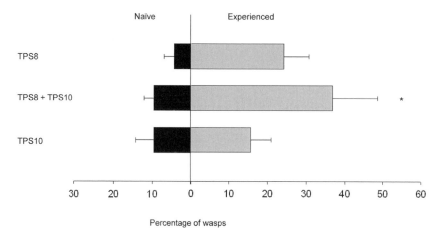

Figure 4.3 Attraction of the parasitic wasp, *Cotesia marginiventris*, in olfactometer experiments to volatile blends from maize, a food plant of its larval hosts (Turlings *et al.*, 2004). Wasps were more attracted to a blend of constitutive and induced sesquiterpenes (asterisk), mimicking the natural blend produced by the caterpillar–plant system, than to either constitutive or induced sesquiterpenes separately (GLM, $P = 0.009$; see Turlings *et al.*, 2004 for details). Volatiles were produced by transgenic *A. thaliana* plants over-expressing maize terpene synthases. Terpene synthase 10 (TPS10) produces most of the herbivore-induced sesquiterpenes of maize; terpene synthase 8 (TPS8) produces most of the constitutively emitted sesquiterpenes of maize. Attraction is compared for naïve wasps (no prior oviposition experience) and experienced wasps (prior oviposition experience on hosts on TPS8 or TPS10 plants or combination). There was no difference in attraction among experienced wasps based on their type of experience. Naïve wasps and those with each type of experience were released in groups of six, and each bioassay was replicated eight times. Bars are ± one standard error of the mean. The percentages do not add to 100% since some wasps made no choice.

volatiles of the host tree (Mumm & Hilker, 2005), although it is not clear which constituents of the full blend are necessary.

In order to determine whether complex mixtures are inherently more attractive to parasitoids than simple mixtures, we carried out another bioassay to compare *C. marginiventris* response with products of one (TPS8), two (TPS8 + TPS10) and three (TPS8 + TPS10 + TPS5) terpene synthases. TPS5 produces the sesquiterpene volatiles of old maize leaves and husks. In this experiment, experienced *C. marginiventris* showed no significant preference for any of the blends offered, but insects tended to orient towards the most complex mixture ($F_{2,45} = 2.67$, $P = 0.08$) (Figure 4.4).

Greater complexity in volatile blends may also be a driving force in attraction of mutualists for pollination. Of the 60 components of the floral scent of

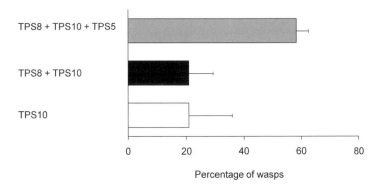

Figure 4.4 Attraction of *Cotesia marginiventris* to blends of increasing diversity. These parasitoid wasps were offered three choices: (1) volatiles of TPS10, (2) volatiles of TPS10 + TPS8 and (3) volatiles of TPS10 + TPS8 + TPS5. Volatiles were produced by *A. thaliana* over-expressing maize terpene synthases, as in Figure 4.3, but here three plant lines were employed, each expressing one of terpene synthase 5, 8 or 10. Terpene synthase 10 (TPS10) produces most of the herbivore-induced sesquiterpenes of maize; terpene synthase 8 (TPS8) produces most of the constitutive sesquiterpenes of maize; terpene synthase 5 (TPS5) produces constitutive sesquiterpenes of old maize leaves. Experiment was performed with experienced wasps (prior oviposition experience on TPS10). Wasps were released in groups of six, with seven replicates. Percentages of wasps indicate the proportion of the total attracted to each blend. There was no statistically significant preference for any of the blends, but wasp preference tended towards the most diverse blend ($F_{2,45}$ = 2.67, P = 0.08). Bars are ± one standard error of the mean.

Datura wrightii, nine elicited a neural response in the pollinator *Manduca sexta*. However, in behavioural assays these nine compounds were only attractive when offered together, not when offered individually (Riffell *et al.*, 2009). The fact that volatile mixtures occur in more than one type of plant-mutualist interaction may indicate a general need for signal diversity in sustaining this type of communication. Alternatively, as seen in Section 4.3.1, mixtures of volatiles from flowers or vegetative organs could also have been selected for as a way to simultaneously target a variety of mutualists, plus different enemies.

4.3.3   Components act synergistically to provide greater protection

The idea that two or more defence components mixed together could provide greater toxicity or deterrence to plant enemies than the equivalent amount of a single defensive substance has been investigated for over 25 years ever since a seminal review article by May Berenbaum (Berenbaum, 1985) appeared. Compounds as diverse as the triterpenoids of *Azadirachta indica* (neem) oil (Stark & Walter, 1995), the furanocoumarins of *Pastinaca sativa* (parsnip)

(Berenbaum *et al.*, 1991; Berenbaum & Zangerl, 1993) and the amides of *Piper* (Dyer *et al.*, 2003) all show enhanced anti-herbivore activity in mixtures that is more than the sum of the respective individual activities. Mixtures of PSMs have also been shown to be more active than individual compounds against pathogens (Kang *et al.*, 1992; Fewell & Roddick, 1993; Segura *et al.*, 1999; Stermitz *et al.*, 2000) and as allelopathic agents against potential competitors (Voukou *et al.*, 2003). Synergy of this type is expected not only to increase the effectiveness of defence, but also to reduce its cost because a lower mass of defensive compound is needed to attain the same degree of protection.

The synergistic effects of mixtures have been observed among compounds of the same chemical class (Berenbaum *et al.*, 1991; Berenbaum & Zangerl, 1993; Fewell & Roddick, 1993; Stark & Walter, 1995; Segura *et al.*, 1999; Dyer *et al.*, 2003; Voukou *et al.*, 2003; Amirhusin *et al.*, 2007) as well as among compounds of different classes (Berenbaum & Neal, 1985; Gunasena *et al.*, 1988; Kang *et al.*, 1992; Castellanos & Espinosa-Garcia, 1997; Guillet *et al.*, 1998; Stermitz *et al.*, 2000; Hummelbrunner & Isman, 2001; Akhtar & Isman, 2003; Steppuhn & Baldwin, 2007). Not all compounds in a mixture are synergistic with respect to activity against attackers (Diawara *et al.*, 1993; Castellanos & Espinosa-Garcia, 1997), but enough examples of synergy are now known that researchers studying the activity of plant compounds against herbivores or pathogens should always test individual compounds against a background of other substances as well as alone. While older studies have mostly assembled mixtures synthetically by combining purified compounds, recent studies, including some described earlier in this chapter, have used transgenic methods to manipulate the diversity of plant defences (Steppuhn & Baldwin, 2007; Mumm *et al.*, 2008).

Synergism of mixtures in direct defence may be attributed to a number of factors, such as one component of the mixture facilitating the transport of another to the target site. For example, the monoterpenes of *Porophyllum*, which synergise the toxicity of the polyacetylene derivative, α-terthienyl to the lepidopteran *Ostrinia nubilalis*, are proposed to act by disrupting membranes, thus increasing access of α-terthienyl to its target (Guillet *et al.*, 1998). A similar phenomenon has been exploited in drug development by using monoterpenes to increase the ability of certain pharmaceuticals to penetrate the skin (Kanikkannan *et al.*, 2000). The same principle applied in reverse explains the synergistic effect of the alkaloid berberine on the antimicrobial activity of a flavonoid derivative from the medicinal plant *Berberis fremontii* (Stermitz *et al.*, 2000). Here the flavonoid inhibits the membranous multi-drug resistance pump of the bacteria, allowing the antibiotic berberine to accumulate to more toxic levels in cells.

Mixtures of defence compounds may also have more value in defence than individual substances if they possess desirable physical properties. For example, the mixing of lower molecular weight monoterpenes (C_{10}) with higher molecular weight diterpenes (C_{20}) in conifer resin ducts makes the resin more fluid, enabling rapid transport of this anti-herbivore defence to the site of enemy attack (Phillips & Croteau, 1999). The sticky blend of monoterpenes and diterpenes in conifer resin is thought to serve as a physical barrier to herbivore invasion. Once exposed to the atmosphere, the monoterpenes rapidly volatilise leaving behind the diterpenes, which may also serve as a chemical barrier to invasion. A mixture of monoterpenes with higher molecular weight terpenes has also been suggested to have defensive value in leaf glands of *Myrica gale* (Carlton *et al.*, 1992). Here the fluidity of the monoterpenes helps the more viscous sesquiterpenes (C_{15}) spread out and cover the leaf surface when glands are broken, thus deploying the antifungal sesquiterpenes over a wider area.

Another example of terpene defence mixtures that have different physical properties comes from species of *Bursera*, which like conifers also secrete terpenes in resin canals in foliage and stems. *Bursera* spp. whose mixtures are dominated by monoterpenes store these compounds under pressure and squirt them out when leaves are damaged by herbivory, thus killing or deterring herbivorous insects in the vicinity (Becerra *et al.*, 2001). On the other hand, resins dominated by mixtures of sesquiterpenes and diterpenes found in other *Bursera* species are not squirted, presumably because of their higher viscosity. These resins are thought to exert their defensive roles in other ways. The differences between *Bursera* with squirting or non-squirting resin are critical for a group of specialised chrysomelid beetles of the genus *Blepharida*, each of which associates only with a small subset of the more than 100 *Bursera* species (Becerra, 1997). Beetles feeding on squirting *Bursera* cut the leaf ducts before feeding on the leaves and use the resin for their own defence as part of a faecal shield, two behaviours not found in beetle species feeding on non-squirting *Bursera* (Becerra *et al.*, 2001).

A common reason given for why defence mixtures are more active than pure compounds is that the components of a mixture could interfere with each other's metabolism by herbivores resulting in greater persistence of defences and correspondingly greater toxicity to the herbivore. This proposal has been well supported by an investigation of furanocoumarin metabolism in black swallowtail larvae (*Papilio polyxenes*) in which mixtures reduced the ability of swallowtail cytochrome P450 monooxygenase enzymes to oxidise furanocoumarins *in vitro*, and reduced growth *in vivo* when fed to larvae (Berenbaum & Zangerl, 1993). Enhancement of the toxic effects of one defence compound by a close structural analogue has been referred to as analogue synergism (McKey, 1979), and may be due to the analogue competing for binding to a detoxification enzyme or irreversibly binding and inactivating the enzyme (Berenbaum & Zangerl, 1996). Interference with metabolism may also occur in mixtures of chemically unrelated compounds. For instance, monoterpenes have been proposed to synergise the toxicity of α-terthienyl by inhibiting cytochrome P450 enzymes involved in α-terthienyl degradation (Guillet *et al.*, 1998).

Mixtures may act synergistically to defend plants in still other ways. Herbivores sometimes overcome plant defences that reduce nutritional quality by simply consuming more tissue. This is the case for larvae of the lepidopteran *Spodoptera exigua* when feeding on the wild tobacco *Nicotiana attenuata* containing trypsin protease inhibitor (Steppuhn & Baldwin, 2007). However, the presence of the alkaloid nicotine synergises the defensive effect of the protease inhibitor by preventing such compensatory feeding.

When defence compounds act as feeding deterrents to herbivores, deterrence sometimes decreases with prolonged exposure. Such a decrease occurred when purified deterrents were added singly to leaf discs and offered to larvae of the lepidopteran *Trichoplusia ni*, but did not occur when the deterrents were added in mixtures (Akhtar & Isman, 2003). This case of synergism was suggested to be a result of effects on sensory neurons, but the mechanism is not clear.

Mixtures may not only help defend plants in ecological time, but also impede the ability of their herbivores to acquire resistance in evolutionary time. After 40 generations of exposure, the aphid *Myzus persicae* had evolved resistance to azadirachtin, a well-known botanical insecticide extracted from *Azadirachta indica*. But, over the same time span, this insect did not develop resistance to a neem seed extract containing azadirachtin plus a mixture of other limonoid triterpenoids (Feng & Isman, 1995). More recently, the acquisition of resistance to *Bacillus thuringienis* (Bt) toxin by the lepidopteran *Plutella xylostella* was shown to be slower on plants having two different Bt toxins than on plants having only a single Bt toxin (Zhao *et al.*, 2003).

Another advantage of having mixtures lies in the way this allows individual plants to have a defence composition different from that of their neighbours. For example, in *Pinus sylvestris* forests, individual trees may possess different mixtures of monoterpenes (Pakeman *et al.*, 2006). When herbivores or pathogens have already developed an adaptation to circumvent some of the defence compounds present in a host–plant population, a novel defence mixture may prove advantageous (Feeny, 1992), especially for long-lived plants whose short-lived herbivores or pathogens go through many more generations on which natural selection might act than their hosts.

A very different view of the value of PSM mixtures in defence has been put forth in a series of papers by Richard Firn and Clive Jones (Jones & Firn, 1991;

Firn & Jones, 1996, 2003). Rather than attributing any direct defensive value to metabolite mixtures, these authors propose that mixtures represent a reservoir of diversity maintained to respond to future defence challenges. Firn and Jones hypothesise that only a few of the many purported defensive metabolites in a plant are likely to be active against an enemy, so having a mixture gives a better chance of having at least one compound that works. These ideas have been extensively critiqued (Berenbaum & Zangerl, 1996), but final judgement on their validity requires much more knowledge of the biological activities of individual mixture components and their mode of action than we have at present.

4.4   Conclusions and future directions

Plant chemical defences are so often deployed as mixtures containing numerous substances from a single class of compounds that understanding how such mixtures are produced and why they are produced should give important insights into how plants defend themselves. In this contribution, we have attempted a brief survey of the biosynthesis of secondary metabolite mixtures and their cost in relation to the possible benefits of possessing mixtures for defence.

   The pathways of plant secondary metabolism associated with the biosynthesis of defence compounds ensure a high diversity of products within a single class. The repeated addition of basic building blocks, the exploitation of enzymes generating a variety of carbon skeletons, the many possibilities for skeletal modification and the low specificity of later pathway steps all help promote mixture formation at a minimal cost of extra biosynthetic machinery.

   For mixtures of defences within a single compound class to persist in evolution, any costs incurred in their manufacture must be exceeded by the benefits obtained from deploying them rather than single defence compounds. The possible benefits of such mixtures in direct defence range from synergistic activities providing protection against individual enemies to broadened effectiveness against multiple enemies. Mixtures may be a practical way for a plant to survive intense pest pressure without investing heavily in defence, and an evolutionary stable strategy to combat the counteradaptation of its enemies. Mixtures may also be valuable in indirect defence, since they provide a framework for transmitting more informative messages about the presence of specific prey or host species. However, all of these explanations need further validation, and our abilities to assess the costs and benefits of a mixture in a single currency must be improved (Schuman & Baldwin, Chapter 15).

   Ideally, the entire arsenal of potential defence compounds in a plant should be tested singly and in mixtures in order to determine exactly which substances are active against specific attackers or in attracting specific enemies of attackers. This is an enormous task when one considers the large number of defence compounds present in even a single plant species and the range of potential enemies that must be considered. Nevertheless, an extensive range of bioassays must be undertaken. These should be combined with other firmly established methods in plant–herbivore research, such as genetic transformation to increase or decrease the complexity of mixtures, phylogenetic comparison among species with and without defence mixtures and quantitative genetics approaches (Rasmann & Agrawal, 2009). In addition, to understand the potential value of synergism better it is important to carry out more research on the mode of action of mixtures. Very little is known about the molecular mode of action of most plant defences at the individual level, and this is even more true for mixtures. With so much to accomplish, progress in understanding the role of mixtures in plant chemical defence may not be rapid in the near future, but an increased awareness of how widespread mixtures are should encourage researchers to study plant defences in this context.

References

Agrawal, A. A. and Kurashige, N. S. (2003) A role for isothiocyanates in plant resistance against the specialist herbivore *Pieris rapae*. *Journal of Chemical Ecology*, **29**, 1403–1415.

Akhtar, Y. and Isman, M. B. (2003) Binary mixtures of feeding deterrents mitigate the decrease in feeding deterrent response to antifeedants following prolonged exposure in the cabbage looper, *Trichoplusia ni* (Lepidoptera: Noctuidae). *Chemoecology*, **13**, 177–182.

Amirhusin, B., Shade, R. E., Koiwa, H. *et al.* (2007) Protease inhibitors from several classes work synergistically against *Callosobruchus maculatus*. *Journal of Insect Physiology*, **53**, 734–740.

Ashihara, H., Deng, W.-W., Mullen, W. and Crozier, A. (2010) Distribution and biosynthesis of flavan-3-ols in *Camellia sinensis* seedlings and expression of genes encoding biosynthetic enzymes. *Phytochemistry*, **71**, 559–566.

Austin, M. B. and Noel, J. P. (2003) The chalcone synthase superfamily of type III polyketide synthases. *Natural Products Reports*, **20**, 79–110.

Becerra, J. X. (1997) Insects on plants: macroevolutionary chemical trends in host use. *Science*, **276**, 253–256.

Becerra, J. X., Venable, D. L., Evans, P. H. and Bowers, W. S. (2001) Interactions between chemical and mechanical defenses in the plant genus *Bursera* and their implications for herbivores. *American Zoologist*, **41**, 865–876.

Bednarek, P., Pislewska-Bednarek, M., Svatos, A. *et al.* (2009) A glucosinolate metabolism pathway in living plant cells mediates broad-spectrum antifungal defense. *Science*, **323**, 101–106.

Berenbaum, M. R. (1985) Brementown revisited: interactions among allelochemicals in plants. In G. A. Cooper-Driver, T. Swain and E. E. Conn (eds.) *Chemically Mediated Interactions between Plants and Other Organisms. Recent Advances in Phytochemistry* Volume 19. New York: Plenum Press, 139–169.

Berenbaum, M. R. and Neal, J. J. (1985) Synergism between myristicin and xanthotoxin, a naturally co-occurring plant toxicant. *Journal of Chemical Ecology*, **11**, 1349–1358.

Berenbaum, M. R., Nitao, J. K. and Zangerl, A. R. (1991) Adaptive significance of furanocoumarin diversity in *Pastinaca sativa* (Apiaceae). *Journal of Chemical Ecology*, **17**, 207–215.

Berenbaum, M. R. and Zangerl, A. R. (1993) Furanocoumarin metabolism in *Papilio polyxenes*: biochemistry, genetic variability, and ecological significance. *Oecologia*, **95**, 370–375.

Berenbaum, M. R. and Zangerl, A. R. (1996) Phytochemical diversity: adaptation or random variation? In J. T. Romeo, J. A. Saunders and P. Barbosa (eds.) *Phytochemical Diversity and Redundancy in Ecological Interactions. Recent Advances in Phytochemistry* Volume 30. New York: Plenum Press, 1–24.

Bolwell, G. P., Bozak, K. and Zimmerlin, A. (1994) Plant cytochrome P450. *Phytochemistry*, **37**, 1491–1506.

Bones, A. M. and Rossiter, J. T. (2006) The enzymic and chemically induced decomposition of glucosinolates. *Phytochemistry*, **67**, 1053–1067.

Brattsten, L. B. (1992) Metabolic defenses against plant allelochemicals. In G. A. Rosenthal and M. R. Berenbaum (eds.) *Herbivores, Their Interactions with Secondary Plant Metabolites*, 2nd edn Volume II. San Diego, CA: Academic Press, 176–242.

Brown, P. D., Tokuhisa, J. G., Reichelt, M. and Gershenzon, J. (2003) Variation of glucosinolate accumulation among different organs and developmental stages of *Arabidopsis thaliana*. *Phytochemistry*, **62**, 471–481.

Buchanan, B. B., Gruissem, W. and Jones, R. L. (eds.) (2000) *Biochemistry and Molecular Biology of Plants*. Rockville, MD: American Society of Plant Physiologists.

Burow, M., Müller, R., Gershenzon, J. and Wittstock, U. (2006) Altered glucosinolate hydrolysis in genetically engineered *Arabidopsis thaliana* and its influence on the larval development of *Spodoptera littoralis*. *Journal of Chemical Ecology*, **32**, 2333–2349.

Burow, M., Bergner, A., Gershenzon, J. and Wittstock, U. (2007) Glucosinolate hydrolysis in *Lepidium sativum* – identification of the thiocyanate-forming protein. *Plant Molecular Biology*, **63**, 49–61.

Burow, M., Losansky, A., Müller, R. *et al.* (2009) The genetic basis of constitutive and herbivore-induced ESP-independent nitrile formation in Arabidopsis. *Plant Physiology*, **149**, 561–574.

Carlton, R. R., Waterman, P. G., Gray, A. I. and Deans, S. G. (1992) The antifungal activity of the leaf gland volatile oil of sweet gale (*Myrica gale*) (Myricaceae). *Chemoecology*, **3**, 55–59.

Castellanos, I. and Espinosa-Garcia, F. J. (1997) Plant secondary metabolite diversity as a resistance trait against insects: a test with *Sitophilus granarius* (Coleoptera: Curculionidae) and seed secondary metabolites. *Biochemical Systematics and Ecology*, **25**, 591–602.

Cole, R. A. (1976) Isothiocyanates, nitriles and thiocyanates as products of autolysis of glucosinolates in Cruciferae. *Phytochemistry*, **15**, 759–762.

D'Auria, J. C. (2006) Acyltransferases in plants: a good time to be BAHD. *Current Opinion in Plant Biology*, **9**, 331–340.

de Vos, M., Kriksunov, K. L. and Jander, G. (2008) Indole-3-acetonitrile production from indole glucosinolates deters oviposition by *Pieris rapae*. *Plant Physiology*, **146**, 916–926.

Degenhardt, J., Köllner, T. G. and Gershenzon, J. (2009) Monoterpene and sesquiterpene synthases and the origin of terpene skeletal diversity in plants. *Phytochemistry*, **70**, 1621–1637.

Diawara, M. M., Trumble, J. T., White, K. K., Carson, W. G. and Martinez, L. A. (1993) Toxicity of linear furanocoumarins to *Spodoptera exigua*: evidence for antagonistic interactions. *Journal of Chemical Ecology*, **19**, 2473–2484.

Dyer, L. A., Dodson, C. D., Stireman, J. O. III *et al.* (2003) Synergistic effects of three *Piper* amides on generalist and specialist herbivores. *Journal of Chemical Ecology*, **29**, 2499–2514.

Engelen-Eigles, G., Holden, G., Cohen, J. D. and Gardner, G. (2006) The effect of temperature, photoperiod, and light quality on gluconasturtiin concentration in watercress (*Nasturtium officinale* R. Br.). *Journal of Agricultural and Food Chemistry*, **54**, 328–334.

Fahey, J. W., Zalcmann, A. T. and Talalay, P. (2001) The chemical diversity and distribution of glucosinolates and isothiocyanates among plants. *Phytochemistry*, **56**, 5–51.

Falk, K. L. and Gershenzon, J. (2007) The desert locust, *Schistocerca gregaria*, detoxifies the glucosinolates of *Schouwia purpurea* by desulfation. *Journal of Chemical Ecology*, **33**, 1542–1555.

Feeny, P. (1992) The evolution of chemical ecology: contributions from the study of herbivorous insects. In G. A. Rosenthal and M. R. Berenbaum (eds.) *Herbivores: Their Interactions with Secondary Plant Metabolites*, 2nd edn Volume II. San Diego, CA: Academic Press, 1–44.

Feng, R. and Isman, M. B. (1995) Selection for resistance to azadirachtin in the green peach aphid, *Myzus persicae*. *Experientia*, **51**, 831–833.

Fewell, A. M. and Roddick, J. G. (1993) Interactive antifungal activity of the glycoalkaloids alpha-solanine and alpha-chaconine. *Phytochemistry*, **33**, 323–328.

Firn, R. D. and Jones, C. G. (1996) An explanation of secondary product 'redundancy'. In J. T. Romeo, J. A. Saunders and P. Barbosa (eds.) *Phytochemical Diversity and Redundancy in Ecological Interactions*, *Recent Advances in Phytochemistry* Volume 30. New York: Plenum Press, 295–312.

Firn, R. D. and Jones, C. G. (2003) Natural products – a simple model to explain chemical diversity. *Natural Products Reports*, **20**, 382–391.

Geervliet, J. B. F., Vet, L. E. M. and Dicke, M. (1994) Volatiles from damaged plants as major cues in long-range host-searching by the specialist parasitoid *Cotesia rubecula*. *Entomologia Experientalis et Applicata*, **73**, 289–297.

Gershenzon, J. and Kreis, W. (1999) Biosynthesis of monoterpenes, sesquiterpenes, diterpenes, sterols, cardiac glycosides and steroid saponins. In M. Wink (ed.) *Biochemistry of Plant Secondary Metabolism* Volume 2. Sheffield: Sheffield Academic Press, 222–299.

Graser, G., Schneider, B., Oldham, N. J. and Gershenzon, J. (2000) The methionine chain elongation pathway in the biosynthesis of glucosinolates in *Eruca sativa* (Brassicaceae). *Archives of Biochemistry and Biophysics*, **378**, 411–419.

Grootwassink, J. W. D., Balsevich, J. J. and Kolenovsky, A. D. (1990) Formation of sulfatoglucosides from exogenous aldoximes in plant cell cultures and organs. *Plant Science*, **66**, 11–20.

Grubb, C. D., Zipp, B. J., Ludwig-Müller, J. *et al.* (2004) Arabidopsis glucosyltransferase UGT74B1 functions in glucosinolate biosynthesis and auxin homeostasis. *Plant Journal*, **40**, 893–908.

Guillet, G., Belanger, A. and Arnason, J. T. (1998) Volatile monoterpenes in *Porophyllum gracile* and *P. ruderale* (Asteraceae): identification, localization and insecticidal synergism with alpha-terthienyl. *Phytochemistry*, **49**, 423–429.

Gunasena, G. H., Vinson, S. B., Williams, H. J. and Stipanovic, R. D. (1988) Effects of caryophyllene, caryophyllene oxide, and their interaction with gossypol on the growth and development of *Heliothis virescens* (F.) (Lepidoptera: Noctuidae). *Journal of Economic Entomology*, **81**, 93–97.

Halkier, B. A. and Gershenzon, J. (2006) Biology and biochemistry of glucosinolates. *Annual Review of Plant Biology*, **57**, 303–333.

Harvey, J. A., Witjes, L. M. A., Benkirane, M., Duyts, H. and Wagenaar, R. (2007) Nutritional suitability and ecological relevance of *Arabidopsis thaliana* and *Brassica oleracea* as foodplants for the cabbage butterfly, *Pieris rapae*. *Plant Ecology*, **189**, 117–126.

Hertweck, C. (2009) The biosynthetic logic of polyketide diversity. *Angewandte Chemie, International Edition*, **48**, 4688–4716.

Howe, G. A. and Jander, G. (2008) Plant immunity to insect herbivores. *Annual Review of Plant Biology*, **59**, 41–66.

Hummelbrunner. L. A. and Isman, M. B. (2001) Acute, sublethal, antifeedant, and synergistic effects of monoterpenoid essential oil compounds on the tobacco cutworm, *Spodoptera litura* (Lep., Noctuidae). *Journal of Agricultural and Food Chemistry*, **49**, 715–720.

Ibrahim, R. K., Bruneau, A. and Bantignies, B. (1998) Plant O-methyltransferases: molecular analysis, common signature and classification. *Plant Molecular Biology*, **36**, 1–10.

Jones, C. G. and Firn, R. D. (1991) On the evolution of plant secondary chemical diversity. *Philosophical Transactions of the Royal Society of London B*, **333**, 273–280.

Kang, R., Helms, R., Stout, M. J. *et al.* (1992) Antimicrobial activity of the volatile constituents of *Perilla frutescens* and its synergistic effects with polygodial. *Journal of Agricultural and Food Chemistry*, **40**, 2328–2330.

Kanikkannan, N., Kandimalla, K., Lamba, S. S. and Singh, M. (2000) Structure-activity relationship of chemical penetration enhancers in transdermal drug delivery. *Current Medicinal Chemistry*, **7**, 593–608.

Kim, J. H., Lee, B. W., Schroeder, F. C. and Jander, G. (2008) Identification of indole glucosinolate breakdown products with antifeedant effects on *Myzus persicae* (green peach aphid). *Plant Journal*, **54**, 1015–1026.

Klein, M. and Papenbrock, J. (2009) Kinetics and substrate specificities of desulfo-glucosinolate sulfotransferases in *Arabidopsis thaliana. Physiologia Plantarum*, **135**, 140–149.

Kliebenstein, D. J., Kroymann, J., Brown, P. *et al.* (2001a) Genetic control of natural variation in Arabidopsis glucosinolate accumulation. *Plant Physiology*, **126**, 811–825.

Kliebenstein, D. J., Lambrix, V. M., Reichelt, M., Gershenzon, J. and Mitchell-Olds, T. (2001b) Gene duplication in the diversification of secondary metabolism: tandem 2-oxoglutarate-dependent dioxygenases control glucosinolate biosynthesis in Arabidopsis. *Plant Cell*, **13**, 681–693.

Köllner, T. G., Schnee, C., Gershenzon, J. and Degenhardt, J. (2004) The sesquiterpene hydrocarbons of maize (*Zea mays*) form five groups with distinct developmental and organ-specific distributions. *Phytochemistry*, **65**, 1895–1902.

Lambdon, P. W., Hassall, M., Boar, R. R. and Mithen, R. (2003) Asynchrony in the nitrogen and glucosinolate leaf-age profiles of *Brassica*: is this a defensive strategy against generalist herbivores? *Agriculture, Ecosystems and Environment*, **97**, 205–214.

Lambrix, V. M., Reichelt, M., Mitchell-Olds, T., Kliebenstein, D. J. and Gershenzon, J. (2001) The Arabidopsis epithiospecifier protein promotes the hydrolysis of glucosinolates to nitriles and influences *Trichoplusia ni* herbivory. *Plant Cell*, **13**, 2793–2807.

McKey, D. (1979) The distribution of secondary compounds within plants. In G. A. Rosenthal and D. H. Janzen (eds.) *Herbivores: Their Interactions with Plant Secondary Metabolites*, 1st edn. New York: Academic Press, 56–133.

Morant, A. V., Jørgensen, K., Jørgensen, C. *et al.* (2008) β-Glucosidases as detonators of plant chemical defense. *Phytochemistry*, **69**, 1795–1813.

Morrissey, J. P. (2009) Biological activity of defence-related plant secondary metabolites. In A. E. Osbourn and V. Lanzotti (eds.) *Plant-derived Natural Products, Synthesis, Function and Application*. Dordrecht: Springer, 283–299.

Müller, R., de Vos, M., Sun, J. Y. *et al.* (2010) Differential effects of indole and aliphatic glucosinolates on lepidopteran herbivores. *Journal of Chemical Ecology*, **36**, 905–913.

Mumm, R. and Hilker, M. (2005) The significance of background odour for an egg parasitoid to detect plants with host eggs. *Chemical Senses*, **30**, 337–343.

Mumm, R., Burow, M., Bukovinszkine'Kiss, G. et al. (2008) Formation of simple nitriles upon glucosinolate hydrolysis affects direct and indirect defense against the specialist herbivore, *Pieris rapae*. *Journal of Chemical Ecology*, **34**, 1311–1321.

Nomura, T., Quesada, A. L. and Kutchan, T. M. (2008) The new beta-D-glucosidase in terpenoid-isoquinoline alkaloid biosynthesis in *Psychotria ipecacuanha*. *Journal of Biological Chemistry*, **283**, 34650–34659.

Pakeman, R. J., Beaton, J. K., Thoss, V. et al. (2006) The extended phenotype of Scots pine *Pinus sylvestris* structures the understorey assemblage. *Ecography*, **29**, 451–457.

Pfalz, M., Vogel, H. and Kroymann, J. (2009) The gene controlling the *Indole Glucosinolate Modifier1* quantitative trait locus alters indole glucosinolate structures and aphid resistance in *Arabidopsis*. *Plant Cell*, **21**, 985–999.

Phillips, M. A. and Croteau, R. B. (1999) Resin-based defenses in conifers. *Trends in Plant Science*, **4**, 184–190.

Prescott, A. G. and Lloyd, M. D. (2000) The iron(II) and 2-oxoacid-dependent dioxygenases and their role in metabolism. *Natural Products Reports*, **17**, 367–383.

Rasmann, S. and Agrawal, A. A. (2009) Plant defense against herbivory: progress in identifying synergism, redundancy, and antagonism between resistance traits. *Current Opinion in Plant Biology*, **12**, 473–478.

Reichelt, M., Brown, P. D., Schneider, B. et al. (2002) Benzoic acid glucosinolate esters and other glucosinolates from *Arabidopsis thaliana*. *Phytochemistry*, **59**, 663–671.

Riffell, J. A., Lei, H., Christensen, T. A. and Hildebrand, J. G. (2009) Characterization and coding of behaviorally significant odor mixtures. *Current Biology*, **19**, 335–340.

Schnee, C., Köllner, T. G., Held, M. et al. (2006) The products of a single maize sesquiterpene synthase form a volatile defense signal that attracts natural enemies of maize herbivores. *Proceedings of the National Academy of Sciences USA*, **103**, 1129–1134.

Schoonhoven, L. M., van Loon, J. J. A. and Dicke, M, (2005) *Insect–Plant Biology*, 2nd edn. Oxford: Oxford University Press.

Schwab, W. (2003) Metabolome diversity: too few genes, too many metabolites? *Phytochemistry*, **62**, 837–849.

Segura, A., Moreno, M., Madueno, F., Molina, A. and Garcia-Olmedo, F. (1999) Snakin-1, a peptide from potato that is active against plant pathogens. *Molecular Plant–Microbe Interactions*, **12**, 16–23.

Seo, S. T. and Tang, C. S. (1982) Hawaiian fruit-flies (Diptera: Tephritidae) – toxicity of benzyl isothiocyanate against eggs or 1st instars of three species. *Journal of Economic Entomology*, **75**, 1132–1135.

Springob, K. and Kutchan, T. M. (2009) Introduction to the different classes of natural products. In A. E. Osbourn and V. Lanzotti (eds.) *Plant-derived Natural Products, Synthesis, Function and Application*. Dordrecht: Springer, 3–50.

Stark, J. D. and Walter, J. F. (1995) Neem oil and neem oil components affect the efficacy of commercial neem insecticides. *Journal of Agricultural and Food Chemistry*, **43**, 507–512.

Steppuhn, A. and Baldwin, I. T. (2007) Resistance management in a native plant: nicotine prevents herbivores from compensating for plant protease inhibitors. *Ecology Letters*, **10**, 499–511.

Stermitz, F. R., Lorenz, P., Tawara, J. N., Zenewicz, L. A. and Lewis, K. (2000) Synergy in a medicinal plant: antimicrobial action of berberine potentiated by 5´-methoxyhydnocarpin, a multidrug pump inhibitor. *Proceedings of the National Academy of Sciences USA*, **97**, 1433–1437.

Taiz, L. and Zeiger, E. (eds.) (2002) *Plant Physiology*, 3rd edn. Sunderland, MA: Sinauer Associates.

Tholl, D. (2006) Terpene synthases and the regulation, diversity and biological roles of terpene metabolism. *Current Opinion in Plant Biology*, **9**, 297–304.

Turlings, T. C. J., Davison, A. C. and Tamo, C. (2004) A six-arm olfactometer permitting simultaneous observation of insect attraction and odour trapping. *Physiological Entomology*, **29**, 45–55.

Voukou, D., Douvli, P., Blionis, G. J. and Halley, J. M. (2003) Effects of monoterpenoids, acting alone or in pairs, on seed germination and subsequent seedling growth. *Journal of Chemical Ecology*, **29**, 2281–2301.

Waffo, A. F. K., Coombes, P. H., Crouch, N. R. *et al.* (2007) Acridone and furoquinoline alklaoids from *Teclea gerrardii* (Rutaceae: Toddalioideae) of southern Africa. *Phytochemistry*, **68**, 663–667.

Williams, D. J., Critchley, C., Pun, S., Chaliha, M. and O'Hare, T. J. (2009) Differing mechanisms of simple nitrile formation on glucosinolate degradation in *Lepidium sativum* and *Nasturtium officinale* seeds. *Phytochemistry*, **70**, 1401–1409.

Wink, M. (2003) Evolution of secondary metabolites from an ecological and molecular phylogenetic perspective. *Phytochemistry*, **64**, 3–19.

Wittstock, U., Kliebenstein, D. J., Lambrix, V. M., Reichelt, M. and Gershenzon, J. (2003) Glucosinolate hydrolysis and its impact on generalist and specialist insect herbivores. In J. T. Romeo (ed.) *Integrative Phytochemistry: From Ethnobotany to Molecular Ecology*, *Recent Advances in Phytochemistry* Volume 37. Amsterdam: Pergamon, 101–125.

Wittstock, U., Agerbirk, N., Stauber, E. J. *et al.* (2004) Successful herbivore attack due to metabolic diversion of a plant chemical defense. *Proceedings of the National Academy of Sciences USA*, **101**, 4859–4864.

Zhao, J.-Z., Cao, J., Li, Y. *et al.* (2003) Transgenic plants expressing two *Bacillus thuringiensis* toxins delay insect resistance evolution. *Nature Biotechnology*, **21**, 1493–1497.

CHAPTER FIVE

The herbivore's prescription: a pharm-ecological perspective on host-plant use by vertebrate and invertebrate herbivores

JENNIFER SORENSEN FORBEY
Boise State University
MARK D. HUNTER
University of Michigan

5.1   Introduction

Plants, and the organisms that eat them, constitute the majority of terrestrial multicellular diversity (Speight *et al.*, 2008). Indeed, co-evolutionary interactions between herbivores and plants are thought by some to be 'the major zone of interaction responsible for generating terrestrial organic diversity' with plant secondary metabolites (PSMs) playing a central role in co-evolutionary processes (Ehrlich & Raven, 1964). As typically described, plants gain fitness advantages and the potential for evolutionary radiation from mutation or recombination events that generate novel PSMs that deter herbivores (or other attackers and competitors, e.g. pathogens). In turn, counter-adaptations, or offences (Karban & Agrawal, 2002; Sorensen & Dearing, 2006), by herbivore populations favour cladogenesis in the consumers and exert further selection pressure for novel PSMs (Janzen, 1980). Antagonistic interactions between plants and herbivores are, therefore, seen as a driving force behind the great diversity of PSMs that occur in plant populations (Rosenthal & Berenbaum, 1992; Gershenzon *et al.*, Chapter 4).

The broad acceptance of a co-evolutionary arms race between plants and herbivores, with antagonism as the pivotal interaction, has led to the general view that PSMs are toxins that must be avoided, tolerated or overcome by consumers (Speight *et al.*, 2008). Perhaps not surprisingly, many ecologists and evolutionary biologists have simply come to regard PSMs as barriers to consumption, with those barriers overcome to varying degrees by the general-ist and specialist herbivore populations that consume plants (Shipley *et al.*, 2009). However, the development of a tri-trophic perspective of plant–herbivore–enemy interactions (Price *et al.*, 1980) paved the way for a deeper understanding of the role of PSMs in the ecology and evolutionary biology of

The Ecology of Plant Secondary Metabolites: From Genes to Global Processes, eds. Glenn R. Iason, Marcel Dicke and Susan E. Hartley. Published by Cambridge University Press. © British Ecological Society 2012.

herbivores. Variation within and among plant populations is now seen to provide the potential for 'enemy free space' for herbivores (Jeffries & Lawton, 1984; Bernays & Graham, 1988) and a template upon which interactions between herbivores, higher trophic levels and the abiotic environment can occur (Hunter & Price, 1992). Moreover, we recognise a variety of external stressors, including predation, disease and abiotic conditions, that may be ameliorated by some level of PSM consumption (Calvert *et al.*, 1979; Hunter & Schultz, 1993; De Roode *et al.*, 2008; Forbey *et al.*, 2009), whereas the same PSMs can impose fitness costs if consumption rates are too high (Rossiter *et al.*, 1988; van Zandt & Agrawal, 2004). These examples demonstrate that PSMs are neither inherently good nor inherently bad for herbivores. Rather, their net effects on herbivore performance should reflect dose, environmental contingency and the complex biotic and abiotic interactions within which consumption takes place (Hunter *et al.*, 1992)

We propose that the field of pharmacology offers the insight needed to explore the interaction between PSM dose and environmental conditions (level of external stress) and establish the cost–benefit function in these interactions. We begin by describing the criteria needed to demonstrate that PSMs are actually exploited by herbivores for therapeutic effects (e.g. self-medication). The dose–response relationship, the therapeutic window and dose regulation are core concepts in pharmacological research that offer useful methods for assessing the balance between costs and benefits of PSMs. The remainder of the chapter provides examples from vertebrate and invertebrate systems in which the behaviour and performance of herbivores to external stressors and PSMs are viewed from the perspective of pharmacology. We conclude with some suggestions for integrating pharmacology and ecology, termed pharm-ecology (Forbey & Foley, 2009), into future areas of research.

5.2   Pharmacological perspective

Humans have long recognised that drugs, including PSMs, have both therapeutic and toxic properties. Given the potential for toxicity, commercially marketed drugs undergo rigorous tests that require decades of research and considerable financial resources to establish the 'safe' and effective dose (Amir-Aslani & Mangematin, 2010). As such, every prescription comes with the knowledge that the benefit of that drug to mitigate an external stress outweighs the potential toxicity. This balance between benefit and toxicity allows humans to select the best therapeutic dose for a specific ailment.

Is it possible that ecological and evolutionary interactions between plants, herbivores and external stressors can lead to the selection of therapeutic doses of PSMs in natural systems as they have in human–drug interactions? To answer this question we need first to establish criteria that promote self-medication (Singer *et al.*, 2009), or the intentional use of PSMs for therapeutic purposes, by herbivores. The first criterion is that the external stress has a fitness cost. Although a decrease in lifetime reproductive fitness is the best measure of the cost of a stressor, this is often difficult to measure in long-lived species. Measurements that indicate a compromised state of homeostasis may provide valid proxies for fitness. Changes in body mass, metabolism, concentrations of stress hormones (e.g. cortisol), individual survival, growth rate, reproductive output etc. can provide useful endpoints of stress responses. However, these endpoints should eventually be validated to determine if they reflect true changes in fitness. The second criterion is that intake of the PSM decreases the fitness of the consumer (i.e. is toxic) at some dose or in the absence of the stressor. The third criterion is that the consumer intentionally consumes PSMs when the external stressor is present and experiences increased fitness as a result.

5.2.1   Dose–response relationship

Central to these criteria are the pharmacological concepts of dose–response, the therapeutic window and dose regulation. The dose–response relationship describes what PSMs do to the body and is generally referred to as pharmacodynamics (Gibaldi & Perrier, 1982; Tozer & Rowland, 2006). Systemic concentration (i.e. concentration of PSM in the body) is the true link to response. However, we assume linearity between the consumed dose and resultant systemic concentrations and, therefore, often use dose in terms of intake for simplicity. The therapeutic view is represented by a sigmoidal curve, wherein an increase in dose (and therefore systemic concentration) decreases external stress (benefit) through therapeutic properties of the PSM (Figure 5.1a). The toxic view of the dose–response relationship is often represented by the inverse of this curve, wherein an increase in dose (and therefore systemic concentration) elicits an adverse response (cost) through toxic properties of the PSM (Figure 5.1b). Efficacy is represented by the maximum response (percentage of maximum, positive or negative), and potency is represented by the slope of the curve and defines the size of the dose needed to produce a response. A steeper curve indicates a more potent, biologically active PSM, where small changes in dose have rapid effects. The curves together illustrate the concept of chemical hormesis (Calabrese, 2005; Hayes, 2007; Raubenheimer & Simpson, 2009), wherein small doses of a PSM have opposite effects (beneficial) than larger doses of the same PSM (costly).

5.2.2   Therapeutic window

The second concept that is critical for assessing self-medication in herbivores is the therapeutic window (TW). This window describes the relative safety of a potentially therapeutic PSM and is represented by the difference between

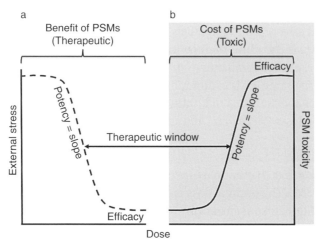

Figure 5.1 Hypothetical dose–response relationships demonstrating the chemical hormetic characteristics of a PSM, wherein increases in dose correspond to increases in systemic concentration, which elicit a therapeutic or toxic response. (a) The left hand, clear panel, dashed line, illustrates the therapeutic benefit of PSMs to reduce external stress at lower doses. (b) The right hand, shaded panel, solid line, illustrates the costs of toxicity at higher doses of a PSM. Efficacy is represented by the maximum response (% of maximum) for the therapeutic effect (bottom of curve in (a)) and toxic effect (top of curve in (b)). The slope of the curve represents therapeutic and toxic potency. Steep slopes indicate more potent therapeutic (a) or toxic (b) responses. The difference between the therapeutic dose and toxic dose is the therapeutic window. Adapted from Calabrese (2005), Hayes (2007) and Raubenheimer and Simpson (2009).

the dose (or corresponding systemic concentrations) alleviating an external stress and the dose resulting in some threshold of toxicity (Figure 5.1). Exceptionally wide TWs may indicate PSMs that are not potent for either reducing stress or increasing toxicity. In other words, they require large increases in dose and subsequent systemic concentration before having an effect. In medicine, the best drugs are those that are potent therapeutics with high efficacy (i.e. maximum response often less than 100%) in patients, but are not potent toxins and do not reach full toxic efficacy. For herbivores, the widest TWs will probably not facilitate self-medication, but may allow herbivores to consume large quantities of PSMs whilst meeting nutritional needs without toxic responses. The therapeutic window is influenced by the biological activity of the PSM and also the physiology of the herbivore of interest. For example, both vertebrate and invertebrate specialist herbivores with mechanisms that minimise PSM absorption and maximise PSM metabolism and elimination (Sorensen & Dearing, 2003; Li *et al.*, 2004; Sorensen *et al.*, 2004; Mao *et al.*, 2006) may experience very low concentrations of systemic PSMs, which may reduce toxicity even when the intake dose is high. Physiological mechanisms that limit systemic concentrations may also reduce efficacy and potency of a PSM to alleviate external stress. Thus, specialists may consume much higher amounts of PSMs to receive the same therapeutic benefit as their generalist counterparts. It is important that comparative studies between different PSMs in the same herbivore or the same PSMs in different herbivores are conducted to adequately investigate the mechanisms that influence the dose–response relationship across a range of doses (McLean & Duncan, 2006; Sorensen *et al.*, 2006).

5.2.3   Dose regulation

Lastly, self-medication requires that animals are capable of regulating feeding to optimise the dose of PSMs. The optimal dose is dependent on what the body does to that dose to influence concentration over time (pharmacokinetics) and what that concentration does to the body (pharmacodynamics) (Figure 5.2). There are many factors that influence pharmacokinetics and pharmacodynamics, and these are described elsewhere (Gibaldi & Perrier, 1982; Tozer & Rowland, 2006). The key point is that there exists an optimal dose that results in a concentration above the therapeutic threshold but below the toxic threshold. In medicine, drugs are formulated and administered specifically to regulate systemic concentrations within the therapeutic window. For example, some drugs are administered intravenously to provide

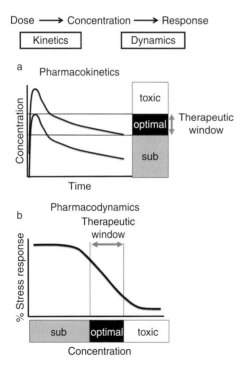

Figure 5.2 The relationship between dose, systemic concentration and response to an ingested PSM. (a) The concentration–time course of an ingested PSM is called pharmacokinetics (PK). The curves represent two doses, one larger than the other, resulting in a larger maximum concentration, but both doses having a similar rate of elimination of PSM (terminal slope). (b) The concentration–response curve is called pharmacodynamics. It represents the effect of a given concentration on the body. Concentrations that are too low are suboptimal (sub) and are therefore not therapeutic, and concentrations that are too high become toxic. The concentration range between suboptimal and toxic is the therapeutic window. Herbivores can theoretically regulate the dose and frequency of doses to maintain an optimal range of concentrations to alleviate a stress response. See McLean and Duncan (2006) and Sorensen *et al.* (2006) for additional details.

maximal systemic concentrations immediately. Others are formulated to allow sustained release following oral administration or are prescribed over various time intervals (much like herbivores might consume plant meals) to sustain systemic concentrations over longer periods of time. The best dosing regime is that which provides the greatest benefit and lowest cost of PSMs in response to an external stress.

There is increasing evidence that herbivores can regulate doses of PSMs to maintain optimal concentrations (Torregrossa & Dearing, 2009). As concentrations of PSMs in the diet increase, many herbivores decrease meal size and increase the intervals between meals as a way to maximise daily intake, but minimise doses of PSMs per meal (Wiggins *et al.*, 2003; Sorensen *et al.*, 2005a; Marsh *et al.*, 2007). Convincing evidence comes from brushtail possums, which, by regulating PSM dose through changes in meal size and frequency, regulate blood concentrations of PSMs below a threshold (approx. $10 \ \mu g \ mL^{-1}$; Boyle *et al.*, 2005; McLean *et al.*, 2008). In these studies, PSM concentration in the blood was maintained even as concentrations of PSMs in the diet changed and as tolerance to them increased. An increase in tolerance due to an increase in activity of detoxification enzymes that reduces systemic concentration of PSMs allowed possums to consume larger doses, but did not increase average blood concentrations (i.e. area under the blood-concentration curve). In other words, tolerant possums eliminated PSMs faster from the circulation, but instead of reducing blood concentrations below some suboptimal level, those possums regulated the dose via increased intake to maintain concentrations of PSMs at a constant level. In this case, possums were not exposed to external stresses and voluntarily consumed the 'optimal' dose, most probably to gain the benefits of maximising nutrient intake while minimising the costs of toxins. Although the mechanisms for dose regulation remain unknown, these results offer the first documentation of self-regulation of PSM concentrations in vertebrate herbivores.

Several studies suggest that invertebrate herbivores also regulate PSM dose relative to its concentration in the diet (Glendinning & Gonzalez, 1995; Simpson & Raubenheimer, 2001; Singer *et al.*, 2002). For example, the presence of PSMs in leaves results in reduced meal number and meal duration in weevils (Wright *et al.*, 2003). Diet mixing may minimise the accumulation of any one PSM in some caterpillar species (Singer *et al.*, 2002) but the selective advantage of diet mixing by insects remains enigmatic (Johns *et al.*, 2009; Karban *et al.*, 2010). Other studies show that PSMs influence feeding by insects in a dose-dependent manner (Senthil-Nathan *et al.*, 2008; Cardinal-Aucoin *et al.*, 2009). However, a critical component missing from these studies is whether changes in feeding behaviour influence systemic concentrations in the body. Additional studies are needed in both vertebrate and invertebrate species that directly measure the amount and frequency of feeding and resulting concentrations of PSMs in the bodies of animals exposed to a range of concentrations of PSMs in the diet.

5.3   Evidence for self-medication with PSMs in herbivores

Is there any evidence that herbivores can regulate dose of PSMs to mitigate the negative consequences of external stress? We propose that the answer is yes. Herbivores meet many of the criteria established for self-medication. External stresses such as predators and parasites have direct and indirect fitness consequences for herbivores (Combes, 2001; Irvine *et al.*, 2006). In addition, there is substantial evidence that PSMs are detrimental and costly to herbivores and are, therefore, generally avoided (Freeland, 1991; Rosenthal & Berenbaum, 1992; Foley & McArthur, 1994; Sorensen *et al.*, 2005c). There is increasing evidence that herbivores choose particular plants when external stresses are elevated (see examples below). There is also evidence that herbivores can regulate intake (Bernays & Chapman, 2000; Wiggins *et al.*, 2003; Sorensen *et al.*, 2005a; Torregrossa & Dearing, 2009) and that doing so can maintain concentrations of PSMs in the body at specific thresholds (Boyle *et al.*, 2005; McLean *et al.*, 2007). Finally, there is evidence that the intentional intake of PSMs can alleviate fitness costs associated with external stresses (see examples below). We now describe some of the more relevant examples that illustrate most, if not all, of the criteria required for demonstrating self-medication in herbivores (see Forbey *et al.*, 2009 for additional examples in vertebrates).

5.3.1   Predators

That PSMs can have both positive and negative effects on herbivores gained general acceptance with pioneering studies of PSM sequestration by insect herbivores as defence against vertebrate predators (Brower *et al.*, 1975; Seiber *et al.*, 1975). Clearly, predators represent a significant source of external stress to herbivores both by eliciting non-consumptive stress and causing mortality (Preisser *et al.*, 2005). The high body concentrations of PSMs sequestered by some herbivores (Wink & Witte, 1991) inevitably impose metabolic costs (if they are metabolised) and storage costs (Rowell-Rahier & Pasteels, 1986; Opitz & Muller, 2009). Sequestered PSMs or their metabolites may also impose toxicity costs and compromise the health of herbivores (Smilanich *et al.*, 2009). However, the costs of sequestering PSMs can be offset if sequestered PSMs provide the benefit of reduced predation (Opitz *et al.*, 2010). Diet mixing by herbivores may represent a foraging strategy to further offset toxicity and opportunity costs associated with sequestration of PSMs (Singer *et al.*, 2004b), thus increasing the benefit-to-cost ratio of sequestration. Herbivores that sequester illustrate elegantly the costs and benefits of pharmacological exploitation of PSMs because there exists a broad range of mimetic insects, grading from toxic to palatable, that represent points along a cost–benefit continuum based upon external conditions (predation pressure, frequency of unpalatable models etc.; Ritland & Brower, 1993). It has been proposed that sequestration may be less common in mobile vertebrate species, because other predator-avoidance strategies, such as fight or flight, are less costly than the costs required to process and sequester high concentrations of PSMs (Forbey *et al.*, 2009). However, studies are needed to determine whether potential direct benefits of sequestration such as immobilisation of PSMs (rather than excretion) are indeed secondary to reduced predation. Future studies should investigate the mechanisms that dictate higher or lower accumulation of PSM concentrations in tissues (i.e. biodistribution), the costs of these mechanisms; and how various levels of sequestered PSMs influence the fitness of herbivores under varying levels of predation risk.

5.3.2   Endoparasites

More recently, ecologists and evolutionary biologists have come to recognise a much broader range of therapeutic exploitation of PSMs beyond those used to mitigate vertebrate predation (Cory & Hoover, 2006). PSMs are exploited for use against invertebrate predators, parasitoids, parasites and agents of disease (Hunter & Schultz, 1993; Huffman *et al.*, 1998; Singer *et al.*, 2004b). Underlying this body of work is the idea that some PSMs are generally 'antibiotic', with a range of negative effects against a diversity of organisms. For example, some phenolic and polyphenolic compounds in plants are active against bacteria, viruses, fungi, nematodes, insects and vertebrates (Schultz *et al.*, 1992). For a herbivore to exploit a polyphenolic pharmacologically requires that the negative impact on the natural enemy more than balances any negative fitness effect of consuming PSMs on the herbivore itself (Foster *et al.*, 1992). Here, we provide several examples of apparent therapeutic use of PSMs by insect and mammalian herbivores. The examples range from generalists to specialist herbivores with stresses ranging from viruses to parasites.

5.3.3   Gypsy moth–host-plant–virus interactions

The gypsy moth, *Lymantria dispar* (Lepidoptera: Lymantriidae), is a polyphagous defoliator of temperate forest trees in Europe, Asia and eastern North America (Doane & Mcmanus, 1981). Despite its wide dietary breadth, the gypsy moth exhibits preferences among host-plant species, and those preferences appear related, in part, to the risk of viral infection (Rossiter, 1987). Plant polyphenols have been implicated as PSMs that may reduce the susceptibility of gypsy moth larvae to a nuclear polyhedrosis virus (NPV), the external stress (Keating *et al.*, 1988). When larvae are dosed with virus on artificial diet, it requires only 800 virus particles to kill 50% of the larvae. In striking contrast, it takes closer to 60 000 virus particles to kill 50% of larvae when

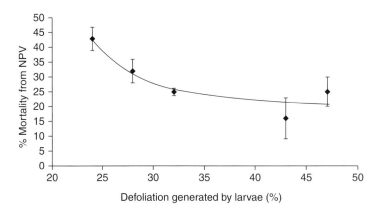

Figure 5.3 Effects of defoliation-induced changes in red oak (*Quercus rubra*) foliage quality on the susceptibility of gypsy moth (*Lymantria dispar*) larvae to a nuclear polyhedrosis virus. Data are from Hunter and Schultz (1993).

virus is administered on foliage containing PSMs. Moreover, virus-induced mortality of gypsy moth among host-plant species is negatively correlated with the polyphenol concentration of the foliage. Experimental application of polyphenols to artificial diet or viral inoculum reduces the susceptibility of larvae to viral infection (Keating *et al.*, 1988, 1990).

In an interesting twist to this story, gypsy moth larvae are known to induce foliar polyphenolic concentrations when they feed on some host-plant species (Rossiter *et al.*, 1988; Hunter & Schultz, 1995). As a consequence, larval susceptibility to virus may decrease as defoliation levels increase. An experimental test of this hypothesis demonstrates that defoliation-induced increases in foliar polyphenols are associated with subsequent inhibition of NPV and higher larval survivorship (Hunter & Schultz, 1993). Simply put, the more foliage that larvae eat, the more pharmaceutically active their diet becomes (Figure 5.3).

As we have stressed previously, we should expect PSMs to have both positive and negative impacts on herbivores, and this is the case with the gypsy moth. In the absence of NPV, foliar polyphenolics have negative impacts on both the growth and fecundity of gypsy moths (Rossiter *et al.*, 1988). The consequences of consuming a polyphenolic diet are, therefore, a balance between negative effects on growth and fecundity at high doses and positive effects on viral inhibition at low doses, illustrating chemical hormesis. The costs and benefits have been modelled analytically (Foster *et al.*, 1992), with results that define an optimal polyphenolic dose for gypsy moth larvae that live in the presence of NPV. Interestingly, that optimal polyphenolic dose is typical of upland oak communities (Kleiner & Montgomery, 1994) where gypsy moth outbreaks typically develop in eastern North America (Doane & McManus, 1981).

5.3.4 Monarch–milkweed–protozoan interactions

If the gypsy moth provides an example of pharmaceutical PSM use by a generalist herbivore, the monarch butterfly, *Danaus plexippus* (Lepidoptera: Danaiidae), provides an example of pharmaceutical PSM use by a dietary specialist. Monarch larvae feed on a subset of species in the genus *Asclepias* (the milkweeds) and closely related plants in the family Apocynaceae. Most milkweed species contain cardenolides, toxic steroids that compromise ion channels in cell membranes (Benson *et al.*, 1977; Mebs *et al.*, 2000). Specialist insects of milkweed, like the monarch, generally have mutations in the genes coding for ion channel proteins and often sequester cardenolides as defences against their own enemies (Seiber *et al.*, 1975, 1980). Although monarchs suffer some fitness costs associated with cardenolide consumption (Zalucki *et al.*, 2001a, 2001b), they are known to gain protection from vertebrate herbivores from sequestered cardenolides (Brower *et al.*, 1975). There is some evidence that cardenolides are also active against parasitic flies in the family Tachinidae that attack monarch larvae (Hunter *et al.*, 1996). More recently, it has been shown that monarchs also gain protection from a protozoan parasite by exploiting milkweeds (De Roode *et al.*, 2008), and we focus on this example below.

Ophryocystis elektroscirrha (McLaughlin & Myers, 1970) (phylum Apicomplexa) is a protozoan parasite that infects monarch populations worldwide (Leong *et al.*, 1997). Transmission generally occurs when females that are laying eggs scatter parasite spores onto leaf surfaces or directly onto the eggs themselves (Altizer *et al.*, 2004). Infection proceeds when caterpillars ingest parasite spores from contaminated eggs or host-plant leaves. Spores lyse in the larval gut, and parasites penetrate the intestinal wall to undergo asexual and sexual replication. Parasite infection imposes a significant fitness cost on hosts, and reduces monarch survival, body mass, ability to fly, mating success and fecundity (De Roode *et al.*, 2007). Recent studies indicate that PSMs in milkweed, specifically cardenolides, can mitigate the costs of parasite infection in monarchs.

The pharmaceutical effects of milkweed on monarchs has been tested by exposing larvae to controlled parasite doses while they are feeding on milkweeds that vary in quality and quantity of cardenolides (De Roode *et al.*, 2008). Larvae consuming parasites on a milkweed species (*Asclepias curassavica*) with higher total concentration and diversity of cardenolides are less likely to be infected than are larvae infected by feeding on a low cardenolide species (*A. incarnata*) (Figure 5.4a). Moreover, the spore load carried by monarchs is lower and adult longevity increases when larvae consume the high cardenolide species (Figure 5.4b, c). In other words, the high cardenolide milkweed, *A. curassavica*, appears to ameliorate some of the fitness costs associated with parasite infection. Consistent with the criteria for self-medication, infected female butterflies preferentially lay their eggs on high cardenolide milkweeds

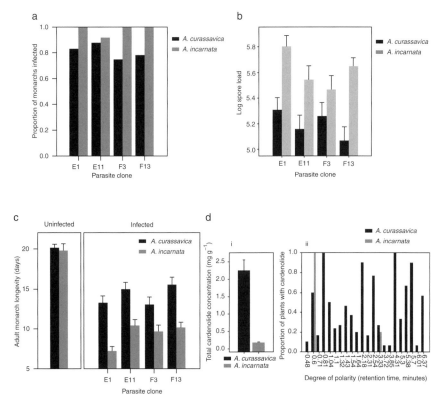

Figure 5.4 A comparison of the effects of milkweeds containing high (*Asclepias curassavica*) and low (*A. incarnata*) cardinolide on the susceptibility of monarch butterfly (*Danaus plexippus*) larvae to a protozoan parasite. Infection levels (a) and spore loads (b) are lower on *A. curassavica* whereas adult longevity (c) is higher. *A curassavica* expresses a greater total concentration (d, i) and diversity (d, ii) of cardenolides in its foliage than does *A. incarnata*. Data are from De Roode *et al.* (2008).

whereas uninfected females do not (Lefèvre *et al.*, 2010). The precise biochemical mechanism underlying the effect of milkweed PSMs on monarch parasites is not yet established. Cardenolides are implicated as potential pharmaceuticals, in part because they are much more concentrated and diverse in *A. curassavica* than in *A. incarnata* (De Roode *et al.*, 2008; Figure 5.4d). Currently, studies are investigating whether *Ophryocystis elektroscirrha* has membrane ion channels that are susceptible to cardenolides (Felibertt *et al.*, 1995) as a potential mechanism for parasite inhibition in monarchs.

5.3.5   Caterpillars–host-plant–parasitoid interactions
The 'holy grail' of pharm-ecology research is evidence of therapeutic self-medication – a change in PSM use by organisms in direct response to an

external stressor that subsequently reduces the fitness costs imposed by that stressor. Although the stressors of interest in self-medication studies are traditionally agents of disease (Clayton & Wolfe, 1993), we consider therapeutic self-medication to include behavioural plasticity in PSM use that alleviates the fitness burden of any external stressor, including predators, parasites and the abiotic environment (Forbey *et al.*, 2009). Studies of therapeutic self-medication in any wild animal are rare, and most provide only anecdotal and equivocal evidence (Clayton & Wolfe, 1993). In insects, most pharmaceutical use of PSMs appears to be prophylactic, in which PSM use is behaviourally fixed and unaffected by the presence or absence of the putative stressor (Keating *et al.*, 1988; Chapuisat *et al.*, 2007). However, in at least two systems (monarchs on milkweed, above, and arctiid caterpillars, below), insects change their use of PSMs depending upon the external conditions that they face, thus demonstrating a major criterion for true self-medication.

In addition to monarch butterflies (Lefèvre *et al.*, 2010), caterpillars in the family Arctiidae provide excellent evidence for therapeutic self-medication in insects (Singer *et al.*, 2004b, 2009). Some arctiid caterpillars are generalists and highly mobile. As a result, they encounter and exploit a variety of host-plant species, with variable quality and quantity of PSMs, during development. For example, the caterpillar *Estigmene acrea* includes both *Senecio longilobus* and *Viguiera dentata* in its diet, the former containing pyrrolizidine alkaloids and the latter lacking them. Laboratory studies indicate that *E. acrea* performs better on *V. dentata* than on *S. longilobus* in the absence of parasitoids, yet older larvae prefer *S. longilobus* under field conditions, where the mortality risk from parasitoids is about 28%. Do the alkaloids in *S. longilobus* help to protect larvae from their parasitoids? It turns out that a mixed diet that includes both plant species confers the greatest protection from parasitism. The combination of high growth (from *V. dentata*) and pharmaceutical defence (from *S. longilobus*) mediates the greatest reduction in parasitism (Singer *et al.*, 2004b).

The pattern of incorporating PSM-rich host plants in their diet occurs in other arctiids also. For example, woolly bear caterpillars, *Grammia incorrupta* (= *geneura*), sacrifice superior growth on alkaloid-free host plants to include plants like *S. longilobus* that contain pyrrolizidine alkaloids in their mixed diets. Mixed diets that include PSM-rich plants provide *G. incorrupta* with some level of protection from at least three species of parasitoids across two insect orders, Braconid wasps and Tachinid flies (Singer *et al.*, 2004a). However, a key question remains: do arctiid caterpillars increase the PSM content of their diets in response to parasitism? In other words, are they capable of therapeutic self-medication?

In at least *G. incorrupta*, the answer appears to be yes. In the absence of parasitoids, excessive ingestion of alkaloids causes a reduction in the survival of *G. incorrupta*, indicating a clear cost associated with alkaloid consumption. However, if larvae are already parasitised, larval survival is greater if caterpillars consume pyrrolizidine alkaloids. Critically, parasitised caterpillars are more likely than unparasitised caterpillars to consume large quantities of alkaloids, providing strong evidence for therapeutic self-medication in *G. incorrupta* (Singer *et al.*, 2009). Future studies should investigate choice of diets in addition to regulation of dose by varying intake of a single diet under both stressed (parasitised) and unstressed (unparasitised) conditions.

5.3.6   Ungulates–tannins–parasite interactions

There are a number of vertebrate animals, including humans and domestic herbivores, that select particular plants and PSMs to mitigate the costs of parasites. Parasites negatively affect the body condition and fecundity of animals (Irvine *et al.*, 2006; Pullan & Brooker, 2008) and therefore represent a large external stressor for vertebrates. Although the majority of evidence for exploiting PSMs to reduce parasite loads is found in the literature on humans (Sneader, 1996), humans and vertebrate herbivores often share the same parasites and, therefore, should both benefit from the antiparasitic properties of PSMs (Huffman *et al.*, 1998).

Studies on grazing livestock provide the most thorough evidence for self-medication by vertebrates to combat parasites. Parasites compromise survival, reproduction and growth rates (Hutchings *et al.*, 2003; Min *et al.*, 2003) in ruminants. Given the economic role of ruminants in food production, substantial effort is aimed at identifying cost-effective methods to control parasite infections. One interesting hypothesis is that ruminants can learn to consume natural pharmaceuticals in plants to treat parasite infection (Lisonbee *et al.*, 2009; Villalba *et al.*, 2010). In support, parasitised lambs consume the optimal dose of tannins in artificial diet to reduce helminthoses (Lisonbee *et al.*, 2009). In addition, sheep voluntarily consume more tannins when infected with parasites than when they are not infected, and this intake results in a decrease in faecal egg counts (Villalba *et al.*, 2010). The voluntary intake of tannins to reduce parasite burdens (0.9 g kg^{-1}; Villalba *et al.*, 2010) is below toxic doses (<1.5 g kg^{-1}) in sheep (Hervas *et al.*, 2003). These results suggest that sheep can detect parasite loads and can voluntarily regulate intake of tannins to reduce the costs of parasites, while minimising the toxicity of tannins.

5.4   Methodological advances in detecting self-medication behaviour

5.4.1   Controlled conditions

To fully understand the costs and benefits of PSM consumption by herbivores and reveal mechanisms that drive diet selection, the exploitation of PSMs for therapeutic use should be investigated under conditions of controlled variation of stress and PSM diversity. First, diet selection should be conducted under conditions that carefully control levels of stress such as predators or disease. Valid therapeutic and toxic endpoints are also needed to assess response. Endpoint measurements should be developed that are closely linked to fitness such as survival, mating success or fecundity (De Roode *et al.*, 2007) and ultimately rates of population growth. Since PSMs often influence herbivores on an individual basis, lifetime reproductive success may be the most relevant fitness parameter to investigate. Pharmacology offers well-established assays for assessing toxicity (Boyd, 1968; Caldwell *et al.*, 2009) and these should be integrated with biological measures of fitness. We recognise that such experimental protocols are more difficult to implement in studies of long-lived species, but we urge those studying vertebrate herbivores to move beyond measurements of food intake as the endpoint response to PSMs.

Researchers should also aim to establish a true dose–response relationship by administering herbivores with a range of doses of PSMs, rather than relying on voluntary consumption, and they should also measure the link between dose, concentrations in the body and various responses. Once dose–response relationships have been established, studies should be conducted that allow herbivores to select diets from a range of food that varies in PSM quality and quantity available in their natural habitat. Because parasites can evolve resistance to a single PSM, and herbivores are likely to be exposed to multiple external stressors simultaneously, the best self-medication practice may be a mixture of PSMs with different biological activities (Villalba & Provenza, 2007). The maximum intake dose in the field of any single PSM should rarely exceed thresholds for toxicity established in controlled dose–response studies. However, dose of any PSM may be higher in the field than that established in laboratory studies, owing to complementary interactions between PSMs, wherein one PSM can increase tolerance to another (Marsh *et al.*, 2006; Mody *et al.*, 2007). Moreover, the dose for PSMs consumed in the field is expected to increase in natural conditions above typical levels when herbivores are exposed to external stress that can be mitigated by PSMs. Field approaches that simultaneously measure shifts in foraging behaviour (i.e. dose regulation and mixed diets) and stress levels may provide further insight into the advantages of mixed diets beyond the traditional hypotheses of nutrient balance and detoxification limitation (Bernays *et al.*, 1994; Dearing *et al.*, 2000; Raubenheimer & Jones, 2006; Mody *et al.*, 2007)

5.4.2   Field conditions

The ultimate goal should be to apply pharm-ecology to understanding host-plant use in the field. This requires developing biomarkers that can be used to assess natural variation in external stressors in herbivores, measuring PSM quality and quantity, and tracking the responses of herbivores to changes in stress and PSMs. This is a major challenge. However, several techniques are currently available and are ripe for broader use in ecological studies. For example, faecal cortisol metabolites (FCMs) are being increasingly used as an effective biomarker of stress in vertebrates (Sheriff *et al.*, 2010). Metabolomics is an emerging pharmacological approach used to monitor and detect physio-logical responses to both external stress from diseases and PSM consumption (Griffin, 2003; Lindon *et al.*, 2004). Infrared thermography provides a non-invasive technique that has been used to detect viral infection in animals (Schaefer *et al.*, 2004; Dunbar *et al.*, 2009) and has the potential to be used to detect parasites in wild animals. Any biomarker for stress should first be vali-dated in the laboratory to ensure that it reflects an accurate assessment of the effects of the external stressor of interest.

Near-infrared reflectance spectrometry (NIRS) offers an emerging tool for assessing the quality and quantity of PSM within and among landscapes (Foley *et al.*, 1998; Wallis & Foley, 2003; Stolter *et al.*, 2006; Wiedower *et al.*, 2009; Moore *et al.*, 2010). For example, it was recently used to describe spatial variation in nitrogen and PSMs in eucalyptus foliage within a habitat and to establish a link between PSM variation and tree use by koalas (Moore *et al.*, 2010). Once NIRS calibrations are established for a given PSM, quantities of PSM can then be determined in the field using hand-held devices (Walsh *et al.*, 2000). Such technology is currently used in the agricultural sector to determine PSMs in fruits (Sinelli *et al.*, 2008). Visible and near-infrared reflectance techniques have also been used to detect stress in plants from pesticides and infection (Delalieux *et al.*, 2007; Luedeling *et al.*, 2009). NIRS is expected to work best for specialist herbivores that have narrow feeding niches because mapping PSM concentration in the habitat can be tran-scribed directly to the quality of food consumed. However, spectral techni-ques applied to a variety of plants could be coupled with telemetry technology that tracks animal location and movement to measure the spa-tial and temporal shifts in foraging in both specialist and generalist herbi-vores in response to changes in parasite loads or other environmental stressors.

Finally, future work should explore the relationship between specialist and generalist diets and the evolution of self-medication (Singer *et al.*,

2004a, 2004b). There is little doubt that escape from stressors can drive, in part, the evolution of host choice in herbivores (Bernays & Graham, 1988; Lill *et al.*, 2002). However, an open question is the degree to which current breadth of diet should influence the prevalence of self-medication. Should we expect dietary generalists to exhibit prophylactic or therapeutic self-medication? How will this compare with dietary specialists? It is tempting to assume that specialist herbivores may have responded evolutionarily to pharmaceutical needs but, as a result, have little flexibility in their current choices among plants. However, even specialists such as the monarch butterfly may exhibit preferences among congeneric plants based on their pharmaceutical properties (Lefèvre *et al.*, 2010). Some specialists with narrow tolerance to plants beyond the preferred host (Futuyma & Moreno, 1988; Mackenzie, 1996; Sorensen *et al.*, 2005b) may have diminished opportunities to self-medicate. It has been suggested that prophylactic self-medication should evolve when parasite risk is predictably high in space or time and that therapeutic self-medication is more beneficial when parasite risk is unpredictable and low (Carrai *et al.*, 2003; Hart, 2005; Castella *et al.*, 2008). However, we cannot assume that dietary specialisation reflects evolution for prophylaxis (Schultz, 1988) nor that generalists are more likely to engage in therapeutic self-medication. Both empirical and theoretical studies of the ecology and evolution of self-medicating behaviours that continue to draw on advances in pharmacology (Forbey & Foley, 2009) are required to answer these questions.

5.5 Conclusions and future directions

Although the notion of pharmaceutical use of PSMs by herbivores has broad appeal, there is increasing recognition that the discipline needs to develop firm criteria for establishing that a behaviour is indeed 'self-medication' (Clayton & Wolfe, 1993; Hart, 2005; Singer *et al.*, 2009). We suggest that the establishment and use of such criteria is of fundamental importance in future research. Just as the term 'co-evolution' came to be used without sufficient care or evidence (Janzen, 1980), so too can the term self-medication be applied without appropriate rigour (Sapolsky, 1994). For a behaviour to be considered self-medication requires that (a) the consumer exhibits increased fitness as a result of the behaviour when the stressor is present; (b) the consumer exhibits decreased fitness as a result of the behaviour when the stressor is absent; and (c) the stressor induces the behaviour (Singer *et al.*, 2009). To this list, we should add that the behaviour should not increase the fitness of any stressor organism (parasite, predator, agent of disease), otherwise it might represent parasite manipulation of host behaviour (Lefèvre *et al.*, 2009). In other words, future studies of self-medication should take a hypothesis-driven, experimental approach based on a priori criteria.

We urge ecologists to use pharmacological insight and tools to establish necessary dose–response relationships. They should test the costs and benefits of stressors, and the intake and systemic concentrations of PSMs, in both laboratory and field conditions. Pharmacological knowledge should also be integrated with advances in metabolomics, remote sensing and telemetry in field studies to fully understand the balance between the positive and negative effects of consuming PSMs under varying environmental conditions. These novel approaches will shed new light on the factors that shape evolutionary trends in host-plant use by herbivores in natural systems.

References

Altizer, S. M., Oberhauser, K. S. and Geurts, K. A. (2004) Transmission of the protozoan parasite, *Ophryocystis elektroscirrha*, in monarch butterfly populations: implications for prevalence and population-level impacts. In Oberhauser, K. S. and Solensky, M. (eds.) *The Monarch Butterfly: Biology and Conservation.* Ithaca, NY, Cornell University Press, 203–218.

Amir-Aslani, A. and Mangematin, V. (2010) The future of drug discovery and development: shifting emphasis towards personalized medicine. *Technological Forecasting and Social Change*, **77**, 203–217.

Benson, J. M., Seiber, J. N., Keeler, R. F. and Johnson, A. E. (1977) Comparative toxicology of cardiac glycosides from milkweed, *Asclepias eriocarpa*. *Toxicology and Applied Pharmacology*, **41**, 131–132.

Bernays, E. A. and Graham, M. (1988) On the evolution of host specificity in phytophagous arthropods. *Ecology*, **69**, 886–892.

Bernays, E. A. and Chapman, R. F. (2000) Plant secondary compounds and grasshoppers: beyond plant defenses. *Journal of Chemical Ecology*, **26**, 1773–1794.

Bernays, E. A., Bright, K. L., Gonzalez, N. and Angel, J. (1994) Dietary mixing in a generalist herbivore – tests of 2 hypotheses. *Ecology*, **75**, 1997–2006.

Boyd, E. M. (1968) Predictive drug toxicity – assessment of drug safety before human use. *Canadian Medical Association Journal*, **98**, 278–293.

Boyle, R. R., McLean, S., Brandon, S. and Wiggins, N. (2005) Rapid absorption of dietary 1,8-cineole results in critical blood concentration of cineole and immediate cessation of eating in the common brushtail possum (*Trichosurus vulpecula*). *Journal of Chemical Ecology*, **31**, 2775–2790.

Brower, L. P., Edmunds, M. and Moffitt, C. M. (1975) Cardenolide content and palatability of a population of *Danaus chrysippus* butterflies from West Africa. *Journal of Entomology*, **49**, 183–196.

Calabrese, E. J. (2005) Paradigm lost, paradigm found: the re-emergence of hormesis as a fundamental dose response model in the toxicological sciences. *Environmental Pollution*, **138**, 378–411.

Caldwell, G. W., Yan, Z. Y., Tang, W. M., Dasgupta, M. and Hasting, B. (2009) ADME optimization and toxicity assessment in early- and late-phase drug discovery. *Current Topics in Medicinal Chemistry*, **9**, 965–980.

Calvert, W. H., Hedrick, L. E. and Brower, L. P. (1979) Mortality of the monarch butterfly (*Danaus plexippus* L.): avian predation at five overwintering sites in Mexico. *Science*, **204**, 847–851.

Cardinal-Aucoin, M., Bauce, E. and Albert, P. J. (2009) Preingestive detection of tannins by *Choristoneura fumiferana* (Lepidoptera: Tortricidae). *Annals of the Entomological Society of America*, **102**, 717–726.

Carrai, V., Borgognini-Tarli, S. M., Huffman, M. A. and Bardi, M. (2003) Increase in tannin consumption by sifaka (*Propithecus verreauxi verreauxi*) females during the birth season: a case for self-medication in prosimians? *Primates*, **44**, 61–66.

Castella, G., Chapuisat, M. and Christe, P. (2008) Prophylaxis with resin in wood ants. *Animal Behaviour*, **75**, 1591–1596.

Chapuisat, M., Oppliger, A., Magliano, P. and Christe, P. (2007) Wood ants use resin to protect themselves against pathogens. *Proceedings of the Royal Society B: Biological Sciences*, **274**, 2013–2017.

Clayton, D. H. and Wolfe, N. D. (1993) The adaptive significance of self-medication. *Trends in Ecology and Evolution*, **8**, 60–63.

Combes, C. (2001) *Parasitism: The Ecology and Evolution of Intimate Interactions*. Chicago, IL: University of Chicago Press.

Cory, J. S. and Hoover, K. (2006) Plant-mediated effects in insect–pathogen interactions. *Trends in Ecology and Evolution*, **21**, 278–286.

De Roode, J. C., Gold, L. R. and Altizer, S. (2007) Virulence determinants in a natural butterfly–parasite system. *Parasitology*, **134**, 657–668.

De Roode, J. C., Pedersen, A. B., Hunter, M. D. and Altizer, S. (2008) Host plant species affects virulence in monarch butterfly parasites. *Journal of Animal Ecology*, **77**, 120–126.

Dearing, M. D., Mangione, A. M. and Karasov, W. H. (2000) Diet breadth of mammalian herbivores: nutrient versus detoxification constraints. *Oecologia*, **123**, 397–405.

Delalieux, S., Van Aardt, J., Keulemans, W., Schrevens, E. and Coppin, P. (2007) Detection of biotic stress (*Venturia inaequalis*) in apple trees using hyperspectral data: non-parametric statistical approaches and physiological implications. *European Journal of Agronomy*, **27**, 130–143.

Doane, C. C. and McManus, M. L. (1981) *The Gypsy Moth: Research Toward Integrated Pest Management. USDA Technical Bulletin 1584.* USDA Forest Service.

Dunbar, M. R., Johnson, S. R., Rhyan, J. C. and Mccollum, M. (2009) Use of infrared thermography to detect thermographic changes in mule deer (*Odocoileus hemionus*) experimentally infected with foot-and-mouth disease. *Journal of Zoo and Wildlife Medicine*, **40**, 296–301.

Ehrlich, P. R. and Raven, P. H. (1964) Butterflies and plants: a study in coevolution. *Evolution*, **18**, 586–608.

Felibertt, P., Bermudez, R., Cervino, V. *et al.* (1995) Ouabain-sensitive Na+, K+-ATPase in the plasma membrane of *Leishmania mexicana*. *Molecular and Biochemical Parasitology*, **74**, 179–187.

Foley, W. J. and McArthur, C. (1994) The effects and costs of allelochemicals for mammalian herbivores: an ecological perspective. In D. J. Chivers and P. Langer (eds.) *The Digestive System in Mammals: Food, Form and Function*. Cambridge: Cambridge University Press.

Foley, W. J., Mcilwee, A., Lawler, I. *et al.* (1998) Ecological applications of near infrared reflectance spectroscopy: a tool for rapid, cost-effective prediction of the composition of plant and animal tissues and aspects of animal performance. *Oecologia*, **116**, 293–305.

Forbey, J. S., and Foley, W. J. (2009) PharmEcology: a pharmacological approach to understanding plant–herbivore interactions. An introduction to the symposium. *Integrative and Comparative Biology*, **49**, 267–273.

Forbey, J. S., Harvey, A. L., Huffman, M. A. *et al.* (2009) Exploitation of secondary metabolites by animals: a response to homeostatic challenges. *Integrative and Comparative Biology*, **49**, 314–328.

Foster, M. A., Schultz, J. C. and Hunter, M. D. (1992) Modeling gypsy moth–virus–leaf chemistry interactions: implications of plant quality for pest and pathogen dynamics. *Journal of Animal Ecology*, **61**, 509–520.

Freeland, W. J. (1991) Plant secondary metabolites. Biochemical evolution with herbivores. In R. Palo and C. T. Robbins (eds.) *Plant Defenses Against Mammalian Herbivory.* Boca Raton, FL: CRC Press.

Futuyma, D. J. and Moreno, G. (1988) The evolution of ecological specialization. *Annual Review of Ecology and Systematics*, **19**, 207–233.

Gibaldi, M. and Perrier, D. (1982) *Pharmacokinetics.* New York: Marcel Dekker.

Glendinning, J. I. and Gonzalez, N. A. (1995) Gustatory habituation to deterrent allelochemicals in a herbivore – concentration and compound specificity. *Animal Behaviour*, **50**, 915–927.

Griffin, J. L. (2003) Metabonomics: NMR spectroscopy and pattern recognition analysis of body fluids and tissues for characterisation of xenobiotic toxicity and disease diagnosis. *Current Opinion in Chemical Biology*, **7**, 648–654.

Hart, B. L. (2005) The evolution of herbal medicine: behavioural perspectives. *Animal Behaviour*, **70**, 975–989.

Hayes, D. P. (2007) Nutritional hormesis. *European Journal of Clinical Nutrition*, **61**, 147–159.

Hervas, G., Perez, V., Giraldez, F. J. *et al.* (2003) Intoxication of sheep with quebracho tannin extract. *Journal of Comparative Pathology*, **129**, 44–54.

Huffman, M. A., Ohigashi, H., Kawanaka, M. *et al.* (1998) African great ape self-medication: a new paradigm for treating parasite disease with natural medicines? In Y. Ebizuka (ed.) *Towards Natural Medicine Research in the 21st Century.* Amsterdam: Elsevier Science, 113–123.

Hunter, M. D., Ohgushi, T. and Price, P. W. (1992) *The Effects of Resource Distribution on Animal-Plant Interactions.* San Diego, CA: Academic Press.

Hunter, M. D., Malcolm, S. B. and Hartley, S. E. (1996) Population-level variation in plant secondary chemistry and the population biology of herbivores. *Chemoecology*, **7**, 45–56.

Hunter, M. D. and Price, P. W. (1992) Playing chutes and ladders: bottom-up and top-down forces in natural communities. *Ecology*, **73**, 724–732.

Hunter, M. D. and Schultz, J. C. (1993) Induced plant defenses breached? Phytochemical induction protects an herbivore from disease. *Oecologia*, **94**, 195–203.

Hunter, M. D. and Schultz, J. C. (1995) Fertilization mitigates chemical induction and herbivore responses within damaged oak trees. *Ecology*, **76**, 1226–1232.

Hutchings, M. R., Athanasiadou, S., Kyriazakis, I. and Gordon, I. J. (2003) Can animals use foraging behaviour to combat parasites? *Proceedings of the Nutrition Society*, **62**, 361–370.

Irvine, R. J., Corbishley, H., Pilkington, J. G. and Albon, S. D. (2006) Low-level parasitic worm burdens may reduce body condition in free-ranging red deer (*Cervus elaphus*). *Parasitology*, **133**, 465–475.

Janzen, D. H. (1980) When is it coevolution? *Evolution*, **34**, 611–612.

Jeffries, M. J. and Lawton, J. H. (1984) Enemy free space and the structure of ecological communities. *Biological Journal of the Linnaean Society*, **23**, 269–286.

Johns, R., Quiring, D. T., Lapointe, R. and Lucarotti, C. J. (2009) Foliage-age mixing within balsam fir increases the fitness of a generalist caterpillar. *Ecological Entomology*, **34**, 624–631.

Karban, R. and Agrawal, A. A. (2002) Herbivore offense. *Annual Review of Ecology and Systematics*, **33**, 641–664.

Karban, R., Karban, C., Huntzinger, M., Pearse, I. and Crutsinger, G. (2010) Diet mixing enhances the performance of a generalist caterpillar, *Platyprepia virginalis*. *Ecological Entomology*, **35**, 92–99.

Keating, S. T., Yendol, W. G. and Schultz, J. C. (1988) Relationship between susceptibility of gypsy moth larvae (Lepidoptera: Lymantriidae) to a baculovirus and host plant foliage constituents. *Environmental Entomology*, **17**, 952–958.

Keating, S. T., Hunter, M. D. and Schultz, J. C. (1990) Leaf phenolic inhibition of gypsy moth nuclear polyhedrosis virus: role of polyhedral inclusion body aggregation. *Journal of Chemical Ecology*, **16**, 1445–1457.

Kleiner, K. W. and Montgomery, M. E. (1994) Forest stand susceptibility to the gypsy moth (Lepidoptera: Lymantriidae): species and site effects on foliage quality to larvae. *Environmental Entomology*, **23**, 699–711.

Lefèvre, T., Adamo, S. A., Biron, D. G. *et al.* (2009) Invasion of the body snatchers: the diversity and evolution of manipulative strategies in host-parasite interactions. *Advances in Parasitology*, **68**, 45–83.

Lefèvre, T., Oliver, L., Hunter, M. D. and De Roode, J. C. (2010) Evidence for trans-generational medication in nature. *Ecology Letters*, 1485–1493.

Leong, K. L. H., Yoshimura, M. A. and Kaya, H. K. (1997) Occurrence of a neogregarine protozoan, *Ophryocystis elektroscirrha* McLaughlin and Myers, in populations of monarch and queen butterflies. *Pan-Pacific Entomologist*, **73**, 49–51.

Li, X., Baudry, J., Berenbaum, M. R. and Schuler, M. A. (2004) Structural and functional divergence of insect CYP6B proteins: from specialist to generalist cytochrome P450. *Proceedings of the National Academy of Sciences USA*, **101**, 2939–2944.

Lill, J. T., Marquis, R. J. and Ricklefs, R. E. (2002) Host plants influence parasitism of forest caterpillars. *Nature*, **417**, 170–173.

Lindon, J. C., Holmes, E., Bollard, M. E., Stanley, E. G. and Nicholson, J. K. (2004) Metabonomics technologies and their applications in physiological monitoring, drug safety assessment and disease diagnosis. *Biomarkers*, **9**, 1–31.

Lisonbee, L. D., Villalba, J. J., Provenza, F. D. and Hall, J. O. (2009) Tannins and self-medication: implications for sustainable parasite control in herbivores. *Behavioural Processes*, **82**, 184–189.

Luedeling, E., Hale, A., Zhang, M. H., Bentley, W. J. and Dharmasri, L. C. (2009) Remote sensing of spider mite damage in California peach orchards. *International Journal of Applied Earth Observation and Geoinformation*, **11**, 244–255.

Mackenzie, A. (1996) A trade-off for host plant utilization in the black bean aphid, *Aphis fabae*. *Evolution*, **50**, 155–162.

Mao, W. F., Berhow, M. A., Zangerl, A. R., McGovern, J. and Berenbaum, M. R. (2006) Cytochrome P450-mediated metabolism of xanthotoxin by *Papilio multicaudatus*. *Journal of Chemical Ecology*, **32**, 523–536.

Marsh, K. J., Wallis, I. R., McLean, S., Sorensen, J. S. and Foley, W. J. (2006) Conflicting demands on detoxification pathways influence how common brushtail possums choose their diets. *Ecology*, **87**, 2103–2112.

Marsh, K. J., Wallis, I. R. and Foley, W. J. (2007) Behavioural contributions to the regulated intake of plant secondary metabolites in koalas. *Oecologia*, **154**, 283–290.

McLaughlin, R. E. and Myers, J. (1970) *Ophryocystis elektroscirrha* sp. n., a neogregarine pathogen of monarch butterfly *Danaus plexippus* (L.) and the Florida queen butterfly *D. gilippus berenice* Cramer. *Journal of Protozoology*, **17**, 300–305.

McLean, S. and Duncan, A. J. (2006) Pharmacological perspectives on the detoxification of plant secondary metabolites: implications for ingestive behavior of herbivores. *Journal of Chemical Ecology*, **32**, 1213–1228.

McLean, S., Boyle, R. R., Brandon, S., Davies, N. W. and Sorensen, J. S. (2007) Pharmacokinetics of 1,8-cineole, a dietary toxin, in the brushtail possum (*Trichosurus vulpecula*): significance for feeding. *Xenobiotica*, **37**, 903–922.

McLean, S., Brandon, S., Boyle, R. R. and Wiggins, N. L. (2008) Development of tolerance to the dietary plant secondary metabolite 1,8-cineole by the brushtail possum (*Trichosurus vulpecula*). *Journal of Chemical Ecology*, **34**, 672–680.

Mebs, D., Zehner, R. and Schneider, M. (2000) Molecular studies on the ouabain binding site of the Na+, K+-ATPase in milkweed butterflies. *Chemoecology*, **10**, 201–203.

Min, B. R., Barry, T. N., Attwood, G. T. and Mcnabb, W. C. (2003) The effect of condensed tannins on the nutrition and health of ruminants fed fresh temperate forages: a review. *Animal Feed Science and Technology*, **106**, 3–19.

Mody, K., Unsicker, S. B. and Linsenmair, K. E. (2007) Fitness related diet-mixing by intraspecific host-plant-switching of specialist insect herbivores. *Ecology*, **88**, 1012–1020.

Moore, B. D., Lawler, I. R., Wallis, I. R., Beale, C. and Foley, W. J. (2010) Palatability mapping: a koala's eye view of spatial variation in habitat quality. *Ecology*, **91**, 3165–3176.

Opitz, S. E. W. and Muller, C. (2009) Plant chemistry and insect sequestration. *Chemoecology*, **19**, 117–154.

Opitz, S. E. W., Jensen, S. R. and Muller, C. (2010) Sequestration of glucosinolates and iridoid glucosides in sawfly species of the genus Athalia and their role in defense against ants. *Journal of Chemical Ecology*, **36**, 148–157.

Preisser, E. L., Bolnick, D. I. and Benard, M. F. (2005) Scared to death? The effects of intimidation and consumption in predator–prey interactions. *Ecology*, **86**, 501–509.

Price, P. W., Bouton, C. E., Gross, P. *et al.* (1980) Interactions among three tropic levels: influence of plants on interactions between insect herbivores and natural enemies. *Annual Review of Ecology and Systematics*, **11**, 41–65.

Pullan, R. and Brooker, S. (2008) The health impact of polyparasitism in humans: are we under-estimating the burden of parasitic diseases? *Parasitology*, **135**, 783–794.

Raubenheimer, D. and Jones, S. A. (2006) Nutritional imbalance in an extreme generalist omnivore: tolerance and recovery through complementary food selection. *Animal Behaviour*, **71**, 1253–1262.

Raubenheimer, D. and Simpson, S. J. (2009) Nutritional pharmecology: doses, nutrients, toxins, and medicines. *Integrative and Comparative Biology*, **49**, 329–337.

Ritland, D. B. and Brower, L. P. (1993) A reassessment of the mimicry relationship among viceroys, queens, and monarchs in Florida. *Natural History Museum of Los Angeles County Science Series*, **38**, 129–139.

Rosenthal, G. and Berenbaum, M. (1992) *Herbivores: Their Interaction with Secondary Plant Metabolites*. New York: Academic Press.

Rossiter, M. C. (1987) Use of a secondary host, pitch pine, by non-outbreak populations of the gypsy moth. *Ecology*, **68**, 857–868.

Rossiter, M. C., Schultz, J. C. and Baldwin, I. T. (1988) Relationships among defoliation, red oak phenolics, and gypsy moth growth and reproduction. *Ecology*, **69**, 267–277.

Rowell-Rahier, M. and Pasteels, J. M. (1986) Economics of chemical defense in chrysomelinae. *Journal of Chemical Ecology*, **12**, 1189–1203.

Sapolsky, R. M. (1994) Fallible instinct – a dose of skepticism about the medicinal knowledge of animals. *Sciences New York*, **34**, 13–15.

Schaefer, A. L., Cook, N., Tessaro, S. V. *et al.* (2004) Early detection and prediction of infection using infrared thermography. *Canadian Journal of Animal Science*, **84**, 73–80.

Schultz, J. C. (1988) Many factors influence the evolution of herbivore diets, but plant chemistry is central. *Ecology*, **69**, 896–897.

Schultz, J. C., Hunter, M. D. and Appel, H. M. (1992) Antimicrobial activity of polyphenols mediates plant–herbivore interactions. In R. W. Hemingway and P. E. Laks (eds.) *Plant Polyphenols: Biogenesis, Chemical Properties, and Significance*. New York: Plenum Press.

Seiber, J. N., Benson, J. M., Roeske, C. A. and Brower, L. P. (1975) Qualitative and quantitative aspects of milkweed cardenolide sequestering by monarch butterflies. *Abstracts of Papers of the American Chemical Society*, **170**, 103.

Seiber, J. N., Tuskes, P. M., Brower, L. P. and Nelson, C. J. (1980) Pharmacodynamics of some individual milkweed cardenolides fed to larvae of the monarch butterfly (*Danaus plexippus* L). *Journal of Chemical Ecology*, **6**, 321–339.

Senthil-Nathan, S., Choi, M. Y., Paik, C. H. and Kalaivani, K. (2008) The toxicity and physiological effect of goniothalamin, a styryl-pyrone, on the generalist herbivore, *Spodoptera exigua* Hubner. *Chemosphere*, **72**, 1393–1400.

Sheriff, M. J., Krebs, C. J. and Boonstra, R. (2010) Assessing stress in animal populations: do fecal and plasma glucocorticoids tell the same story? *General and Comparative Endocrinology*, **166**, 614–619.

Shipley, L. A., Forbey, J. S. and Moore, B. D. (2009) Revisiting the dietary niche: when is a mammalian herbivore a specialist? *Integrative and Comparative Biology*, **49**, 274–290.

Simpson, S. J. and Raubenheimer, D. (2001) The geometric analysis of nutrient–allelochemical interactions: a case study using locusts. *Ecology*, **82**, 422–439.

Sinelli, N., Spinardi, A., Di Egidio, V., Mignani, I. and Casiraghi, E. (2008) Evaluation of quality and nutraceutical content of blueberries (*Vaccinium corymbosum* L.) by near and mid-infrared spectroscopy. *Postharvest Biology and Technology*, **50**, 31–36.

Singer, M. S., Bernays, E. A. and Carriere, Y. (2002) The interplay between nutrient balancing and toxin dilution in foraging by a generalist insect herbivore. *Animal Behaviour*, **64**, 629–643.

Singer, M. S., Carriere, Y., Theuring, C. and Hartmann, T. (2004a) Disentangling food quality from resistance against parasitoids: diet choice by a generalist caterpillar. *American Naturalist*, **164**, 423–429.

Singer, M. S., Rodrigues, D., Stireman, J. O. and Carriere, Y. (2004b) Roles of food quality and enemy-free space in host use by a generalist insect herbivore. *Ecology*, **85**, 2747–2753.

Singer, M. S., Mace, K. C. and Bernays, E. A. (2009) Self-medication as adaptive plasticity: increased ingestion of plant toxins by parasitized caterpillars. *PLoS ONE*, **4**, e4796.

Smilanich, A. M., Dyer, L. A., Chambers, J. Q. and Bowers, M. D. (2009) Immunological cost of chemical defence and the evolution of herbivore diet breadth. *Ecology Letters*, **12**, 612–621.

Sneader, W. (1996) *Drug Prototypes and Their Exploitation*. New York: Wiley.

Sorensen, J. S. and Dearing, M. D. (2003) Elimination of plant toxins: an explanation for dietary specialization in mammalian herbivores. *Oecologia*, **134**, 88–94.

Sorensen, J. S. and Dearing, M. D. (2006) Efflux transporters as a novel herbivore countermechanism to plant chemical defenses. *Journal of Chemical Ecology*, **32**, 1181–1196.

Sorensen, J. S., Turnbull, C. A. and Dearing, M. D. (2004) A specialist herbivore (*Neotoma stephensi*) absorbs fewer plant toxins than a generalist (*Neotoma albigula*). *Physiological and Biochemical Zoology*, **77**, 139–148.

Sorensen, J. S., Heward, E. and Dearing, M. D. (2005a) Plant secondary metabolites alter the feeding patterns of a mammalian herbivore (*Neotoma lepida*). *Oecologia*, **146**, 415–422.

Sorensen, J. S., Mclister, J. D. and Dearing, M. D. (2005b) Novel plant secondary metabolites impact dietary specialists more than generalists (Neotoma spp.). *Ecology*, **86**, 140–154.

Sorensen, J. S., Mclister, J. D. and Dearing, M. D. (2005c) Plant secondary metabolites compromise the energy budgets of specialist and generalist mammalian herbivores. *Ecology*, **86**, 125–139.

Sorensen, J. S., Skopec, M. M. and Dearing, M. D. (2006) Application of pharmacological approaches to plant–mammal interactions. *Journal of Chemical Ecology*, **32**, 1229–1246.

Speight, M. R., Hunter, M. D. and Watt, A. D. (2008) *The Ecology of Insects: Concepts and Applications*. Oxford: Wiley-Blackwell.

Stolter, C., Julkunen-Tiitto, R. and Ganzhorn, J. U. (2006) Application of near infrared reflectance spectroscopy (NIRS) to assess some properties of a sub-arctic ecosystem. *Basic and Applied Ecology*, **7**, 167–187.

Torregrossa, A. M. and Dearing, M. D. (2009) Nutritional toxicology of mammals: regulated intake of plant secondary compounds. *Functional Ecology*, **23**, 48–56.

Tozer, T. N. and Rowland, M. (2006) *Introduction to Pharmacokinetics and Pharmacodynamics (The Quantitative Basis of Drug Therapy)*. Baltimore, MD: Lippincott William & Wilkins.

Van Zandt, P. A. and Agrawal, A. A. (2004) Community-wide impacts of herbivore-induced plant responses in milkweed (*Asclepias syriaca*). *Ecology*, **85**, 2616–2629.

Villalba, J. J. and Provenza, F. D. (2007) Self-medication and homeostatic endeavor in herbivores: learning about the benefits of nature's pharmacy. *Animal*, **1**, 1360–1370.

Villalba, J. J., Provenza, F. D., Hall, J. O. and Lisonbee, L. D. (2010) Selection of tannins by sheep in response to a gastro-intestinal nematode infection. *Journal of Animal Science*, **88**, 2189–2198.

Wallis, I. R. and Foley, W. J. (2003) Validation of near-infrared reflectance spectroscopy to estimate the potential intake of Eucalyptus foliage by folivorous marsupials. *Australian Journal of Zoology*, **51**, 95–98.

Walsh, K. B., Guthrie, J. A. and Burney, J. W. (2000) Application of commercially available, low-cost, miniaturised NIR spectrometers to the assessment of the sugar content of intact fruit. *Australian Journal of Plant Physiology*, **27**, 1175–1186.

Wiedower, E., Hansen, R., Bissell, H. *et al.* (2009) Use of near infrared spectroscopy to discriminate between and predict the nutrient composition of different species and parts of bamboo: application for studying giant panda foraging ecology. *Journal of Near Infrared Spectroscopy*, **17**, 265–273.

Wiggins, N. L., McArthur, C., McLean, S. and Boyle, R. (2003) Effects of two plant secondary metabolites, cineole and gallic acid, on nightly feeding patterns of the common brushtail possum. *Journal of Chemical Ecology*, **29**, 1447–1464.

Wink, M. and Witte, L. (1991) Storage of quinolizidine alkaloids in *Macrosiphum albifrons* and *Aphis genistae* (Homoptera: Aphididae). *Entomologia Generalis*, **15**, 237–254.

Wright, G. A., Simpson, S. J., Raubenheimer, D. and Stevenson, P. C. (2003) The feeding behavior of the weevil, *Exophthalmus jekelianus*, with respect to the nutrients and allelochemicals in host plant leaves. *Oikos*, **100**, 172–184.

Zalucki, M. P., Brower, L. P. and Alonso, A. (2001a) Detrimental effects of latex and cardiac glycosides on survival and growth of first-instar monarch butterfly larvae *Danaus plexippus* feeding on the sandhill milkweed *Asclepias humistrata*. *Ecological Entomology*, **26**, 212–224.

Zalucki, M. P., Malcolm, S. B., Paine, T. D. *et al.* (2001b) It's the first bites that count: survival of first-instar monarchs on milkweeds. *Australian Ecology*, **26**, 547–555.

CHAPTER SIX

Volatile isoprenoids and abiotic stresses

FRANCESCA BAGNOLI, SILVIA FINESCHI and
FRANCESCO LORETO

Consiglio Nazionale delle Ricerche, Istituto per la Protezione delle Piante

6.1 Introduction

Plants produce thousands of chemicals that are not recognised as primary or basic metabolites (i.e. necessary for the survival of the cells). These secondary metabolites usually only occur in special, differentiated cells and are not necessary for the cells themselves, but may be useful for the plant as a whole. Plants at different taxonomic levels (family, genus, species) produce a characteristic mix of secondary metabolites that can be utilised as characters in classifying plants. Both primary and secondary metabolism overlap (Gershenzon *et al.*, Chapter 4) and it is often not understood why a certain compound is produced.

Secondary metabolites can be classified on the basis of their chemical structure, composition, solubility in various solvents or the pathway by which they are synthesised. Three main groups are recognised: isoprenoids (composed almost entirely of carbon and hydrogen); phenolics (made from simple sugars, containing benzene rings, hydrogen and oxygen); and nitrogen-containing compounds (extremely diverse, may also contain sulfur).

Isoprene and monoterpenes, also known as volatile isoprenoids (VIPs; Vickers *et al.*, 2009a), represent a part of the biogenic volatile organic compounds. They play a critical role in interactions between the biosphere and atmosphere, and are key constraints of the physical and chemical properties of the atmosphere and climate. In plants, volatile isoprenoids are predominantly made by photosynthesis intermediates via a chloroplastidic pathway (Lichtenthaler *et al.*, 1997) and are emitted by vegetative organs as well as from flowers (Knudsen *et al.*, 1993) and roots (Steeghs *et al.*, 2004). Isoprenoids are also emitted by animals, but the emission rates by plants are several orders of magnitude higher than by animals, and account for a significant amount of carbon fixed by photosynthesis (Sharkey & Loreto, 1993). Isoprenoids can be emitted by plants constitutively, or the emission may be induced in response to biotic and abiotic stress factors. Both constitutive and induced VIPs act as defensive compounds and are often crucial for plant protection in stressful environments, as well as for plant communication with other organisms

The Ecology of Plant Secondary Metabolites: From Genes to Global Processes, eds. Glenn R. Iason, Marcel Dicke and Susan E. Hartley. Published by Cambridge University Press. © British Ecological Society 2012.

(Schuman & Baldwin, Chapter 15; Dicke *et al.*, Chapter 16). This chapter will review the current understanding of how VIPs contribute to interactions between plants and the surrounding environment, focusing in particular on the possible ecological roles of these compounds in plant protection against abiotic stresses.

6.2   VIPs in plant–atmosphere interactions

In the atmosphere, VIPs perform different actions, depending on the presence of anthropogenic pollutants. In the presence of high levels of nitrogen oxides (NO_x), which are emitted by combustion processes and are therefore largely anthropogenic, VIPs are the main precursors of photochemical ozone (O_3) production in the troposphere (Chameides *et al.*, 1988). In addition to being an important greenhouse gas (Denman *et al.*, 2007), O_3 is also a toxic pollutant that influences ecological interactions (Lindroth, Chapter 7), significantly reduces crop and forestry yield worldwide, and is responsible for health problems in humans and animals during acute pollution episodes (Bernstein *et al.*, 2004).

Volatile isoprenoids are rapidly reacting molecules. As the basic VIP chemical structure is deprived of oxygen atoms (only present in oxygenated isoprenoids as one alcoholic group), these compounds are rapidly oxidised. Once emitted into the atmosphere, VIP oxidation may even affect the levels of radicals (singlet oxygen ($^1O_2$), superoxide (O_2^-), hydrogen peroxide (H_2O_2) and hydroxyl radicals (OH^-)) and hence the overall oxidising capacity of the atmosphere. Isoprenoid scavenging of OH radicals, which is the cleansing agent for many atmospheric pollutants, has been demonstrated over large forested areas (Di Carlo *et al.*, 2004). Volatile isoprenoids, particularly monoterpenes and sesquiterpenes, also affect atmospheric composition via their oxidation products which contribute to the growth of secondary organic aerosol particles (Kavouras *et al.*, 1998). Such particles are climatically important: they scatter and absorb solar and thermal radiation; they act as cloud condensation nuclei, thereby affecting cloud properties and precipitation; they serve as reaction sites for heterogeneous chemistry in the whole troposphere.

6.3   Role of VIPs in plant defence against biotic stressors

When VIPs accumulate in specialised organs of leaves, stems or trunks, such as in most conifers, significant quantities are frequently released after wounding. At such high concentrations, VIPs are phytotoxic and are also toxic to pathogens. They inhibit the infection of wounded tissues and seal the wound by their rapid oxidation and polymerisation (thus solidification) when they come into contact with the atmosphere (Pasqua *et al.*, 2002).

Isoprenoids emitted by flowers attract pollinators and are important for cross-fertilisation. Many studies have shown how constitutive and herbivore-induced VIPs and other volatiles may attract parasitoids and predators

(Gershenzon & Dudareva, 2007). This is a growing area of research with important applied consequences in the fields of sustainable plant growth and biotechnological plant defence against biotic stresses. Indeed, *Arabidopsis thaliana* plants that were engineered to over-express a terpene synthase significantly repelled aphids in a choice test, thus suggesting that emitted monoterpenes could function in plant defence (Aharoni *et al.*, 2003; Chen *et al.*, 2003). Moreover, transgenic *A. thaliana* plants, obtained by switching the subcellular localisation of sesquiterpene synthase to the mitochondria, produced new isoprenoids that indirectly increased their defence by attracting carnivorous predatory mites (Kappers *et al.*, 2005).

Until recently, isoprene, the most simple and abundantly emitted VIP, was not reported to have any function in plant communication with other organisms. However, recent studies (Laothawornkitkul *et al.*, 2008; Loivamäki *et al.*, 2008) demonstrated a repellent action of isoprene emission on herbivorous insects by comparing non-emitting wild types and plants genetically transformed to induce isoprene synthesis. Moreover, it was suggested that the attraction of 'helpful' insects (such as pollinators or carnivores that attack herbivores) could be hampered by neighbouring plants that strongly emit isoprene (Loivamäki *et al.*, 2008). The adaptive value of herbivore-induced plant volatiles (HIPVs) was recently reviewed (Dicke & Baldwin, 2010) and is covered elsewhere in this book (Schuman & Baldwin, Chapter 15; Dicke *et al.*, Chapter 16).

6.4 Role of VIPs in plant protection againstabiotic stressors

The role of VIPs in protecting plants against abiotic stressors has now been demonstrated in many experimental systems, including different plants and different stressors (Loreto & Schnitzler, 2010). However, the mechanism(s) by which such a protective action is accomplished are not fully understood, and the question of whether VIPs have evolved to protect plants against environmental constraints, as well as to defend them against herbivores and pathogens, remains unresolved. We will review the often circumstantial and indirect evidence collected so far, leading to the hypothesis that VIPs exert such a functional role. A key ecological objective is to understand the role of VIPs in protecting plants against abiotic stress, considering the need for plants to develop strategies to adapt to possible rapid climatic changes.

6.4.1 VIP biosynthesis/emission is a process resistant to stress

Volatile isoprenoids are synthesised in the chloroplast from photosynthetic carbon shunted from the Calvin cycle, and a small contribution of extrachloroplastidic carbon sources. This is called the MEP pathway, after its intermediate, methylerythritol phosphate (Lichtenthaler *et al.*, 1997). It is a virtually independent pathway with respect to the mevalonic acid (MVA) pathway that

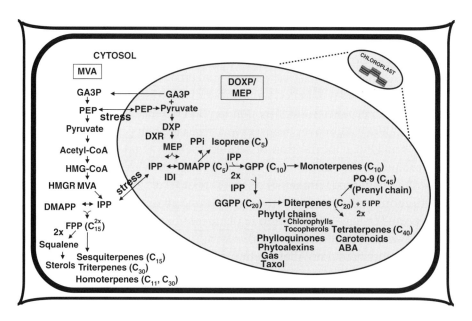

Figure 6.1 Scheme of the chloroplastic (MEP) and extrachloroplastic (MVA) pathways of biosynthesis of plant isoprenoids. Volatile isoprenoids are mainly made of carbon directly shunted from photosynthetic carbon fixation. However, at least two compounds, phosphoenolpyruvate (PEP) and isopentenyl pyrophosphate (IPP), are able to cross membranes and to allow cross-talk between chloroplastic and extrachloroplastic routes, and this cross-talk is suggested to be activated particularly under stress conditions, sustaining chloroplastic VIP biosynthesis and emission. Figure redrawn from Vickers *et al.* (2009a). See plate section for colour version.

occurs in the cytosol and produces, among others, sesquiterpenes and fatty acids (Figure 6.1). Discovery of the MEP pathway explained the dependence of VIP emission on photosynthesis, at least in plants that do not store VIPs in large pools (Loreto & Sharkey, 1990; Loreto *et al.*, 1996). However, many environmental stresses that strongly and rapidly reduce photosynthesis do not affect VIP emission. Most significant is the case of plants undergoing drought stress. Uncoupling of photosynthesis and isoprene emission in drought-stressed plants was first seen by Sharkey and Loreto (1993), but was clearly demonstrated in more recent experiments (Brilli *et al.*, 2007; Fortunati *et al.*, 2008). Labelling experiments revealed that ^{13}C incorporation in the isoprene molecule is low in stressed leaves. Because the labelled carbon is immediately fixed by photosynthesis in the chloroplast, it is now clear that isoprene continues to be formed with a growing contribution of extrachloroplastidic carbon sources when photosynthesis is impaired by the stress. Two compounds, phosphoenolpyruvate (PEP) and isopentenyl pyrophosphate (IPP), have been reported to be shuttled in and out of the chloroplasts, allowing cross-talk between the MVA and MEP pathways of isoprenoid biosynthesis (Vickers *et al.*, 2009a). We hypothesise that such cross-talk is activated when photosynthesis is not a strong enough source of carbon for VIP formation in the chloroplasts, such as in stressed leaves (Figure 6.1). The cytosolic IPP pool is reported to be largely dominant over the chloroplastic IPP pool, and is not rapidly labelled by ^{13}C (Loreto *et al.*, 2004a), which may thus explain why isoprene emitted under stress also remains unlabelled.

6.4.2 VIP biosynthesis/emission increases in plants recovering from abiotic stresses

Stressed plants generally do not die but recover from stresses multiple times during their lifespan (at least in nature; they often die in laboratory experiments!). Besides being resistant to abiotic stresses, VIP emission is also stimulated in plants recovering from the stress. VIP biosynthesis implies significant energy costs for the plant; and thus this increase in emissions suggests a vital function exerted by these compounds.

The seminal experiment by Sharkey and Loreto (1993) showed a four-fold stimulation of isoprene emission in kudzu plants recovering from drought, which was sustained for several days before returning to the original emission level. Brilli *et al.* (2007) showed a similar trend in poplar plants recovering from drought stress, and similar stimulations of VIP emissions occur when plants are mechanically damaged (Loreto & Sharkey, 1993), salt-stressed (Loreto & Delfine, 2000) or exposed to pollutants. In particular, exposure to O_3 may induce larger emission rates of isoprene (Fares *et al.*, 2006) and monoterpenes (Loreto *et al.*, 2004b) over several weeks. Fares *et al.* (2006) also reported that the expression of the gene (*IspS*) encoding for isoprene synthase (ISPS), the enzyme that mediates isoprene synthesis, is enhanced in leaves that have not been directly exposed to O_3 stress; rather, they sense the stress that other leaves of the same plant were exposed to. However, Fortunati *et al.* (2008) showed that the large temperature dependence of VIPs (with a Q_{10} = 2–4, typical of reactions driven enzymatically) is suppressed in plants recovering from stress, a finding that may indicate stress-induced impairment of the enzyme ISPS or of other enzymes of the biosynthetic pathway of VIP. This finding may also have an important consequence when considering that climate change is likely to expose plants to a combination of recurrent droughts under increasing temperatures. Global warming has been thought to induce 'a more fragrant world' (Peñuelas & Staudt, 2010), but the combined effect of drought and rising temperature may somehow reduce or even counterbalance altogether such an expected scenario, and its consequences for plant interactions with organisms and the environment.

6.4.3 VIPs protect plants against temperature stress

Isoprene has lipophilic properties, and its emission strongly increases at rising temperatures. Consequently, it may specifically enhance thermotolerance of leaves by reducing membrane denaturation under heat stress (Sharkey & Singsaas, 1995; Singsaas *et al.*, 1997). Sharkey and collaborators demonstrated that photosynthesis becomes more resistant to high temperatures in isoprene-emitting leaves, and that the protection of photosynthesis depends on the amount of isoprene present in the leaves (Singsaas *et al.*, 1997). Furthermore, evidence was collected that the protective effect was strongest when heat stress occurred rapidly and transiently, such as under heat waves (Sharkey *et al.*, 2001), but recent experiments indicate that such protection is achieved independently of the type of heat stress. Sasaki *et al.* (2007) used a rather unrealistic stress (60 °C) to test thermotolerance, but we now have evidence that thermotolerance is also enhanced in isoprene-emitting plants when exposed to moderate heat (38–40 °C) for several days (Fortunati *et al.*, in preparation). Further evidence of isoprene-enhanced thermotolerance was provided in different isoprene-emitting plant species (Sharkey *et al.*, 2001; Velikova & Loreto, 2005; Velikova *et al.*, 2006) in which isoprene synthesis was blocked by fosmidomycin. The same pattern of protection of photosynthesis against high temperature was detected in *Quercus ilex*, a monoterpene-emitting oak whose emission, suppressed by fosmidomycin, was restored with exogenous monoterpene fumigation (Loreto *et al.*, 1998). The idea that chemical inhibition by fosmidomycin could be an effective and specific way to test isoprene thermal protection has been recently challenged, as Possell *et al.* (2010) reported that fosmidomycin may have important side effects on carbon metabolism. However, an additional opportunity to study the effect of VIPs in thermotolerance induction was offered by the availability of genetically modified plants. Specifically, the use of genetically modified plants in which isoprene biosynthesis has been overexpressed (*A. thaliana*) or silenced (*Populus* spp.) is a powerful tool for understanding biochemical pathways and functions. Recent results showed that isoprene-emitting transgenic plants are more resistant to heat stress than wild types, thus confirming results obtained by chemical inhibition of the emission. Loivamäki *et al.* (2007) used isoprene-emitting *Arabidopsis* genotypes that express *IspS* from *Populus canescens* and verified that transformed plants have significantly enhanced growth rates under high temperature. This indicates the effect of isoprene on enhanced growth. However, this study does not clarify whether the enhanced growth is directly due to a protective role of isoprene. A comparison of *IspS*-overexpressing *Arabidopsis* and wild-type plants under heat stress treatment showed that the transformed plants had higher heat tolerance than the wild types; moreover, high isoprene emission and a decrease in the leaf surface temperature were observed in transgenic plants (Sasaki *et al.*, 2007). This result supports the hypothesis that isoprene plays a specific and crucial role in heat protection because no difference was observed between wild types and isoprene-emitting transformed plants under strong light and drought treatments. Finally, Behnke *et al.* (2007) tested the function of isoprene in the protection against rapid and transient heat stress by using engineered poplars (*P. canescens*) in which isoprene emission was knocked down. Poplar trees lacking the capacity for isoprene emission showed increased damage to photosynthesis by heat spikes relative to control, isoprene-emitting trees. Thus, the observation that VIPs induce thermotolerance and specifically protect photosynthesis against high temperature stress has now been convincingly demonstrated.

6.4.4   VIPs protect plants against oxidant stress factors

Evidence that VIP and reactive oxygen species (ROS) interact can be extrapolated from studies at the atmospheric level (Chameides *et al.*, 1988) down to the leaf mesophyll level, where ROS are known to build up and cause oxidant damage under a number of stress conditions (Loreto & Velikova, 2001). However, a protective function of VIPs against ROS damage has been shown. Acute and short exposure to O_3 does not injure leaves if they are fumigated with exogenous isoprene (Loreto *et al.*, 2001). A contrasting approach tested the antioxidant functions of isoprene by blocking its biosynthesis in a strong emitter, the common reed (*Phragmites australis*) (Loreto & Velikova, 2001). In this case the negative effects of O_3 on the photosynthetic characteristics (photosynthesis, stomatal conductance and fluorescence parameters) were significantly enhanced after blocking isoprene emission. Leaves that were unable to emit isoprene also accumulated larger quantities of ROS and membrane peroxidation markers, indicating a larger degree of oxidative damage at the membrane level.

The discovery that isoprene protects leaves against ROS has been successfully tested in several other systems since the release of the two papers by Loreto and collaborators (Loreto & Velikova, 2001; Loreto *et al.*, 2001). Among others, Affek and Yakir (2002) and Velikova *et al.* (2004) successfully tested the capacity of isoprene to protect against singlet oxygen, another ROS that is produced inside plant leaves by the interaction of molecular oxygen with triplet state chlorophyll under high light intensities.

Isoprene and monoterpenes substantially increase O_3 uptake by leaves whereas O_3-dependent damage is consistently reduced in plants that emit these VIPs (Loreto & Fares, 2007). This finding indicates that O_3 flux inside leaves does not always sufficiently indicate plant damage because O_3 flux may be driven by its efficient detoxification by reacting with VIPs.

The availability of poplar clones characterised by different levels of O_3 sensitivity offered another experimental opportunity to test the claim that

VIPs are efficient antioxidants. The interactive effect of elevated O_3 and CO_2 concentrations in different aspen clones, one O_3-tolerant and the other O_3-sensitive, grown for several years under controlled conditions, was measured (Calfapietra *et al.*, 2008). In the O_3-sensitive clone both elevated CO_2 and O_3 induced a significant decrease in isoprene emission, whereas a significant reduction was not measured for the O_3-tolerant clone. These results suggest that the O_3-sensitivity of poplar clones is inversely related to their capacity to emit isoprene.

Finally, the antioxidant properties of VIPs could be tested on novel experimental systems now available via plant transformation. Vickers *et al.* (2009b) demonstrated that tobacco genotypes engineered to enhance isoprene emission were more resistant to O_3 than non-emitting wild types. In contrast, an experiment with transgenic poplars, in which isoprene emission had been repressed and virtually knocked down, indicated that these genotypes were less sensitive to O_3 (Behnke *et al.*, 2010). However, the genetic transformation of poplars induced the onset of several epigenetic responses, some of which might compensate isoprene functions. For example, poplars with repressed isoprene synthesis showed higher levels of carotenoids and enzymatic antioxidants that are known to cope efficiently with ROS and protect leaves from oxidative damage (Behnke *et al.*, 2010). The discovery that plants may adjust their metabolism to replace compounds that are unavailable for plant protection is important in understanding the mechanisms of acclimation and adaptation to environmental constraints.

6.5   Mechanisms of action of VIPs in abiotic stress

Evidence that VIPs are involved in protection against abiotic stress is accompanied by ambiguous knowledge of the mechanism(s) by which the protective action is exerted. Here, we summarise and comment on the hypothesised mechanisms.

6.5.1   Metabolic hypothesis

The metabolic hypothesis particularly focuses on isoprene and suggests that this compound allows quenching of energy or metabolites, therefore acting as a 'safety valve'. It has been suggested that isoprene emission may dissipate excess carbon under stress conditions imposed by restricting CO_2 acceptors in photosynthesis (Sanadze, 2004). Another study (Rosenstiel *et al.*, 2003) proposed isoprene biosynthesis as a pathway to prevent an excess of chloroplastic dimethylallyl pyrophosphate (DMAPP): the authors observed that there was a significant positive correlation between DMAPP content and isoprene emission, both in whole leaves of *Populus deltoides* and isolated protoplasts, thus indicating that the controls of DMAPP synthesis underlie the suppression of isoprene emission. However, Nogues *et al.* (2006) found that the

DMAPP pool is often inversely associated to isoprene emission, inferring that low emissions are not caused by substrate limitation and that the pool size of the precursor builds up when isoprene is not emitted at high rates.

The 'opportunistic hypothesis' (Owen & Peñuelas, 2005) may also be considered as a metabolic hypothesis. It states that VIPs are formed by DMAPP only after the requirements of higher isoprenoid biosynthesis are satisfied, or that the same pool of carbon may generate volatile and /or essential isoprenoids (namely carotenoids) depending on the need to face different constraints. This is a sound hypothesis based on the acclaimed importance of non-volatile isoprenoids as protective compounds in plants and animals. However, so far there is little supporting experimental evidence. Interestingly, the 'opportunistic hypothesis' may also explain the switch of emissions between different classes of VIP. Brilli *et al.* (2009) detected such a switch between isoprene and monoterpenes in poplar leaves attacked by the beetle *Chrysomela populi*, and demonstrated that induced monoterpene emission by adult leaves attracts insects that would otherwise feed on young leaves constitutively emitting the same monoterpene blend. This seems to be a very sophisticated example of co-evolution, based on the 'opportunistic' use of the different volatiles that can be synthesised within the same biochemical pathway.

6.5.2   Membrane stabilisation

A second group of hypotheses formulate a more distinctive functional role for volatile organic compounds (VOCs) in plant protection against abiotic stressors. The first one derives from the observation that isoprene, as well as monoterpenes, protects leaves against heat-induced damage (Sharkey & Yeh, 2001). To explain this observation, Sharkey and Singsaas (1995) hypothesised that isoprene, and other volatile molecules characterised by the presence of double bounds in the molecule and a high lipophilicity, stabilise chloroplast membranes during high temperature events, thus protecting the photosynthetic apparatus. Under heat stress conditions, membranes become more fluid and the efficiency of photosynthetic processes decreases. Therefore, it was proposed that every lipophilic molecule may partition into the centre of phospholipid bilayer membranes, stabilising lipid–lipid interactions, lipid–protein interactions or protein–protein interactions (Singsaas *et al.*, 1997).

Several experiments support the membrane stabilisation theory: (i) in plants fumigated with terpenes or naturally emitting VIPs, the photosynthetic apparatus is better protected against heat stress than in non-emitters or in plants in which VIP biosynthesis has been chemically blocked (Singsaas *et al.*, 1997; Velikova & Loreto, 2005; Sharkey *et al.*, 2008); (ii) molecular dynamic simulations of phospholipid bilayers with and without isoprene suggest that isoprene actually increases the packing of lipid tails into the free volume at the centre of the membrane bilayer, counteracting in part the effect of a higher temperature (Siwko *et al.*, 2007); (iii) transgenic plants engineered to emit isoprene are more resistant to heat stress than wild types (Behnke *et al.*, 2007; Loivamäki *et al.*, 2007; Sasaki *et al.*, 2007), and heat resistance is reduced in poplar plants that have been engineered to knock down ISPS, thus inhibiting the naturally high isoprene emission in this species; (iv) new experimental evidence is being collected that shows, by the use of circular dichroism, that the photosystems embedded in the photosynthetic membranes are indeed more stable in heat-stressed leaves in which isoprene is also present (A. Fortunati *et al.*, unpublished data).

Experimental evidence shows that isoprene protection is not achieved equally in all plant species and is absent in reconstructed artificial membranes (Logan & Monson, 1999). If the effect of isoprene was only to physically stabilise membranes, then species-specific differences in membrane properties (e.g. fluidity) might explain why protection is not always found. More studies are also needed to establish clearly how isoprene physically stabilises natural membranes under stress conditions.

6.5.3   Antioxidant action

A second functional hypothesis is that VIPs are part of the non-enzymatic oxidative defence system, thus acting as an antioxidant and scavenging ROS, possibly by reactions using the conjugated double bond system (Loreto & Velikova, 2001; Loreto *et al.*, 2001; Affek & Yakir, 2002; Velikova *et al.*, 2004). It is known that non-volatile isoprenoids (namely carotenoids) may directly scavenge ROS and may quench triplet chlorophyll and singlet oxygen (Edreva, 2005). The following findings support the hypothesis that volatile isoprenoids also form part of the non-enzymatic oxidative defence systems and may be explained by the physiological actions outlined below:

1   Under similar abiotic stress conditions, plants that cannot produce isoprene show a greater accumulation of hydrogen peroxide and increased levels of lipid peroxidation and antioxidant enzyme activities, compared with isoprene-emitting plants. This is especially true of, but not exclusively related to, oxidative stresses (Loreto & Velikova, 2001; Velikova & Loreto, 2005; Velikova *et al.*, 2005b, 2006).

2   Isoprene may react directly with O_3, but, because O_3 is almost completely and instantaneously transformed into ROS upon its stomatal uptake by leaves, isoprene and other VIPs are more likely to react with ROS. As already mentioned, VIP–ROS interaction has a series of interesting effects: first, it causes a higher uptake of O_3 into leaves (Loreto & Fares, 2007); second, it quenches the ROS produced by O_3 transformation (Loreto *et al.*, 2001); and third, it reduces physiological, anatomical and ultrastructural damage by ROS (Loreto *et al.*, 2001).

3 VIPs may also have other indirect effects on the oxidative state; for example, they interact with the level of reactive nitrogen species. Endogenous isoprene quenches the NO present in leaves under O_3 exposure (Velikova et al., 2005a, 2008). As H_2O_2 and NO are the two signalling molecules that initiate the hypersensitive stress response leading to cell death (Delledonne et al., 1998; Durner & Klessig, 1999), VIPs may directly or indirectly modulate this vital defensive mechanism. The low formation of NO may be an indirect consequence of the fact that isoprene reduces the oxidative stress at the primary metabolism level (see 1, 2). However, direct reactions with reactive peroxynitrites, formed by oxidation of NO in leaves, cannot be excluded, although such reactions have never been demonstrated in plants. Irrespective of the mechanism, it can be assumed that a further ecological function of isoprene is to remove the excess NO produced under stress conditions, and to prevent NO from reaching cellular concentrations that are damaging in leaves.

4 Isoprene primes the defence system in plants through induction of higher H_2O_2 levels in non-stressed conditions. This effect was discovered by using transformed lines of A. thaliana that produce and emit isoprene or sesquiterpenes (A. Fortunati et al., unpublished data). It is well known that the accumulation of non-toxic levels of H_2O_2 can stimulate a range of defence-related mechanisms, thus activating the enzymatic antioxidant system, namely: ascorbate peroxidase (Karpinski et al., 1997; Panchuk et al., 2002; Devletova et al., 2005); glutathione S transferase; phenylanine ammonia lyase; and heat-shock factors and proteins (Levine et al., 1994; Desikan et al., 1998; Neill et al., 1999; Grant et al., 2000; Vandenabeele et al., 2003; Volkov et al., 2006).

It was observed in Drosophila melanogaster that increased H_2O_2 levels induced the activation of heat-shock factor trimerisation (Zhong et al., 1998), and in Arabidopsis that H_2O_2 regulated the oxidative stress response of ascorbate peroxidase by the activation of a specific heat-shock factor (Devletova et al., 2005). We can therefore assume that the activation of specific heat-shock factors is the crucial mechanism by which H_2O_2 acts as signalling molecule.

6.6 Evolutionary perspectives

Isoprene is the principal VIP and the most abundant VOC released from plants into the atmosphere, although it is only one component of a diverse range of secondary plant metabolites. Since the concept of 'secondary metabolites' was created (Kössel, 1891), it has been assumed that these molecules are generally less important than the 'primary metabolites', mostly because we do not know why plants produce them (Pichersky et al., 2006; Firn & Jones, 2009). Regarding the specific roles of volatile plant secondary metabolites, some authors theorised that many of these molecules have no function for the emitting plant (Firn & Jones, 2006; Owen & Peñuelas, 2006). This suggests that they make no contribution to the fitness of the plant and infers that these molecules may be produced for future benefits. On the other hand, if the metabolite possesses properties that confer benefits on the organism, then selection operates on the fitness consequences thereof and not on future potential. Moreover, in a population, the frequency of new alleles, arising by random mutations, would be unlikely to increase and reach fixation if they did not confer a selective advantage and particularly if they unnecessarily cost the plant energy (Pichersky *et al.*, 2006). Despite there being no definitive answer to the question of the evolutionary and biological significance of isoprene, in the past decade a substantial number of studies have added support to the 'functional role' hypothesis (Sharkey & Singsaas, 1995; Vickers *et al.*, 2009a).

It is worth emphasising that since the first VIP measurements performed in plants, significant progress has been made in the ability to detect volatile compounds (Tholl *et al.*, 2006). Therefore, we would like to draw attention to the following critical observations that may further feed back on evolutionary considerations: (i) many plant species previously classified as non-emitting with respect to isoprene might in fact be low emitters when monitored with more sensitive detection methods; (ii) with the exception of cork oak (Staudt *et al.*, 2004; Loreto *et al.*, 2009), no study has been reported on isoprene emission at the intra-specific level.

Emission of VIPs takes place in most of the major groups of land plants, including dicots, monocots, gymnosperms, pteridophytes and mosses (Kesselmeier & Staudt, 1999), but the scattered distribution of VIP-emitting plants across the phylogeny of land plants suggests multiple gains and/or losses of such metabolites (Figure 6.2). Several hypotheses have been proposed by the scientific community regarding isoprene evolution: the key enzyme ISPS may have evolved many times (Harley *et al.*, 1999), but Hanson *et al.* (1999) proposed that the isoprene emission trait was instead lost many times. More recently, Lerdau and Gray (2003) suggested an independent origin of isoprene emission in gymnosperms and angiosperms, with this occurring only once in the angiosperms, and multiple losses of the trait accounting for the distribution of isoprene emission within this phylum.

Sharkey *et al.* (2005) explained that angiosperm isoprene synthases form a monophyletic group, but it is not yet possible to assess whether this means evolution from a common ancestral gene or convergence based on function. Furthermore, Affek and Yakir (2002) pointed out that isoprene shares a common biochemical production pathway with carotenoids, and they speculated that isoprene was a primitive protection mode against singlet oxygen that evolved into the more dedicated radical scavengers. Vickers *et al.* (2009a) also

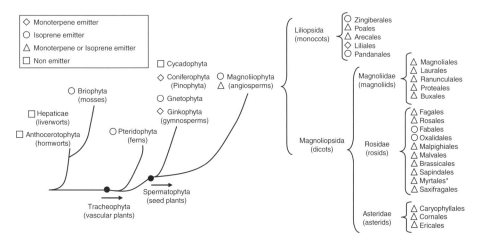

Figure 6.2 The trait isoprene and/or monoterpene emission is assumed to have evolved several times in plants. The trait isoprene emission is present in the most ancient plants, Bryophyta (mosses) and Pteridophyta (ferns). Gymnosperms, the most ancient seed plants, emit isoprene or monoterpenes. Exhaustive information on isoprenoid emission in angiosperms is not yet available, given the large number of angiosperm families and species. Information on VIP inventory studies is given at the order level. In green and blue (see colour plate) are orders for which one or more species that emit isoprene or monoterpenes, respectively, are reported. In red are those orders with species that emit (within the same family or in different families) either isoprene or monoterpenes. Orders for which no studies of VIP emission are available, or no emission has been detected, are not reported.

Within the order Myrtales two genera (*Myrtus* and *Eucalyptus*) contain species that show the unusual feature of emitting simultaneously isoprene and monoterpenes. This figure is based on a picture kindly provided by Professor Thomas Sharkey. See plate section for colour version.

seem to support the idea that isoprene is a primitive trait, possibly evolving when plants became terrestrial in order to help cope with high levels of oxygen and strong oxidative stress, which were previously non-existent in an oxygen-deprived environment such as the aquatic medium (Gross & Bakker, Chapter 8).

Several experiments suggested that isoprene emission characterises mostly hygrophylic (poplars, pedunculate oak) and aquatic plants (common reed), whereas monoterpene emitters are mostly ecologically xerophylic. There is no definite evidence yet about such a correlation between the emission trait and the ecological classification of plants. However, one of the first systematic experiments seems to indicate that hygrophylic tree species of the Italian flora are mainly isoprene emitters (F. Loreto *et al.*, unpublished data).

There is apparently no relationship between VIP emission and phylogeny: for example, genera of the same family, or species within the same genus (e.g. *Quercus*), may have different emission patterns. Even individuals within the same species may show different VIP emission patterns, as is the case for cork oak (*Quercus suber*) (Staudt *et al.*, 2004). It is well known that emission strength and profile could be affected by environmental conditions, or by seasonality (Peñuelas & Llusià, 1999; Niinemets *et al.*, 2004) and, of course, by induction of other biogenic VOCs, e.g. upon the occurrence of biotic stresses. On the other hand, Loreto *et al.* (2009) hypothesised that the emission profiles can be used as markers of intra-specific diversity, thus helping to understand patterns of geographic variations associated with adaptation to the environment. By using plants of different geographical origin that were grown in a 'common garden', Loreto *et al.* (2009) showed that the different VIP emissions of cork oak are not caused by adaptation to the environment, but might rather be associated with past genetic isolation of *Q. suber* populations, whose gene pools have been dated to originate from several million years before present (Magri *et al.*, 2007). Isoprenoids are important in plant stress resistance, permitting the possibility that their geographic diversity might have evolved as a result of environmental pressures. Furthermore, they might drive future adaptations to more severe climatic conditions, such as high temperature and drought. To test this hypothesis, the profile of VIP emission was measured in cork oak plants originating from different European regions and corresponding to the genotypes identified by Magri *et al.* (2007). Individuals from the Portuguese region emitted higher levels of limonene compared with plants originating from all Mediterranean geographic areas. The emission of limonene was especially low in Italian peninsular sites and intermediate in Sardinia and Catalonia. Each of the three major classes of monoterpenes (myrcenes, limonene and pinene) are the products of a distinct terpene synthase. The expression of the genes coding for these enzymes, as well as their regulation, are currently under investigation. The geographic diversity of isoprenoid emission in cork oaks probably reflects the different expression of the corresponding terpene synthase gene. Future phylogeographic studies should focus on the inter- and intra-specific characterisation of terpene synthase genes. However, the trait 'high limonene', which was measured in Portuguese oaks, does not seem to be related to improved resistance to stress factors and is not a result of hybridisation of cork oak with different oak species across the range of this genus. The high limonene emission trait might instead be correlated to selection for a productive trait characteristic of the Portuguese oak, namely cork quality or quantity (Loreto *et al.*, 2009).

6.7   Future directions

Literature on plant VIP emission has steadily increased in the past 15 years, demonstrating the interdisciplinary importance of this field among biologists and chemists. Volatile isoprenoids are currently well characterised in their physiological and ecological functions. On the other hand, more knowledge is needed about the molecular and biochemical mechanisms involved in the role of VIPs in plant protection, particularly in the context of global change, which can directly and indirectly influence biotic and abiotic stress occurrence and mechanisms of resistance to stressors by vegetation. In general, climate and global change will strongly affect induction of VIPs and/or alteration of constitutive VIP emission that may contribute to modifying the communication between plants and other organisms, and plant resistance to environmental constraints. Moreover, VIPs are emerging as fundamental molecules for the signalling of stresses. As one of the most promising prospects to address global change problems lies in the knowledge and use of plant biodiversity to assist the ecosystems to 'adapt' to a rapidly changing climate, a future target might be the manipulation of isoprenoid biosynthesis based, for example, on terpene synthase expression profiles.

Acknowledgements
We thank Megan Beckett for helping with style cleaning and proofreading the final version of this contribution.

References
Affek, H. P. and Yakir, D. (2002) Protection by isoprene against singlet oxygen in leaves. *Plant Physiology*, **129**, 269–277.

Aharoni, A., Giri, A. P., Deuerlein, S. *et al.* (2003) Terpenoid metabolism in wild-type and transgenic *Arabidopsis* plants. *Plant Cell*, **15**, 2866–2884.

Behnke, K., Ehlting, B., Teuber, M. *et al.* (2007) Transgenic, non-isoprene emitting poplars do not like it hot. *Plant Journal*, **51**, 485–499.

Behnke, K., Loivamaki, M., Zimmer, I. *et al.* (2010) Isoprene emission protects photosynthesis in sunfleck exposed grey poplar. *Photosynthesis Research*, **104**, 5–17.

Bernstein, J. A., Alexis, N., Barnes, C. *et al.* (2004) Health effects of air pollution. *Journal of Allergy and Clinical Immunology*, **114**, 1116–1123.

Brilli, F., Barta, C., Fortunati, A. *et al.* (2007) Response of isoprene emission and carbon metabolism to drought in white poplar (*Populus alba*) saplings. *New Phytologist*, **175**, 244–254.

Brilli, F., Ciccioli, P., Frattoni, M. *et al.* (2009) Constitutive and herbivore-induced monoterpenes emitted by Populus x euroamericana leaves are key volatiles that orient *Chrysomela populi* beetles. *Plant Cell and Environment*, **32**, 542–552.

Calfapietra, C., Scarascia Mugnozza, G. S., Karnosky, D. F., Loreto, F. and Sharkey, T. D. (2008) Isoprene emission rates under elevated CO_2 and O_3 in two field-grown aspen clones differing for their sensitivity to O_3. *New Phytologist*, **179**, 55–61.

Chameides, W. L., Lindsay, R. W., Richardson, J. and Kiang, C. S. (1988) The role of biogenic hydrocarbons in urban photochemical smog: Atlanta as a case study. *Science*, **241**, 1473–1475.

Chen, F., Tholl, D., D'Auria, J. C. *et al.* (2003) Biosynthesis and emission of terpenoid volatiles from *Arabidopsis* flowers. *Plant Cell*, **15**, 481–494.

Delledonne, M., Xia, Y., Dixon, R. A. and Lamb, C. (1998) Nitric oxide functions as a signal in plant disease resistance. *Nature*, **394**, 585–588.

Denman, K. L., Brasseur, G., Chidthaisong, A. *et al.* (2007) Couplings between changes in the climate system and biogeochemistry. In S. Solomon, D. Qin, M. Manning *et al.* (eds.) *Climate Change 2007: The Physical Science Basis. Contribution of Working Group I to the Fourth Assessment Report of the Intergovernmental Panel on Climate Change.* Cambridge and New York: Cambridge University Press.

Desikan, R., Burnett, E. C., Hancock, J. T. and Neill, S. J. (1998) Harpin and hydrogen peroxide induce the expression of a homologue of gp91-phox in *Arabidopsis thaliana* suspension cultures. *Journal of Experimental Botany*, **49**, 1767–1771.

Devletova, S., Rizhsky, L., Liang, H. *et al.* (2005) Cytosolic ascorbate peroxidase 1 is a central component of the reactive oxygen gene network of Arabidopsis. *Plant Cell*, **17**, 268–281.

Di Carlo, P., Brune, W. H., Martinez, M. *et al.* (2004) Missing OH reactivity in a forest: evidence for unknown reactive biogenic VOCs. *Science*, **304**, 722–725.

Dicke, M. and Baldwin, I. T. (2010) The evolutionary context for herbivore-induced plant volatiles: beyond the 'cry for help'. *Trends in Plant Science*, **15**, 167–175.

Durner, J. and Klessig, D. (1999) Nitric oxide as a signal in plants. *Current Opinions in Plant Biology*, **2**, 369–374.

Edreva, A. (2005) Generation and scavenging of reactive oxygen species in chloroplasts: a submolecular approach. *Agriculture Ecosystems and Environment*, **106**, 119–133.

Fares, S., Barta, C., Brilli, F. *et al.* (2006) Impact of high ozone on isoprene emission, photosynthesis and histology of developing *Populus alba* leaves directly or indirectly exposed to the pollutant. *Physiologia Plantarum*, **128**, 456–465.

Firn, R. D. and Jones, C. G. (2006) Do we need a new hypothesis to explain plant VOC emissions? *Trends in Plant Science*, **11**, 112–113.

Firn, R. D. and Jones, C. G. (2009) A Darwinian view of metabolism: molecular properties determine fitness. *Journal of Experimental Botany*, **60**, 719–726.

Fortunati, A., Barta, C., Brilli, F. *et al.* (2008) Isoprene emission is not temperature-dependent during and after severe drought-stress: a physiological and biochemical analysis. *Plant Journal*, **55**, 687–697.

Gershenzon, J. and Dudareva, N. (2007) The function of terpene natural products in the natural world. *Nature Chemical Biology*, **3**, 408–414.

Grant, J. J., Yun, B. W. and Loake, G. J. (2000) Oxidative burst and cognate redox signaling reported by luciferase imaging: identification of a signal network that functions independently of ethylene, SA and Me-JA but is dependent on MAPKK activity. *Plant Journal*, **24**, 569–582.

Hanson, D. T., Swanson, S., Graham, L. E. and Sharkey, T. D. (1999) Evolutionary significance of isoprene emission from mosses. *American Journal of Botany*, **86**, 634–639.

Harley, P. C., Monson, R. K. and Lerdau, M. T. (1999) Ecological and evolutionary aspects of isoprene emission from plants. *Oecologia*, **118**, 109–123.

Kappers, I. F., Aharoni, A., van Herpen, T. W. J. M. *et al.* (2005) Genetic engineering of terpenoid metabolism attracts bodyguards to Arabidopsis. *Science*, **309**, 2070–2072.

Karpinski, S., Escobar, C., Karpinska, B., Creissen, G. and Mullineaux, P. M. (1997) Photosynthetic electron transport regulates the expression of cytosolic ascorbate peroxidase genes in Arabidopsis during excess light stress. *Plant Cell*, **9**, 627–640.

Kavouras, I. G., Mihalopoulos, N. and Stephanou, E. G. (1998) Formation of atmospheric particles from organic acids produced by forests. *Nature*, **395**, 683–686.

Kesselmeier, J. and Staudt, M. (1999) Biogenic volatile organic compounds (VOC): an overview on emission, physiology and ecology. *Journal of Atmospheric Chemistry*, **33**, 23–88.

Knudsen, J. T., Tollsten, L. and Bergstorm, L. G. (1993) Floral scents: a checklist of volatile compounds isolated by head-space techniques. *Phytochemistry*, **33**, 253–280.

Kössel, A. (1891) Archives of analytical physiology. *Physiologie Abteilung*, 181–186.

Laothawornkitkul, J., Paul, N. D., Vickers, C. E. *et al.* (2008) Isoprene emissions influence herbivore feeding decisions. *Plant, Cell and Environment*, **31**, 1410–1415.

Lerdau, M. and Gray, D. (2003) Ecology and evolution of light-dependent and light-independent phytogenic volatile organic carbon. *New Phytologist*, **157**, 199–211.

Levine, A., Tenhaken, R., Dixon, R. and Lamb, C. (1994) H_2O_2 from the oxidative burst orchestrates the plant hypersensitive disease resistance response. *Cell*, **79**, 583–593.

Lichtenthaler, H. K., Schwender, J., Disch, A. and Rohmer, M. (1997) Biosynthesis of isoprenoids in higher plant chloroplasts proceeds via a mevalonate-independent pathway. *FEBS Letters*, **400**, 271–274.

Logan, B. A. and Monson, R. K. (1999) Thermotolerance of leaf discs from four isoprene-emitting species is not enhanced by exposure to exogenous isoprene. *Plant Physiology*, **120**, 821–826.

Loivamäki, M., Gilmer, F., Fischbach, R. J. *et al.* (2007) Arabidopsis, a model to study biological functions of isoprene emission? *Plant Physiology*, **144**, 1066–1078.

Loivamäki, M., Mumm, R., Dicke, M. and Schnitzler, J. P. (2008) Isoprene interferes with the attraction of bodyguards by herbaceous plants. *Proceedings of the National Academy of Sciences USA*, **105**, 17430–17435.

Loreto, F. and Delfine, S. (2000) Emission of isoprene from salt-stressed *Eucalyptus globulus* leaves. *Plant Physiology*, **123**, 1605–1610.

Loreto, F. and Fares, S. (2007) Is ozone flux inside leaves only a damage indicator? Clues from volatile isoprenoid studies. *Plant Physiology*, **143**, 1096–1100.

Loreto, F. and Schnitzler, J. P. (2010) Abiotic stresses and induced BVOCs. *Trends in Plant Science*, **15**, 154–166.

Loreto, F. and Sharkey, T. D. (1990) A gas exchange study of photosynthesis and isoprene emission in red oak (*Quercus rubra* L.). *Planta*, **182**, 523–531.

Loreto, F. and Sharkey, T. D. (1993) On the relationship between isoprene emission and photosynthetic metabolites under different environmental-conditions. *Planta*, **189**, 420–424.

Loreto, F. and Velikova, V. (2001) Isoprene produced by leaves protects the photosynthetic apparatus against ozone damage, quenches ozone products, and reduces lipid peroxidation of cellular membranes. *Plant Physiology*, **127**, 1781–1787.

Loreto, F., Ciccioli, P., Brancaleoni, E. *et al.* (1996) Different sources of reduced carbon contribute to form three classes of terpenoid emitted by *Quercus ilex* L. leaves. *Proceedings of the National Academy of Sciences USA*, **93**, 9966–9969.

Loreto, F., Forster, A., Durr, M., Csiky, O. and Seufert, G. (1998) On the monoterpene emission under heat stress and on the increased thermotolerance of leaves of *Quercus ilex* L. fumigated with selected monoterpenes. *Plant Cell and Environment*, **21**, 101–107.

Loreto, F., Mannozzi, M., Maris, C. *et al.* (2001) Ozone quenching properties of isoprene and its antioxidant role in plants. *Plant Physiology*, **126**, 993–1000.

Loreto, F., Pinelli, P., Brancaleoni, E. and Ciccioli, P. (2004a) C-13 labeling reveals chloroplastic and extrachloroplastic pools of dimethylallyl pyrophosphate and their contribution to isoprene formation. *Plant Physiology*, **135**, 1903–1907.

Loreto, F., Pinelli, P., Manes, F. and Kollist, H. (2004b) Impact of ozone on monoterpene emissions and evidence for an isoprene-like antioxidant action of monoterpenes emitted by *Quercus ilex* leaves. *Tree Physiology*, **24**, 361–367.

Loreto, F., Bagnoli, F. and Fineschi, S. (2009) One species, many terpenes: matching chemical and biological diversity. *Trends in Plant Science*, **14**, 416–420.

Magri, D., Fineschi, S., Bellarosa, R. *et al.* (2007) The distribution of *Quercus suber* chloroplast haplotypes matches the palaeogeographical history of the western Mediterranean. *Molecular Ecology*, **16**, 5259–5266.

Neill, S., Desikan, R., Clarke, A. and Hancock, J. (1999) H_2O_2 signaling in plant cells. In M. F. Smallwood, C. M. Calvert and D. J. Bowels (eds.) *Plant Responses to Environmental Stress*. Oxford: BIOS Scientific Publishers, 59–64.

Niinemets, Ü., Loreto, F. and Reichstein, M. (2004) Physiological and physicochemical controls on foliar volatile organic compound emissions. *Trends in Plant Science*, **9**, 180–186.

Nogues, I., Brilli, F. and Loreto, F. (2006) Dimethylallyl diphosphate and geranyl diphosphate pools of plant species characterized by different isoprenoid emissions. *Plant Physiology*, **141**, 721–730.

Owen, S. and Peñuelas, J. (2005) Opportunistic emissions of volatile isoprenoids. *Trends in Plant Science*, **10**, 420–426.

Owen, S. M. and Peñuelas, J. (2006) Response to Firn and Jones: volatile isoprenoids, a special case of secondary metabolism. *Trends in Plant Science*, **11**, 113–114.

Panchuk, I. I., Volkov, R. A. and Schöffl, F. (2002) Heat stress- and heat shock transcription factor-dependent expression and activity of ascorbate peroxidase in Arabidopsis. *Plant Physiology*, **129**, 838–853.

Pasqua, G., Monacelli, B., Manfredini, C., Loreto, F. and Perez, G. (2002) The role of isoprenoid accumulation and oxidation in sealing wounded needles of Mediterranean pines. *Plant Science*, **163**, 355–359.

Peñuelas, J. and Llusià, J. (1999) Seasonal emission of monoterpenes by the Mediterranean tree *Quercus ilex* in field conditions relations with photosynthetic rates, temperature and volatility. *Physiologia Plantarum*, **105**, 641–647.

Peñuelas, J. and Staudt, M. (2010) BVOCs and global change. *Trends in Plant Science*, **15**, 133–144.

Pichersky, E., Sharkey, T. D. and Gershenzon, J. (2006) Plant volatiles: a lack of function or a lack of knowledge? *Trends in Plant Science*, **11**, 421.

Possell, M., Ryan, A., Vickers, C. E., Mullineaux, P. M. and Hewitt, C. N. (2010) Effects of fosmidomycin on plant photosynthesis as measured by gas exchange and chlorophyll fluorescence. *Photosynthesis Research*, **104**, 49–59.

Rosenstiel, T. N., Potosnak, M. J., Griffin, K. L., Fall, R. and Monson, R. K. (2003) Increased CO_2 uncouples growth from isoprene emission in an agriforest ecosystem. *Nature*, **421**, 256–259.

Sanadze, G. A. (2004) Biogenic isoprene (a review). *Russian Journal of Plant Physiology*, **51**, 729–741.

Sasaki, K., Saito, T., Lamsa, M. *et al.* (2007) Plants utilize isoprene emission as a thermotolerance mechanism. *Plant Cell Physiology*, **48**, 1254–1262.

Sharkey, T. D. and Loreto, F. (1993) Water-stress, temperature, and light effects on the capacity for isoprene emission and photosynthesis of kudzu leaves. *Oecologia*, **95**, 328–333.

Sharkey, T. D. and Singsaas, E. L. (1995) Why plants emit isoprene. *Nature*, **374**, 769.

Sharkey, T. D. and Yeh, S. S. (2001) Isoprene emission from plants. *Annual Review of Plant Physiology and Plant Molecular Biology*, **52**, 407–436.

Sharkey, T. D., Chen, X. Y. and Yeh, S. (2001) Isoprene increases thermotolerance of fosmidomycin-fed leaves. *Plant Physiology*, **125**, 2001–2006.

Sharkey, T. D., Yeh, S., Wiberley, A. E. *et al.* (2005) Evolution of the isoprene biosynthetic pathway in kudzu. *Plant Physiology*, **137**, 700–712.

Sharkey, T. D., Wiberley, A. E. and Donohue, A. R. (2008) Isoprene emission from plants: why and how. *Annals of Botany*, **101**, 5–18.

Singsaas, E. L., Lerdau, M., Winter, K. and Sharkey, T. D. (1997) Isoprene increases thermotolerance of isoprene-emitting species. *Plant Physiology*, **115**, 1413–1420.

Staudt, M., Mir, C., Joffre, R. *et al.* (2004) Isoprenoid emissions of *Quercus* spp. (*Q. suber* and *Q. ilex*) in mixed stands contrasting in interspecific genetic introgression. *New Phytologist*, **163**, 573–584.

Siwko, M. E., Marrink, S. J., de Vries, A. H. *et al.* (2007) Does isoprene protect plant membranes from thermal shock? A molecular dynamics study. *Biochemical and Biophysical Acta: Biomembranes*, **1768**, 198–206.

Steeghs, M., Bais, H. P., de Gouw, J. *et al.* (2004) Proton-transfer-reaction mass spectrometry (PTR-MS) as a new tool for real time analysis of root-secreted volatile organic compounds (VOCs) in *Arabidopsis thaliana*. *Plant Physiology*, **135**, 47–58.

Tholl, D., Boland, W., Hansel, A. *et al.* (2006) Practical approaches to plant volatile analysis. *Plant Journal*, **45**, 540–560.

Vandenabeele, S., van der Kelen, K., Dat, J. *et al.* (2003) A comprehensive analysis of hydrogen peroxide-induced gene expression in tobacco. *Proceedings of the National Academy of Sciences USA*, **100**, 16113–16118.

Velikova, V. and Loreto, F. (2005) On the relationship between isoprene emission and thermotolerance in *Phragmites australis* leaves exposed to high temperatures and during the recovery from a heat stress. *Plant Cell and Environment*, **28**, 318–327.

Velikova, V., Edreva, A., Loreto, F. *et al.* (2004) Endogenous isoprene protects *Phragmites australis* leaves against singlet oxygen. *Physiologia Plantarum*, **122**, 219–225.

Velikova, V., Pinelli, P., Pasqualini, S. *et al.* (2005a) Isoprene decreases the concentration of nitric oxide in leaves exposed to elevated ozone. *New Phytologist*, **166**, 419–426.

Velikova, V., Pinelli, P. and Loreto, F. (2005b) Consequences of inhibition of isoprene synthesis in *Phragmites australis* leaves exposed to elevated temperature. *Agriculture, Ecosystems and Environment*, **106**, 209–217.

Velikova, V., Loreto, F., Tsonev, T., Brilli, F. and Edreva, A. (2006) Isoprene prevents the negative consequences of high temperature stress in *Platanus orientalis* leaves. *Functional Plant Biology*, **33**, 931–940.

Velikova, V., Fares, S. and Loreto, F. (2008) Isoprene and nitric oxide reduce damages in leaves exposed to oxidative stress. *Plant Cell and Environment*, **31**, 1882–1894.

Vickers, C. E., Gershenzon, J., Lerdau, M. T. and Loreto, F. (2009a) A unified mechanism of action for volatile isoprenoids in plant abiotic stress. *Nature Chemical Biology*, **5**, 283–291.

Vickers, C. E., Possell, M., Cojocariu, C. I. *et al.* (2009b) Isoprene synthesis protects tobacco plants from oxidative stress. *Plant Cell and Environment*, **32**, 520–531.

Volkov, R. A., Panchuk, I. I., Mullineaux, P. M. and Schöffl, F. (2006) Heat stress-induced H_2O_2 is required for effective expression of heat shock genes in *Arabidopsis*. *Plant Molecular Biology*, **61**, 733–746.

Zhong, M., Orosz, A. and Wu, C. (1998) Direct sensing of heat and oxidation by Drosophila heat shock transcription factor. *Molecular Cell*, **2**, 101–108.

CHAPTER SEVEN

Atmospheric change, plant secondary metabolites and ecological interactions

RICHARD L. LINDROTH

Department of Entomology, University of Wisconsin-Madison

7.1   Introduction

Fifty years ago, when Fraenkel (1959) first placed plant secondary metabolites into an ecological context, the myriad of anthropogenic forces that today influence ecosystem processes at a global scale were poorly recognised, if not altogether unknown. We now know that factors such as atmospheric change, climate warming, invasive species, terrestrial and aquatic eutrophication, and land use are having profound and extensive impacts on the Earth's ecosystems. Less well appreciated, however, are the central roles played by PSMs in many of those processes.

Plant secondary metabolites *respond to* global environmental change; *perpetuate*, via interaction networks, the consequences of global change; and *feed back* to influence future global change (Lindroth, 2010). For example, the carbon cycle, which strongly influences climate, is itself influenced by the chemical matrices into which plants deposit carbon. Rates of photosynthesis are affected by atmospheric CO_2 levels, and subsequent allocation of photosynthates to carbohydrate, cellulose, lignin and tannin pools influences long-term carbon sequestration.

This chapter addresses the effects of two major atmospheric changes – elevated CO_2 and O_3 concentrations – on PSMs and ecological interactions in both 'grazing' and 'decomposing' food webs. Although conceptually similar to the Lindroth (2010) review, this text differs in that the focus is specifically on secondary metabolites, rather than broadly on plant chemistry, and in that both herbaceous and woody plant systems, rather than only the latter, will be addressed. The potential subject matter for this review is enormous, so several caveats are in order. First, this chapter will not address the direct effects of CO_2 and O_3 on consumer organisms; rather, it will cover those effects mediated by PSMs. Second, it will not address the impacts of environmental factors associated with CO_2 and O_3 – such as climate change – on organisms and ecological interactions. Third, because of the volume of literature available and space constraints here, this review is comprehensive and representative but not exhaustive.

The Ecology of Plant Secondary Metabolites: From Genes to Global Processes, eds. Glenn R. Iason, Marcel Dicke and Susan E. Hartley. Published by Cambridge University Press. © British Ecological Society 2012.

7.2   Atmospheric carbon dioxide and ozone

Carbon dioxide and ozone are recognised as the two anthropogenic air pollutants with the greatest, and opposing, impacts on plant growth (Saxe *et al.*,
1998; Ashmore, 2005; Karnosky *et al.*, 2005; Wittig *et al.*, 2009). Although both
are naturally occurring constituents of the Earth's atmosphere, their concentrations have increased greatly since the industrial revolution, primarily owing
to human activities such as the combustion of fossil fuels. Both CO_2 and O_3 are
currently exerting substantial direct and indirect effects on the Earth's ecosystems (Figure 7.1), with impacts predicted to escalate in the near future.

Atmospheric CO_2 concentrations are now 387 ppm (Earth System Research
Laboratory, September 2010), approximately 40% higher than the levels that
persisted for millennia prior to the industrial revolution (Petit *et al.*, 1999).

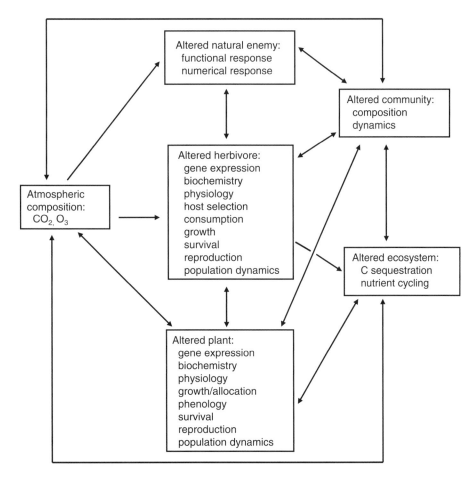

Figure 7.1 Complex of interacting abiotic and biotic factors that interact to influence
food webs, community and ecosystem structure and function, and atmospheric
composition (from Lindroth, 2010).

Concentrations for the end of the twenty-first century are predicted to fall in the range 550–900 ppm (Karl *et al.*, 2009). Chemically unreactive, the half-life of CO_2 in the atmosphere approaches 100 years.

Tropospheric (lower-atmosphere) concentrations of O_3 have increased by 20–50% (average, 38%) since the pre-industrial era (Denman *et al.*, 2007). Although O_3 is a regional pollutant, increasingly large regions are affected and background levels (not due to local origin) are growing worldwide (Vingarzan, 2004). Ozone is a secondary pollutant, produced in the presence of nitrogen oxides (NO_x) and sunlight via the oxidation of hydrocarbons (e.g. volatile organic compounds, VOCs) by OH radicals (Wennberg & Dabdub, 2008). Ozone is highly reactive, so it persists in the atmosphere for only short periods (hours to weeks).

Interestingly, PSMs that are also VOCs (e.g. isoprene, monoterpenes) can contribute to both the formation and degradation of O_3 in the atmosphere (Laothawornkitkul *et al.*, 2009; Yuan *et al.*, 2009; Bagnoli *et al.*, Chapter 6). In the presence of NO_x, oxidation of biogenic VOCs contributes to O_3 production. Alternatively, when NO_x levels are low, oxidation of biogenic VOCs consumes O_3.

7.3   Effects of carbon dioxide and ozone on plant secondary metabolites

Carbon dioxide and ozone directly and indirectly influence both the assimilation of carbon and acquisition of nutrients by plants, which in turn alter pools and fluxes of precursors for the synthesis of PSMs (Figure 7.2). In addition, CO_2 and O_3 may trigger signal transduction pathways that elicit production of PSMs. Consequently, plants grown in environments enriched in CO_2 or O_3 frequently exhibit altered levels of secondary metabolites. Such effects have been summarised in numerous previous reviews (Kangasjärvi *et al.*, 1994; Peñuelas *et al.*, 1997; Bezemer & Jones, 1998; Koricheva *et al.*, 1998; Peñuelas & Estiarte, 1998; Norby *et al.*, 2001; Zvereva & Kozlov, 2006; Stiling & Cornelissen, 2007; Valkama *et al.*, 2007; Bidart-Bouzat & Imeh-Nathaniel, 2008; Lindroth, 2010). Here, I provide updated compilations of the impacts of CO_2 (74 studies, 70 species, Table 7.1) and O_3 (30 studies, 19 species, Table 7.2) on the secondary chemistry of herbaceous and woody plant species.

7.3.1   Carbon dioxide

The vast majority of studies of the effects of enriched CO_2 on PSMs have investigated temperate-zone plants, including agricultural crops, as well as wild herbaceous and forest tree species. The studies have focused largely on 'carbon-based' compounds, especially phenolics and terpenoids. Thus, source–sink balance hypotheses (e.g. carbon–nutrient balance hypothesis, Bryant *et al.*, 1983; growth–differentiation balance hypothesis, Herms & Mattson, 1992)

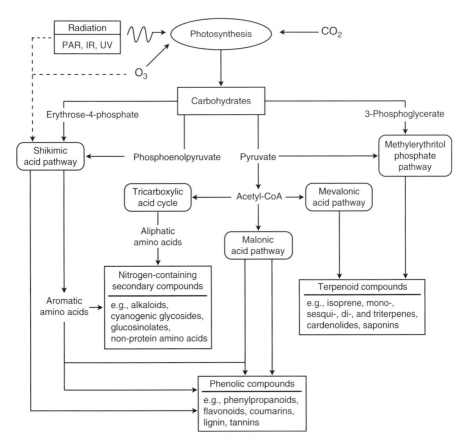

Figure 7.2 Atmospheric CO_2 and O_3 influence major biosynthetic pathways leading to the production of plant secondary metabolites (adapted from Lindroth, 2010). Dashed lines indicate signal transduction pathways. PAR, photosynthetically active radiation.

have been used to provide predictions of plant responses to elevated CO_2. The hypotheses predict that concentrations of carbon-based PSM will increase in high-CO_2 plants, especially if soil nutrient availability is limiting (Peñuelas & Estiarte, 1998). In general, however, empirical results have not matched theoretical predictions: broad classes of PSMs, as well as specific types of PSMs, appear to be differentially influenced by enriched CO_2 concentrations (Koricheva *et al.*, 1998; Zvereva & Kozlov, 2006; Stiling & Cornelissen, 2007; Bidart-Bouzat & Imeh-Nathaniel, 2008; Lindroth, 2010).

Given the documented diversity and importance of nitrogen-containing PSMs (e.g. alkaloids, cyanogenic glycosides, glucosinolates, non-protein amino acids, proteinase inhibitors) to ecological interactions, surprisingly little is known about their response to CO_2-enriched environments (Table 7.1). Responses of alkaloids are species-specific, generally ranging

Table 7.1 *Effects of elevated atmospheric CO_2 concentrations on levels of secondary metabolites in green plant tissues.*

Secondary compound	Species	Change relative to control	Reference
N-containing PSMs			
Alkaloids	*Datura stramonium*	+ or 0	Ziska *et al.*, 2005
Alkaloids	*Lolium perenne*	+ or 0	Hunt *et al.*, 2005
Alkaloids	*Nicotiana tabacum*	– or 0	Ziska *et al.*, 2005; Matros *et al.*, 2006
Alkaloids	*Papaver setigerum*	+	Ziska *et al.*, 2008
Cyanogenic glycosides	*Eucalyptus cladocalyx*	0	Gleadow *et al.*, 1998
Cyanogenic glycosides	*Lotus corniculatus*	–	Goverde *et al.*, 2004
Glucosinolates	*Brassica juncea*	–	Karowe *et al.*, 1997
Glucosinolates	*Brassica napus*	–	Himanen *et al.*, 2008
Glucosinolates	*Arabidopsis thaliana*	+/– or 0	Bidart-Bouzat *et al.*, 2005
Glucosinolates	*Brassica oleracea*	+/– or 0	Reddy *et al.*, 2004; Schonhof *et al.*, 2007
Glucosinolates	*Brassica rapa*	0	Karowe *et al.*, 1997
Glucosinolates	*Raphanus sativus*	0	Karowe *et al.*, 1997
Proteinase inhibitors	*Glycine max*	–	Zavala *et al.*, 2008
Phenolics and derivatives			
Caffeic acid	*Plantago maritima*	+ or 0	Davey *et al.*, 2004
Catechin	*Betula pendula*	+ or 0	Lavola & Julkunen-Tiitto, 1994; Kuokkanen *et al.*, 2001, 2003; Peltonen *et al.*, 2005
Catechin	*Pinus taeda*	+	Booker *et al.*, 1996
Catechin	*Populus tremula × P. tremuloides*	0	Haikio *et al.*, 2009
Chlorogenic acid	*Nicotiana tabacum*	+	Matros *et al.*, 2006
Cinnamoylquinic acids	*Betula pendula*	+	Kuokkanen *et al.*, 2003
Condensed tannins	*Acer saccharum*	+	Lindroth *et al.*, 1993; Kinney *et al.*, 1997

Table 7.1 (*cont.*)

Secondary compound	Species	Change relative to control	Reference
Condensed tannins	*Alphitonia petriei*	0	Kanowski, 2001
Condensed tannins	*Betula pendula*	+	Lavola & Julkunen-Tiitto, 1994; Kuokkanen *et al.*, 2001, 2003; Peltonen *et al.*, 2005
Condensed tannins	*Carpinus betulus*	0	Hättenschwiler & Schafellner, 2004
Condensed tannins	*Eucalyptus tereticornis*	+	Lawler *et al.*, 1997
Condensed tannins	*Fagus sylvatica*	–	Hättenschwiler & Schafellner, 2004
Condensed tannins	*Flindersia brayleyana*	+	Kanowski, 2001
Condensed tannins	*Gossypium hirsutum*	+ or 0	Coviella *et al.*, 2002; Chen *et al.*, 2005; Gao *et al.*, 2008
Condensed tannins	*Lotus corniculatus*	+	Goverde *et al.*, 2004
Condensed tannins	*Picea abies*	+	Hättenschwiler & Schafellner, 1999
Condensed tannins	*Pinus eldarica*	–	Peñuelas *et al.*, 1996
Condensed tannins	*Pinus taeda*	+	Booker *et al.*, 1996; Gebauer *et al.*, 1998
Condensed tannins	*Populus tremula* × *P. tremuloides*	0	Haikio *et al.*, 2009
Condensed tannins	*Quercus petraea*	0	Hättenschwiler & Schafellner, 2004
Condensed tannins	*Quercus rubra*	+ or 0	Lindroth *et al.*, 1993; Kinney *et al.*, 1997
Condensed tannins	*Salix myrsinifolia*	+ or 0	Julkunen-Tiitto *et al.*, 1993; Veteli *et al.*, 2002
Condensed tannins	*Acer mono*	+ or 0	Koike *et al.*, 2006
Condensed tannins	*Alnus hirsuta*	0	Koike *et al.*, 2006
Condensed tannins	*Betula allegheniensis*	+	Traw *et al.*, 1996
Condensed tannins	*Betula papyrifera*	+	Agrell *et al.*, 2000

Table 7.1 (*cont.*)

Secondary compound	Species	Change relative to control	Reference
Condensed tannins	*Betula platyphylla*	–	Koike *et al.*, 2006
Condensed tannins	*Betula populifolia*	+	Traw *et al.*, 1996
Condensed tannins	*Galactia elliottii*	0	Hall *et al.*, 2005
Condensed tannins	*Pinus sylvestris*	+ or 0	Heyworth *et al.*, 1998; Räisänen *et al.*, 2008
Condensed tannins	*Populus tremuloides*	+ or 0	Lindroth *et al.*, 1993; Roth *et al.*, 1998; McDonald *et al.*, 1999; Agrell *et al.*, 2000; Holton *et al.*, 2003; Kopper & Lindroth, 2003a, b
Condensed tannins	*Quercus chapmanii*	0	Hall *et al.*, 2005
Condensed tannins	*Quercus geminata*	0	Hall *et al.*, 2005
Condensed tannins	*Quercus mongolica*	+ or 0	Koike *et al.*, 2006
Condensed tannins	*Quercus myrtifolia*	0	Hall *et al.*, 2005
Coumarins	*Artemisia tridentata*	0	Johnson & Lincoln, 1990
Coumarins	*Nicotiana tabacum*	+	Matros *et al.*, 2006
Ferulic acid	*Plantago maritima*	0	Davey *et al.*, 2004
Flavan-3-ols	*Pinus sylvestris*	0	Räisänen *et al.*, 2008
Flavone aglycones	*Betula pendula*	– or 0	Kuokkanen *et al.*, 2001, 2003; Peltonen *et al.*, 2005
Flavones	*Pinus sylvestris*	0	Räisänen *et al.*, 2008
Flavonoids	*Artemisia tridentata*	0	Johnson & Lincoln, 1990
Flavonoids	*Glycine max*	+ or 0	Kretzschmar *et al.*, 2009; O'Neill *et al.*, 2010
Flavonoids	*Medicago sativa*	+/– or 0	Agrell *et al.*, 2004
Flavonol glycosides	*Betula pendula*	+/– or 0	Lavola & Julkunen-Tiitto, 1994; Kuokkanen *et al.*, 2001, 2003; Peltonen *et al.*, 2005
Flavonol glycosides	*Populus tremula x P. tremuloides*	0	Haikio *et al.*, 2009
Flavonols	*Pinus sylvestris*	0	Räisänen *et al.*, 2008
Furanocoumarins	*Apium graveolens*	0	Reitz *et al.*, 1997

Table 7.1 (*cont.*)

Secondary compound	Species	Change relative to control	Reference
Hydrolysable tannins	*Carpinus betulus*	–	Hättenschwiler & Schafellner, 2004
Hydrolysable tannins	*Fagus sylvatica*	0	Hättenschwiler & Schafellner, 2004
Hydrolysable tannins	*Galactia elliottii*	0	Hall *et al.*, 2005
Hydrolysable tannins	*Quercus chapmanii*	0	Hall *et al.*, 2005
Hydrolysable tannins	*Quercus geminata*	0	Hall *et al.*, 2005
Hydrolysable tannins	*Quercus myrtifolia*	0	Hall *et al.*, 2005
Hydrolysable tannins	*Quercus petraea*	0	Hättenschwiler & Schafellner, 2004
Hydrolysable tannins (ellagitannins)	*Acer saccharum*	+ or 0	Lindroth *et al.*, 1993; Kinney *et al.*, 1997; Agrell *et al.*, 2000
Hydrolysable tannins (ellagitannins)	*Quercus rubra*	+/–	Lindroth *et al.*, 1993; Kinney *et al.*, 1997
Hydrolysable tannins (gallotannins)	*Acer saccharum*	+ or 0	Lindroth *et al.*, 1993; Kinney *et al.*, 1997; Agrell *et al.*, 2000
Hydrolysable tannins (gallotannins)	*Quercus rubra*	0	Lindroth *et al.*, 1993; Kinney *et al.*, 1997
Lignin	*Cardamine hirsuta*	+	Hartley *et al.*, 2000
Lignin	*Eucalyptus tereticornis*	– or 0	Lawler *et al.*, 1997
Lignin	*Poa annua*	0	Hartley *et al.*, 2000
Lignin	*Senecio vulgaris*	0	Hartley *et al.*, 2000
Lignin	*Spergula arvensis*	0	Hartley *et al.*, 2000
Lignin	*Vaccinium myrtillus*	0	Asshoff & Hättenschwiler, 2005
Lignin	*Vaccinium uliginosum*	0	Asshoff & Hättenschwiler, 2005
Lignin	*Nicotiana tabacum*	+	Matros *et al.*, 2006
p-Coumaric acid	*Plantago maritima*	+ or 0	Davey *et al.*, 2004
Phenolic acids	*Betula pendula*	+ or 0	Peltonen *et al.*, 2005

Table 7.1 (*cont.*)

Secondary compound	Species	Change relative to control	Reference
Phenolic acids	*Populus tremula × P. tremuloides*	+ or 0	Haikio *et al.*, 2009
Phenolic acids	*Salix myrsinifolia*	0	Veteli *et al.*, 2002
Phenolic glycosides	*Populus tremula × P. tremuloides*	0	Haikio *et al.*, 2009
Phenolic glycosides	*Populus tremuloides*	+ or 0	Lindroth *et al.*, 1993; McDonald *et al.*, 1999; Agrell *et al.*, 2000; Holton *et al.*, 2003; Kopper & Lindroth, 2003a, b
Phenolic glycosides	*Salix myrsinifolia*	− or 0	Julkunen-Tiitto *et al.*, 1993; Veteli *et al.*, 2002
Phenolics	*Brassica oleracea*	− or 0	Reddy *et al.*, 2004
Protein binding capacity	12 temperate tree species	+ or 0	Knepp *et al.*, 2005
Protein binding capacity	*Quercus myrtifolia*	0	Rossi *et al.*, 2004
Protein binding capacity	*Quercus myrtifolia*	0	Rossi *et al.*, 2004
Total phenolics	*Acer mono*	+	Koike *et al.*, 2006
Total phenolics	*Acer rubrum*	0	Williams *et al.*, 2000, 2003
Total phenolics	*Acer saccharum*	0	Williams *et al.*, 2000
Total phenolics	*Alnus hirsuta*	−	Koike *et al.*, 2006
Total phenolics	*Alphitonia petriei*	0	Kanowski, 2001
Total phenolics	*Betula pendula*	+	Kuokkanen *et al.*, 2003
Total phenolics	*Betula platyphylla*	0	Koike *et al.*, 2006
Total phenolics	*Cardamine hirsuta*	0	Hartley *et al.*, 2000
Total phenolics	*Carpinus betulus*	0	Hättenschwiler & Schafellner, 2004
Total phenolics	*Cercis canadensis*	0	Hamilton *et al.*, 2004
Total phenolics	*Citrus aurantium*	0	Peñuelas *et al.*, 1996
Total phenolics	*Eucalyptus tereticornis*	+	Lawler *et al.*, 1997
Total phenolics	*Fagus sylvatica*	0	Hättenschwiler & Schafellner, 2004
Total phenolics	*Flindersia brayleyana*	0	Kanowski, 2001
Total phenolics	*Galactia elliottii*	0	Hall *et al.*, 2005
Total phenolics	*Liquidambar styraciflua*	0	Hamilton *et al.*, 2004

Table 7.1 (cont.)

Secondary compound	Species	Change relative to control	Reference
Total phenolics	*Liriodendron tulipifera*	0	Hamilton *et al.*, 2004
Total phenolics	*Lotus corniculatus*	+	Goverde *et al.*, 2004
Total phenolics	*Picea abies*	+	Hättenschwiler & Schafellner, 1999
Total phenolics	*Pinus eldarica*	−	Peñuelas *et al.*, 1996
Total phenolics	*Pinus taeda*	+	Booker *et al.*, 1996
Total phenolics	*Quercus alba*	0	Williams *et al.*, 1998
Total phenolics	*Quercus chapmanii*	0	Hall *et al.*, 2005
Total phenolics	*Quercus geminata*	0	Hall *et al.*, 2005
Total phenolics	*Quercus mongolica*	+ or 0	Koike *et al.*, 2006
Total phenolics	*Quercus myrtifolia*	0	Hall *et al.*, 2005
Total phenolics	*Quercus petraea*	0	Hättenschwiler & Schafellner, 2004
Total phenolics	*Senecio vulgaris*	0	Hartley *et al.*, 2000
Total phenolics	*Spergula arvensis*	0	Hartley *et al.*, 2000
Total phenolics	*Ulmus alata*	0	Hamilton *et al.*, 2004
Total phenolics	*Bromus erectus*	+ or 0	Castells *et al.*, 2002
Total phenolics	*Dactylis glomerata*	+	Castells *et al.*, 2002
Total phenolics	*Ficus insipida*	0	Coley *et al.*, 2002
Total phenolics	*Pinus sylvestris*	+ or 0	Kainulainen *et al.*, 1998; Sallas *et al.*, 2001
Total phenolics	*Pinus taeda*	+	Gebauer *et al.*, 1998
Total phenolics	*Poa annua*	+ or 0	Bezemer *et al.*, 2000; Hartley *et al.*, 2000
Total phenolics	*Trifolium repens*	+ or 0	Francini *et al.*, 2007
Total phenolics	Tropical tree community (nine tree species)	+ or 0	Coley *et al.*, 2002
Total phenolics	*Virola surinamensis*	+ or 0	Coley *et al.*, 2002
Verbascoside	*Plantago lanceolata*	+/−	Fajer *et al.*, 1992
Verbascoside	*Plantago maritima*	0	Davey *et al.*, 2004

Table 7.1 (*cont.*)

Secondary compound	Species	Change relative to control	Reference
Polyketides			
Urushiol variants	*Toxicodendron radicans*	+/−	Mohan *et al.*, 2006
Terpenoids and derivatives			
Cardenolides	*Asclepias syriaca*	−	Vannette & Hunter, 2011
Diterpene resin acid (papyriferic)	*Betula pendula*	+ or 0	Lavola & Julkunen-Tiitto, 1994
Diterpene resin acids	*Pinus sylvestris*	0	Kainulainen *et al.*, 1998; Sallas *et al.*, 2001
Iridoid glycosides	*Plantago lanceolata*	−	Fajer *et al.*, 1992
Isoprene	Multiple species	−	Young *et al.*, 2009; Bagnoli *et al.*, 2011
Isoprene	*Populus tremuloides*	− or 0	Calfapietra *et al.*, 2008
Monoterpenes	*Artemisia tridentata*	0	Johnson & Lincoln, 1990
Monoterpenes	*Brassica oleracea*	− or 0	Vuorinen *et al.*, 2004a
Monoterpenes	*Mentha piperita*	0	Lincoln & Couvet, 1989
Monoterpenes	*Pinus ponderosa*	0	Constable *et al.*, 1999
Monoterpenes	*Pinus sylvestris*	+ or 0	Heyworth *et al.*, 1998; Kainulainen *et al.*, 1998; Sallas *et al.*, 2001; Räisänen *et al.*, 2008
Monoterpenes	*Pinus taeda*	− or 0	Williams *et al.*, 1994, 1997
Monoterpenes	*Pseudotsuga menziesii*	0	Constable *et al.*, 1999
Saponins	*Medicago sativa*	+/− or 0	Agrell *et al.*, 2004
Sesquiterpene (gossypol)	*Gossypium hirsutum*	+ or 0	Agrell *et al.*, 2004; Chen *et al.*, 2005; Gao *et al.*, 2008
Sesquiterpenes	*Artemisia tridentata*	0	Johnson & Lincoln, 1990
Sesquiterpenes	*Mentha piperita*	0	Lincoln & Couvet, 1989
Sesquiterpenes	*Nicotiana tabacum*	0	Matros *et al.*, 2006
Sesquiterpenes	*Pinus sylvestris*	0	Sallas *et al.*, 2001
Sesquiterpenes	*Pinus taeda*	0	Williams *et al.*, 1998
Terpenoid aldehydes	*Gossypium hirsutum*	0	Coviella *et al.*, 2002; Agrell *et al.*, 2004
Terpenoids	*Betula pendula*	+	Kuokkanen *et al.*, 2001

+ = increase, − = decrease, 0 = no change.

Table 7.2 *Effects of elevated atmospheric O_3 concentrations on levels of secondary metabolites in green plant tissues.*

Secondary compound class	Species	Change relative to control	Reference
N-containing PSMs			
Alkaloids	*Nicotiana tabacum*	–	Jackson *et al.*, 2000
Glucosinolates	*Brassica napus*	+/–	Himanen *et al.*, 2008
Phenolics and derivatives			
Catechin	*Betula pendula*	+/– or 0	Lavola *et al.*, 1994; Saleem *et al.*, 2001; Peltonen *et al.*, 2005
Catechin	*Picea abies*	0	Heller *et al.*, 1990
Catechin	*Pinus sylvestris*	0	Kainulainen *et al.*, 1994
Catechin	*Pinus taeda*	+	Booker *et al.*, 1996
Chlorogenic acid	*Populus nigra*	–	Fares *et al.*, 2010
Condensed tannins	*Betula pendula*	+/–	Lavola *et al.*, 1994; Peltonen *et al.*, 2005
Condensed tannins	*Picea abies*	0	Heller *et al.*, 1990
Condensed tannins	*Pinus taeda*	+	Booker *et al.*, 1996
Condensed tannins	*Betula papyrifera*	0	Lindroth *et al.*, 2001
Condensed tannins	*Populus tremula × P. tremuloides*	+	Freiwald *et al.*, 2008
Condensed tannins	*Populus tremuloides*	+	Holton *et al.*, 2003; Kopper & Lindroth, 2003a, b
Condensed tannins	*Populus tremuloides*	0	Lindroth *et al.*, 2001
Flavone aglycones	*Betula pendula*	+ or 0	Peltonen *et al.*, 2005
Flavone glycosides	*Petroselinium crispum*	+	Hahlbrock & Scheel, 1989
Flavonoids	*Acer saccharum*	+	Sager *et al.*, 2005
Flavonoids	*Glycine max*	+ or 0	O'Neill *et al.* 2010
Flavonoids	*Gossypiuum hirsutum*	+	Booker, 2000
Flavonoids	*Nicotiana tabacum*	–	Jackson *et al.*, 2000
Flavonoids	*Populus nigra*	+ or 0	Fares *et al.*, 2010

Table 7.2 (cont.)

Secondary compound class	Species	Change relative to control	Reference
Flavonol glycosides	*Betula pendula*	+/– or 0	Lavola *et al.*, 1994; Saleem *et al.*, 2001; Peltonen *et al.*, 2005
Furanocoumarins	*Petroselinium crispum*	+	Hahlbrock & Scheel, 1989
Hydroxycinnamic acid cmpd	*Phaseolus vulgaris*	–	Kanoun *et al.*, 2001
Isoflavonoids	*Glycine max*	+	Keen & Taylor, 1975; O'Neill *et al.* 2010
Isoflavonoids	*Phaseolus vulgaris*	+ or 0	Kanoun *et al.*, 2001
Lignin	*Picea abies*	+	Sandermann, 1996
Lignin	*Trifolium pratense*	+	Muntifering *et al.*, 2006
Lignin	*Trifolium repens*	+	Muntifering *et al.*, 2006
Phenolic acids	*Betula pendula*	+ or 0	Saleem *et al.*, 2001; Peltonen *et al.*, 2005
Phenolic acids	*Pinus strobus*	–	Shadkami *et al.*, 2007
Phenolic glycosides	*Betula pendula*	–	Lavola *et al.*, 1994
Phenolic glycosides	*Picea abies*	0	Heller *et al.*, 1990
Phenolic glycosides	*Populus nigra*	– or 0	Fares *et al.*, 2010
Phenolic glycosides	*Populus tremula* × *P. tremuloides*	+ or 0	Freiwald *et al.*, 2008
Phenolic glycosides	*Populus tremuloides*	–	Holton *et al.*, 2003; Kopper & Lindroth, 2003b
Total phenolics	*Pinus taeda*	+	Booker *et al.*, 1996
Total phenolics	*Pinus sylvestris*	0	Kainulainen *et al.*, 1994, 1998; Manninen *et al.*, 2000; Sallas *et al.*, 2001
Total phenolics	*Trifolium pratense*	0	Muntifering *et al.*, 2006
Total phenolics	*Trifolium repens*	0	Muntifering *et al.*, 2006
Terpenoids and derivatives			
Diterpene resin acid (papyriferic)	*Betula pendula*	+ or 0	Lavola *et al.*, 1994
Diterpene resin acids	*Pinus strobus*	0	Shadkami *et al.*, 2007

Table 7.2 (*cont.*)

Secondary compound class	Species	Change relative to control	Reference
Diterpene resin acids	*Pinus sylvestris*	– or 0	Kainulainen *et al.*, 1994, 1998; Manninen *et al.*, 2000; Sallas *et al.*, 2001
Diterpenes	*Nicotiana tabacum*	–	Jackson *et al.*, 1999
Homoterpenes	*Phaseolus lunatus*	+	Vuorinen *et al.*, 2004b
Isoprene	*Populus alba*	+	Fares *et al.*, 2006
Isoprene	*Populus nigra*	–	Fares *et al.*, 2010
Isoprene	*Populus tremula × P. tremuloides*	0	Blande *et al.*, 2007
Isoprene	*Populus tremuloides*	– or 0	Calfapietra *et al.*, 2008
Monoterpenes	*Brassica oleracea*	–	Pinto *et al.*, 2007
Monoterpenes	*Picea abies*	0	Sandermann, 1996
Monoterpenes	*Picea abies*	–	Heller *et al.*, 1990
Monoterpenes	*Pinus sylvestris*	+ or 0	Kainulainen *et al.*, 1994, 1998; Manninen *et al.*, 2000; Sallas *et al.*, 2001
Monoterpenes	*Quercus ilex*	+	Loreto *et al.*, 2004
Monoterpenes	*Populus tremula × P. tremuloides*	+ or 0	Blande *et al.*, 2007
Sesquiterpenes	*Pinus sylvestris*	0	Sallas *et al.*, 2001
Sesquiterpenes	*Populus tremula × P. tremuloides*	0	Blande *et al.*, 2007

+ = increase, – = decrease, 0 = no change.

from no effect to increased concentrations, with one reported case of decreased concentrations. Levels of cyanogenic glycosides, glucosinolates and proteinase inhibitors generally decrease or are unaffected by CO_2; only in rare instances have increases been reported.

Phenolic compounds have been the most thoroughly studied of all PSMs with respect to responses to elevated CO_2 (Table 7.1). Relatively simple phenolics, including phenolic acids, flavonoids, coumarins and salicylate phenolic glycosides, exhibit the full range of responses, from decreases to increases under enriched CO_2. Condensed tannin levels exhibit a similar range of responses, although levels frequently increase. Indeed, condensed tannins increase under elevated CO_2 more consistently than does any other class of PSMs (60% of cases documented in Table 7.1). In contrast, concentrations of hydrolysable tannins (including ellagitannins and gallotannins) are rarely responsive to CO_2 levels. Similarly, lignin, a major sink for foliar carbon, has rarely exhibited concentration changes in response to CO_2 environment. Numerous studies have assessed the effects of CO_2 on 'total phenolics', typically quantified via the Folin–Ciocalteau assay. Total phenolic levels have rarely declined, and have increased in 38% of the cases examined.

The remaining major class of PSMs widely evaluated in CO_2-enrichment research is the terpenoids and derivatives thereof. Numerous studies have demonstrated that emissions of isoprene, the 5-carbon building block of terpenoids, are reduced under elevated CO_2 (Young et al., 2009; Bagnoli et al., Chapter 6). Nonetheless, several reviews (Stiling & Cornelissen, 2007; Lindroth, 2010) have reported that terpenoids ($\geq C_{10}$) are largely unresponsive to atmospheric CO_2 levels, and that perspective is borne out by data in Table 7.1. In 16 studies of 9 plant species grown under elevated CO_2, levels of monoterpenes rarely increased or decreased, while levels of sesquiterpenes, with the exception of gossypol, never changed. Other terpenoid derivatives, such as cardenolides, iridoid glycosides, resin acids and saponins, have been poorly studied, and exhibit decreases, increases or no change in response to elevated CO_2.

Although at some level frustrating, it should not be surprising that few general patterns exist with respect to responses of PSMs to atmospheric CO_2. The quality and quantity of PSMs expressed in plants are the product of long evolutionary histories of multiple direct, indirect and interacting factors. Variation in evolutionary strategies for optimal allocation of carbon to defence versus growth and reproduction has resulted in differences in CO_2 responses among broad taxonomic groups, species and genotypes (Zvereva & Kozlov, 2006; Stiling & Cornelissen, 2007; Bidart-Bouzat & Imeh-Nathaniel, 2008; Lindroth, 2010). Interactions with abiotic environmental factors, such as temperature or availability of resources (e.g. light, nutrients), can augment or ameliorate the effects of atmospheric CO_2 on PSMs (e.g. Kinney et al., 1997; McDonald et al., 1999; Booker & Maier, 2001; Coley et al., 2002). Interactions with biotic factors, such as herbivores and pathogens, may also influence the effects of CO_2 on PSMs (Bidart-Bouzat & Imeh-Nathaniel, 2008; Lindroth, 2010).

7.3.2   Ozone

As is the case for CO_2, nearly all studies of the effects of elevated O_3 on PSMs have investigated temperate species, especially agricultural crops and forest trees. And again, research has focused on carbon-based phenolic and terpenoid compounds. Atmospheric O_3 can influence concentrations of PSMs by inhibiting carbohydrate production or by eliciting defence signalling pathways (Figure 7.2).

Effects of O_3 on nitrogen-containing PSMs have been evaluated in only two studies, both with agricultural species (Table 7.2). Elevated O_3 levels decreased concentrations of alkaloids (nicotine) and increased or decreased concentrations of individual glucosinolates.

Ozone contributes to oxidative stress and proliferation of oxygen radicals. Thus, like many other stressors, O_3 leads to the up-regulation of genes and enzymes associated with the shikimate–phenylpropanoid pathway (Kangasjärvi et al., 1994, 2005). Consequently, numerous studies, mostly of forest trees, have assessed the impacts of O_3 on antioxidant phenolic metabolites (Table 7.2). Concentrations of simple phenolics (e.g. catechin), phenolic acids, phenolic glycosides and condensed tannins exhibit the full range of negative, positive and nil effects. In contrast, levels of flavonoids and isoflavonoids generally increase under elevated O_3. The meta-analysis of Valkama et al. (2007) revealed that concentrations of phenolic acids and flavonoids significantly increased under elevated O_3, whereas those of tannins did not.

Effects of O_3 on terpenoid compounds have been evaluated in only about a dozen species (Table 7.2). Concentrations of the compounds generally change relatively little under elevated O_3, although some positive and negative changes in concentrations have been reported for monoterpenes and diterpene resin acids. Valkama et al. (2007) concluded that monoterpenes and sesquiterpenes show no general response to O_3, whereas diterpene resin acids tend to increase in concentration.

Of growing interest in recent years has been the influence of atmospheric O_3 on the production and fate of plant volatile organic compounds. VOCs comprise a rich diversity of terpenoids, benzenoids, phenylpropanoids and volatile amino acid and fatty acid derivatives (Pinto et al., 2010). These compounds serve a variety of physiological and ecological functions, including within- and between-plant signalling, defence against herbivores and pathogens, and attraction of pollinators, seed dispersers, predators and parasitoids. Ozone can influence the emission of VOCs by altering their biosynthesis or by eliciting oxidative reactions at the plant surface (Pinto et al., 2010). Interestingly, biogenic VOCs, when oxidised in the presence of NO_x, can increase atmospheric O_3 concentrations, potentially exacerbating O_3 damage to vegetation in the environment. Alternatively, isoprenoid-emitting plants may be less prone to O_3 damage, owing to the oxidant scavenging properties of isoprene (Loreto & Fares, 2007). These and related topics are addressed in several recent reviews (Laothawornkitkul et al., 2009; Yuan et al., 2009; Pinto et al., 2010; Bagnoli et al., Chapter 6).

Surprisingly, exceedingly few studies have evaluated how abiotic (e.g. resource availability) or biotic (e.g. herbivory) factors influence plant chemical responses to elevated O_3 concentrations. The one notable exception is research on the interactive and combined effects of O_3 and CO_2. Atmospheres enriched in CO_2 can ostensibly offset reductions in photosynthesis caused by O_3, as well as increase carbon available for production of defence compounds via pathways up-regulated in response to O_3. The meta-analysis of Valkama et al. (2007) showed that elevated atmospheric CO_2 reduced the effects of O_3 on phenolic concentrations, but increased the effects of O_3 on terpenoid concentrations. Their analysis did not, however, differentiate between simple additive versus truly interactive effects of CO_2 and O_3. Peltonen et al. (2005) reported that CO_2 and O_3 interacted to influence concentrations of about 75% of the over 30 simple phenolics they evaluated in *Betula pendula*; CO_2 tended to reduce the induction of phenolics by O_3. Carbon dioxide and ozone have been reported to interactively influence levels of condensed tannins in *Populus tremuloides* (Holton et al., 2003) and mono- and sesquiterpenes in *Pinus sylvestris* (Sallas et al., 2001).

7.4   Effects of carbon dioxide and ozone on grazing food webs

Plant secondary metabolites strongly influence interactions between plants and herbivores, as well as between herbivores and natural enemies. Thus, elevated levels of atmospheric CO_2 and O_3 are likely to shape trophic dynamics in ecological communities of the future. Such possibilities have been the subject of considerable research and numerous reviews (e.g. Bezemer & Jones, 1998; Coviella & Trumble, 1999; Zvereva & Kozlov, 2006; Stiling & Cornelissen, 2007; Valkama et al., 2007; Bidart-Bouzat & Imeh-Nathaniel, 2008; Lindroth, 2010). Here, I will summarise the main findings of those reviews, and provide examples illustrating the potential for CO_2- and O_3-mediated changes in PSMs to influence trophic dynamics at the levels of individuals, populations and communities.

7.4.1   Carbon dioxide

Enriched CO_2 concentrations, expressed as altered plant chemistry, influence the individual performance of herbivores, in terms of host selection, consumption, growth, survival and reproduction. Typically, herbivores respond to a complex of chemical changes elicited by CO_2, including increases in carbohydrates and decreases in nitrogen (protein) in addition to changes in PSMs.

Several studies have documented that oviposition preferences of insects (Kopper & Lindroth, 2003a; Peltonen et al., 2006), as well as feeding preferences of invertebrates and mammals (e.g. Peters et al., 2000; Kuokkanen et al., 2003; Mattson et al., 2004; Agrell et al., 2005), shift between plants grown under ambient and under elevated CO_2. More relevant, however, is the question of whether genotype or species preferences of herbivores will be re-ordered under universally elevated CO_2 conditions of the future. Minimal research to date has shown that in some cases, host preferences will change significantly (Traw *et al.*, 1996; Agrell *et al.*, 2005), whereas in others, they will not (Traw *et al.*, 1996; Peters *et al.*, 2000).

Individual performance (consumption, growth, survival, reproduction) of herbivores changes under elevated CO_2. The magnitude and direction of changes differ, however, among plant and herbivore species, and in various environmental contexts. Of all response measures, food consumption patterns show the most uniform response to CO_2 enrichment, with most herbivores exhibiting increased consumption of high-CO_2 food. Elevated consumption rates are generally considered to be 'compensatory' because of low protein (N) values in CO_2-enriched plant tissues. Compensatory feeding is probably restricted to below optimum, however, owing to the presence of higher concentrations of PSMs, especially condensed tannins, in host plants (Lindroth *et al.*, 1993). In insects, development times generally increase, while growth rates, final pupal and adult weights, and survivorship typically decline or are unaffected. Effects of CO_2-altered food quality on reproduction have rarely been investigated, although decreases, increases and no change have been reported (Lindroth *et al.*, 1997; Awmack *et al.*, 2004; Chen *et al.*, 2005). Finally, few studies have evaluated the effects of elevated CO_2 on individual performance of mammalian herbivores. C. Habeck and R. L. Lindroth (unpublished data) found that growth rates of young voles were reduced when fed herbaceous vegetation grown under elevated CO_2.

As described previously for PSMs, the effects of CO_2 on herbivore performance are influenced by interactions with environmental factors such as resource availability (e.g. Kinney *et al.*, 1997; Lawler *et al.*, 1997; Roth *et al.*, 1997; Hättenschwiler & Schafellner, 1999; Agrell *et al.*, 2000) and climate (Bale *et al.*, 2002). The meta-analysis of Zvereva and Kozlov (2006) revealed that climate warming will probably ameliorate the predicted negative, food-associated impacts of CO_2 on insect performance.

Carbon dioxide-mediated shifts in plant quality could potentially also alter interactions between herbivores and natural enemies. Chen *et al.* (2005) reported improved performance of ladybird beetle predators feeding on aphids on CO_2-enriched cotton plants, while Stiling *et al.* (1999) found higher parasitoid-caused mortality to leaf-miners on CO_2-enriched oaks. In contrast, however, studies with the parasitoids *Cotesia melanoscela* and *Compsilura concinnata* revealed no effects of CO_2 environment on parasitoid fitness traits (Roth & Lindroth, 1995; Holton *et al.*, 2003). Similarly, elevated CO_2 had no impact on virulence of the gypsy moth pathogen nucleopolyhedrosis virus (Lindroth *et al.*, 1997). In general, CO_2-mediated effects on PSMs are expected to be increasingly muted at higher trophic levels. Much more work is required, however, to ascertain relative degrees of buffering or amplification of the effects of PSMs at the third and higher trophic levels, and the roles of species- and environment-specific factors in those interactions.

Population densities of herbivores in future CO_2-enriched environments will be determined by a host of bottom-up, top-down and environmental factors (Figure 7.2). Although PSMs certainly influence the strength of those interactions, the extent to which they will alter population dynamics in future environments is largely unknown. Putative relationships between atmospheric CO_2, PSMs and herbivore populations have typically been inferred from studies of individual herbivore performance. Not surprisingly, most true population-level studies have explored the effects of CO_2 on aphids. Results indicate that elevated CO_2 can increase, decrease or not alter aphid population densities, and that outcomes are influenced by factors such as plant species, temperature and soil fertility (e.g. Bezemer et al., 1999; Percy et al., 2002; Newman et al., 2003). In one of the more thoroughly studied non-aphid systems to date, Stiling et al. (1999, 2002, 2003, 2009) reported that populations of diverse insect herbivores in a scrub-oak community declined under elevated CO_2, primarily owing to both plant- and parasitoid-induced mortality. Finally, Hillstrom and Lindroth (2008) found that elevated CO_2 tended to reduce population densities of phloem-feeding insects, but increase densities of leaf-chewing insects, at the Aspen FACE facility (aspen and birch forest, Wisconsin, USA).

The effects of enriched CO_2 on the diversity and composition of insect communities have rarely been investigated. Following four years of censuses at the Aspen FACE site, M. L. Hillstrom and R. L. Lindroth (unpublished data) concluded that enriched atmospheric CO_2 has little impact on forest insect diversity, and only transient effects on community composition.

7.4.2 Ozone

Tropospheric ozone concentrations influence plant chemical quality and thus selection for use by herbivores as oviposition substrates or food. Elevated O_3 resulted in increased oviposition on tobacco by tobacco horn-worm moths (Manduca sexta), notwithstanding O_3-mediated reductions in cuticular oviposition stimulants (Jackson et al., 1999). In contrast, O_3 enrich-ment decreased oviposition by leaf beetles on Populus deltoides (Jones & Coleman, 1988) and by leaf-miners on P. tremuloides (Kopper & Lindroth, 2003a). Similarly, insects exhibit a variety of responses to O_3 in terms of food selection. Preferences for O_3-treated foliage increased in leaf beetles fed P. deltoides (Jones & Coleman, 1988) and in common leaf weevils fed hybrid aspen (Freiwald et al., 2008), but decreased in forest tent caterpillars fed P. tremuloides (Agrell et al., 2005). Forest tent caterpillars also exhibited shifts in the rank order of species and genotype preferences when host trees were grown under elevated O_3 (Agrell et al., 2005).

Elevated O_3 can also change the individual fitness performance (consumption, growth, survival, reproduction) of herbivores, although again, effects are species- and context-specific. Based on a meta-analysis of 22 species of trees and 10 species of insects, Valkama *et al.* (2007) reported that under elevated O_3, rates of food consumption, survival and reproduction are generally unaffected, but development times decrease and pupal masses increase. For example, development times decreased and pupal weights increased for forest tent caterpillars reared on high-O_3 *P. tremuloides* (Kopper & Lindroth, 2003b). Research on insects feeding on herbaceous plants has generally provided results similar to those summarised by Valkama *et al.* (2007) for trees. For example, Colorado potato beetles showed no change in growth or fecundity when fed potato, *Solanum tuberosum*, grown in enriched O_3 conditions (Costa *et al.*, 2001). But again, herbivore responses are species-specific. Growth, survival and fecundity of chrysomelid beetles were improved on ozonated *Rumex obtusifolia*, relative to unfumigated controls (Whittaker *et al.*, 1989). Very little research has evaluated the consequences of O_3-mediated changes in plant quality for mammalian herbivores. Muntifering *et al.* (2006) reported that several clover species grown under elevated O_3 had reduced *in vitro* dry-matter digestibility and *in vitro* cell-wall digestibility by ruminal microbes. Habeck and Lindroth (unpublished data) recently found that growth of female (but not male) voles was substantially reduced when weanlings were fed several herbaceous species grown under elevated O_3.

How other environmental factors may influence the impacts of O_3 on plant–herbivore interactions is largely unknown. Most research on this topic has considered interactions between O_3 and CO_2. In their review of the topic, Valkama *et al.* (2007) concluded that the positive effects of O_3 on insect development and growth were generally counteracted by elevated CO_2.

Elevated atmospheric O_3 levels will likely alter the dynamics of interactions between herbivores and their natural enemies. Most research in this area has focused on the effects of ozone on plant- or insect-derived VOCs, which influence the behaviours of insect herbivores, predators and parasitoids (Yuan *et al.*, 2009; Pinto *et al.*, 2010). For example, Mondor *et al.* (2004) found that elevated O_3 strikingly increased pheromone-mediated aphid escape behaviours at the Aspen FACE site (Wisconsin, USA). At the same research site, Holton *et al.* (2003) found that elevated O_3 reduced phenolic glycoside concentrations in *P. tremuloides*, which led to improved performance of forest tent caterpillars and reduced survivorship of the parasitoid *Compsilura concinnata*.

Few studies have evaluated the population-level responses of herbivores to elevated atmospheric O_3 levels. Percy *et al.* (2002) reported that at Aspen FACE, populations of aspen-feeding aphids markedly increased, while populations of aphid predators and parasitoids decreased, in elevated O_3 plots. Awmack *et al.* (2004) found that under elevated O_3, populations of birch-feeding aphids increased when protected from predators and parasitoids, but not when exposed. Several years of pan-trap collections at Aspen FACE revealed that elevated O_3 stands had higher abundances of some phloem-feeding insects, but lower abundances of parasitoids, relative to control stands. In contrast, recent visual surveys of canopy insects at Aspen FACE showed that O_3 fumigation decreased population densities of phloem-feeding species, but increased densities of several leaf-chewing species, on *P. tremuloides* (M. L. Hillstrom and R. L. Lindroth, unpublished data). Given that elevated O_3 appears to affect development times more than survival and reproductive rates (described above), we might expect that population densities of multivoltine insects will respond more strongly to O_3 environments than those of univoltine insects.

To my knowledge, the effects of O_3 on animal community diversity and composition have been evaluated only at the Aspen FACE research site. Hillstrom and Lindroth (2008) found that the composition of insects captured in pan traps differed between ambient and elevated O_3 environments in two out of four years. Visual surveys of canopy insect communities, however, revealed no effects of O_3 on either community diversity or composition (M. L. Hillstrom and R. L. Lindroth, unpublished data). Clearly, additional research on the community-level impacts of tropospheric O_3 is needed.

7.5   Effects of carbon dioxide and ozone on decomposing food webs

Rates of plant litter decomposition and associated nutrient cycling are regulated by litter chemistry, soil organisms and climate. Litter chemistry influences the palatability of litter to invertebrates and the resistance of litter to microbial and physical breakdown. Soil invertebrates comminute and transport litter, while inoculating it with microorganisms that further enhance enzymatic breakdown. Climate, especially temperature and moisture, regulates the rates at which both biological and physical breakdown occur. Each of the three principal factors – chemistry, organisms and climate – is, in turn, influenced directly or indirectly by atmospheric CO_2 and O_3 concentrations (Lindroth, 2010). These topics have been explored in several reviews (e.g. Norby *et al.*, 2001; Lindroth, 2010). Here, the main conclusions of those studies, relevant to the PSM theme of this chapter, will be summarised and illustrated. For both CO_2 and O_3, I will first address the indirect impacts on soil invertebrates (mediated through litter quality) and then consequences for litter decomposition.

7.5.1   Carbon dioxide

Enriched CO_2 atmospheres can change the palatability of plant litter to soil invertebrates, although general patterns of response are difficult to ascertain. Most studies have been conducted with isopods, and results have shown that

CO_2 can decrease, increase or have no effect on consumption of litter (Cotrufo *et al.*, 1998, 2005a; Hättenschwiler *et al.*, 1999; David *et al.*, 2001), and can alter preferences among litter species (Hättenschwiler & Bretscher, 2001). More recent work with earthworms revealed no effect of CO_2 concentration on leaf litter consumption (Meehan *et al.*, 2010).

In terms of effects of CO_2 environment on individual- and population-level performance of soil invertebrates, results are similarly mixed. Individual growth of earthworms decreased when fed high-CO_2 *Betula pendula* (Kasurinen *et al.*, 2007) or *P. tremuloides* litter (Meehan *et. al.*, 2010), and population growth of collembola declined on high-CO_2 *P. tremuloides* litter (Meehan *et al.*, 2010). In contrast, however, survivorship and growth of isopods was not affected by CO_2 in several studies (Hättenschwiler & Bretscher, 2001; Kasurinen *et al.*, 2007). Loranger *et al.* (2004) reported that at Aspen FACE, population densities of collembola and oribatid mites, but not other soil invertebrates, declined in high-CO_2 treatments. As concluded previously for insects feeding on green plant tissues, the effects of CO_2 environment on soil invertebrates are both plant species- and animal species-specific.

The potential for enriched CO_2 concentrations to alter decomposition rates of plant litter has been a matter of primary interest to ecologists. According to the 'litter quality' hypothesis (Strain & Bazzaz, 1983), elevated atmospheric CO_2 can be expected to reduce litter N concentrations while increasing C concentrations (e.g. lignin and tannins), leading to reduced rates of decomposition. Slower decomposition will in turn reduce soil N availability, contributing to 'progressive nitrogen limitation' and negative feedbacks on plant growth (Luo *et al.*, 2004). In fact, plant litter produced under high CO_2 concentrations generally does exhibit reduced N levels and increased C:N ratios (Ceulemans *et al.*, 1999; Norby *et al.*, 2001).

Surprisingly, studies over the past two decades have not revealed consistent effects of elevated CO_2 on rates of litter decomposition. A meta-analysis of studies covering approximately 70 plant species showed that neither litter mass loss nor respiration rates differed systematically between litter produced under ambient versus elevated CO_2 (Norby, 2001). More recent and realistic (long-term; field-oriented) studies have shown similar results. In some cases, litter from enriched-CO_2 sites exhibited reduced decomposition rates (Parsons *et al.*, 2004, 2008; Cotrufo *et al.*, 2005b), whereas in others it did not (Finzi *et al.*, 2001; Hall *et al.*, 2006; Liu *et al.*, 2009). The diversity of effects of CO_2 on litter decomposition is likely to be due to factors such as species-, context- and temporal-specificity (Lindroth, 2010).

7.5.2 Ozone

A small number of studies have explored the impacts of O_3 fumigation on individual- and population-level performance of soil invertebrates. Growth and mortality rates of isopods were unaffected, whereas growth rates of earthworms declined, when fed *Betula pendula* litter from elevated O_3 plots (Kasurinen *et al.*, 2007). In contrast, earthworm growth did not change when individuals were reared on high-O_3 *P. tremuloides*, whereas population growth of collembola increased on the same litter (Meehan *et al.*, 2010). Loranger *et al.* (2004) made a census of soil invertebrates at Aspen FACE and found that numbers of mites declined in high-O_3 plots, whereas numbers of collembola were unaffected.

A similarly small number of studies have evaluated the effects of O_3 concentrations on litter decomposition. Growth under elevated O_3 levels reduced litter decomposition rates in soybean (Booker *et al.*, 2005) and three species of deciduous trees (Findlay *et al.*, 1996; Kasurinen *et al.*, 2006; Parsons *et al.*, 2008). In contrast, litter decomposition was not affected by O_3 in four species of deciduous trees (Boerner & Rebbeck, 1995; Scherzer *et al.*, 1998) and two species of *Pinus* (Scherzer *et al.*, 1998; Kainulainen *et al.*, 2003). Interestingly, at Aspen FACE, elevated O_3 increased the decomposition of *Betula papyrifera* litter from ambient CO_2 treatments, but decreased the decomposition of litter from enriched CO_2 treatments (a significant $O_3 \times CO_2$ interaction; Parsons *et al.*, 2008). As was concluded for CO_2, the effects of elevated O_3 on litter decomposition appear to be species- and context-specific.

7.6   Conclusions and future directions

Clearly, elevated concentrations of atmospheric CO_2 and O_3 have the potential to alter levels of PSMs and, consequently, a host of ecological interactions, ranging from simple pairwise interactions to complex ecosystem processes. Still, an admittedly frustrating conclusion of this and related reviews (e.g. Bidart-Bouzat & Imeh-Nathaniel, 2008; Tylianakis *et al.*, 2008; Lindroth, 2010) is that the effects of CO_2 and O_3 on PSMs are so species-, context- and time-specific that they can appear, if not truly be, idiosyncratic. Moreover, CO_2- and O_3-mediated effects on PSMs, and resulting consequences for species interactions and ecosystem dynamics, may be damped by the large complex of factors that interactively govern ecosystem structure and function. Identification of general patterns, if they exist, will require new, substantive and coordinated research efforts, with an emphasis on the following research priorities (adapted from Lindroth, 2010):

1   Increase the diversity of PSMs evaluated in the context of elevated CO_2 and O_3. To date, most research has emphasised phenolic compounds, with a secondary emphasis on terpenoids. We know little, if anything, about effects of atmospheric change on entire classes of N-containing PSMs, including, for example, alkaloids, cyanogenic glycosides, proteinase inhibitors and non-protein amino acids.

2 Employ emerging molecular genetic/genomics tools to explore the functional significance of PSM genes with respect to atmospheric change, and consequences thereof for shifts in plant metabolomes. Searches for patterns of responses at the molecular level, and how those patterns may be phylogenetically linked, should inform and improve our search for patterns at the whole-plant and species levels.

3 Broaden the representation of species and biomes studied. As of this time, studies of the effects of CO_2 and O_3 on PSMs have each evaluated fewer than 100 of the nearly 300 000 plant species known, and have focused on temperate and boreal biomes. Priority should be given to foundation species (Ellison *et al.*, 2005) and biomes of particular ecological and/or economic importance.

4 Incorporate trophic cascade studies in an ecosystem context. Increasingly, trophic dynamics and their consequences for population and community processes are recognised as deriving from both bottom-up and top-down, as well as direct and indirect, interactions. PSMs play key roles in these processes, influencing not only interactions between plants and herbivores but between herbivores and their natural enemies. Complex multi-species interactions have the potential to amplify or dampen PSM-mediated effects on bivariate interactions. Similarly, the roles of PSMs in other exploitative (e.g. allelopathic) or mutualistic (e.g. fruit dispersal) associations should be evaluated.

5 Address multiple drivers of global change (e.g. CO_2, O_3, warming, invasive species) simultaneously, over sufficiently long time periods, to better understand the context- and time-dependency of ecological responses. Because these drivers are all in play now, a multi-factor approach is essential for accurate predictions of their impacts in environments of the future.

Acknowledgements

I thank Stefan Levay-Young and Erynne Sengele for assistance with preparation of the figures. This work was partially supported by US Department of Energy (Office of Science, OBER) grant DE-FG02-06ER64232 and National Science Foundation (Division of Environmental Biology) grant DEB-0841609.

References

Agrell, J., McDonald, E. P. and Lindroth, R. L. (2000) Effects of CO_2 and light on tree phytochemistry and insect performance. *Oikos*, **88**, 259–272.

Agrell, J., Anderson, P., Oleszek, W., Stochmal, A. and Agrell, C. (2004) Combined effects of elevated CO_2 and herbivore damage on alfalfa and cotton. *Journal of Chemical Ecology*, **30**, 2309–2324.

Agrell, J., Kopper, B., McDonald, E. P. and
Lindroth, R. L. (2005) CO_2 and O_3 effects on
host plant preferences of the forest tent
caterpillar (*Malacosoma disstria*). *Global Change
Biology*, **11**, 588–599.

Ashmore, M. R. (2005) Assessing the future global
impacts of ozone on vegetation. *Plant, Cell
and Environment*, **28**, 949–964.

Asshoff, R. and Hättenschwiler, S. (2005) Growth
and reproduction of the alpine grasshopper
Miramella alpina feeding on CO_2-enriched
dwarf shrubs at treeline. *Oecologia*, **142**,
191–201.

Awmack, C. S., Harrington, R. and Lindroth, R. L.
(2004) Aphid individual performance may
not predict population responses to elevated
CO_2 or O_3. *Global Change Biology*, **10**,
1414–1423.

Bale, J. S., Masters, G. J., Hodkinson, I. D. *et al.*
(2002) Herbivory in global climate change
research: direct effects of rising temperature
on insect herbivores. *Global Change Biology*, **8**,
1–16.

Bezemer, T. M. and Jones, T. H. (1998) Plant–
insect herbivore interactions in elevated
atmospheric CO_2: quantitative analyses and
guild effects. *Oikos*, **82**, 212–222.

Bezemer, T. M., Knight, K. J., Newington, J. E. and
Jones, T. H. (1999) How general are aphid
responses to elevated atmospheric CO_2?
Annals of the Entomological Society of America,
92, 724–730.

Bezemer, T. M., Jones, T. H. and Newington, J. E.
(2000) Effects of carbon dioxide and
nitrogen fertilization on phenolic content
in *Poa annua* L. *Biochemical Systematics and
Ecology*, **28**, 839–846.

Bidart-Bouzat, M. G. and Imeh-Nathaniel, A.
(2008) Global change effects on plant
chemical defenses against insect herbivores.
Journal of Integrative Plant Biology, **50**,
1339–1354.

Bidart-Bouzat, M. G., Mithen, R. and
Berenbaum, M. R. (2005) Elevated CO_2
influences herbivory-induced defense
responses of *Arabidopsis thaliana*. *Oecologia*,
145, 415–424.

Blande, J. D., Tiiva, P., Oksanen, E. and
Holopainen, J. K. (2007) Emission of
herbivore-induced volatile terpenoids from
two hybrid aspen (*Populus tremula ×
tremuloides*) clones under ambient and
elevated ozone concentrations in the field.
Global Change Biology, **13**, 2538–2550.

Boerner, R. E. J. and Rebbeck, J. (1995)
Decomposition and nitrogen release from
leaves of three hardwood species grown
under elevated O_3 and/or CO_2. *Plant and Soil*,
170, 149–157.

Booker, F. L. (2000) Influence of carbon dioxide
enrichment, ozone and nitrogen
fertilization on cotton (*Gossypium hirsutum* L.)
leaf and root composition. *Plant, Cell and
Environment*, **23**, 573–583.

Booker, F. L. and Maier, C. A. (2001) Atmospheric
carbon dioxide, irrigation, and fertilization
effects on phenolic and nitrogen
concentrations in loblolly pine (*Pinus taeda*)
needles. *Tree Physiology*, **21**, 609–616.

Booker, F. L., Anttonen, S. and Heagle, A. S. (1996)
Catechin, proanthocyanidin and lignin
contents of loblolly pine (*Pinus taeda*) needles
after chronic exposure to ozone. *New
Phytologist*, **132**, 483–492.

Booker, F. L., Prior, S. A., Torbert, H. A *et al.* (2005)
Decomposition of soybean grown under
elevated concentrations of CO_2 and O_3.
Global Change Biology, **11**, 685–698.

Bryant, J. P., Chapin, F. S. III and Klein, D. R. (1983)
Carbon/nutrient balance of boreal plants in
relation to vertebrate herbivory. *Oikos*, **40**,
357–368.

Calfapietra, C., Mugnozza, G. S., Karnosky, D. F.,
Loreto, F. and Sharkey, T. D. (2008) Isoprene
emission rates under elevated CO_2 and O_3 in
two field-grown aspen clones differing in
their sensitivity to O_3. *New Phytologist*, **179**,
55–61.

Castells, E., Roumet, C., Peñuelas, J. and Roy, J.
(2002) Intraspecific variability of phenolic
concentrations and their responses to
elevated CO_2 in two Mediterranean
perennial grasses. *Environmental and
Experimental Botany*, **47**, 205–216.

Ceulemans, R., Janssens, I. A. and Jach, M. E. (1999) Effects of CO_2 enrichment on trees and forests: lessons to be learned in view of future ecosystem studies. *Annals of Botany*, **84**, 577–590.

Chen, F. J., Ge, F. and Parajulee, M. N. (2005) Impact of elevated CO_2 on tri-trophic interaction of *Gossypium hirsutum*, *Aphis gossypii*, and *Leis axyridis*. *Environmental Entomology*, **34**, 37–46.

Coley, P. D., Massa, M., Lovelock, C. E. and Winter, K. (2002) Effects of elevated CO_2 on foliar chemistry of saplings of nine species of tropical tree. *Oecologia*, **133**, 62–69.

Constable, J. V. H., Litvak, M. E., Greenberg, J. P. and Monson, R. K. (1999) Monoterpene emission from coniferous trees in response to elevated CO_2 concentration and climate warming. *Global Change Biology*, **5**, 255–267.

Costa, S. D., Kennedy, G. G. and Heagle, A. S. (2001) Effect of host plant ozone stress on Colorado potato beetles. *Environmental Entomology*, **30**, 824–831.

Cotrufo, M. F., Briones, M. J. and Ineson, P. (1998) Elevated CO_2 affects field decomposition rate and palatability of tree leaf litter: importance of changes in substrate quality. *Soil Biology and Biochemistry*, **36**, 1565–1571.

Cotrufo, M. F., Drake, B. and Ehleringer, J. R. (2005a) Palatability trials on hardwood leaf litter grown under elevated CO_2: a stable carbon isotope study. *Soil Biology and Biochemistry*, **37**, 1105–1112.

Cotrufo, M. F., De Angelis, P. and Polle, A. (2005b) Leaf litter production and decomposition in a poplar short-rotation coppice exposed to free air CO_2 enrichment (POPFACE). *Global Change Biology*, **11**, 971–982.

Coviella, C. and Trumble, J. T. (1999) Effects of elevated atmospheric CO_2 on insect–plant interactions. *Conservation Biology*, **13**, 700–712.

Coviella, C., Stipanovic, R. D. and Trumble, J. T. (2002) Plant allocation to defensive compounds: interactions between elevated CO_2 and nitrogen in transgenic cotton plants. *Journal of Experimental Botany*, **53**, 323–331.

Davey, M. P., Bryant, D. N., Cummins, I. *et al.* (2004) Effects of elevated CO_2 on the vasculature and phenolic secondary metabolism of *Plantago maritima*. *Phytochemistry*, **65**, 2197–2204.

David, J.-F., Malet, N., Couteaûx, M.-M. and Roy, J. (2001) Feeding rates of the woodlouse *Armadillidium vulgare* on herb litters produced at two levels of atmospheric CO_2. *Oecologia*, **127**, 343–349.

Denman, K. L., Brasseur, G., Chidthaisong, A. *et al.* (2007) Couplings between changes in the climate system and biogeochemistry. In *Climate Change 2007: The Physical Science Basis*. New York: Cambridge University Press, 499–588.

Earth System Research Laboratory (2010) http://www.esrl.noaa.gov/gmd/ccgg/trends/.

Ellison, A. M., Bank, M. S., Clinton, B. D. *et al.* (2005) Loss of foundation species: consequences for the structure and dynamics of forested ecosystems. *Frontiers in Ecology and Environment*, **3**, 479–486.

Fajer, E. D., Bowers, M. D. and Bazzaz, F. A. (1992) The effect of nutrients and enriched CO_2 environments on production of carbon-based allelochemicals in *Plantago*: a test of the carbon/nutrient balance hypothesis. *American Naturalist*, **140**, 707–723.

Fares, S., Bartaa, C., Brillia, F. *et al.* (2006) Impact of high ozone on isoprene emission, photosynthesis and histology of developing *Populus alba* leaves directly or indirectly exposed to the pollutant. *Physiologia Plantarum*, **128**, 456–465.

Fares, S., Oksanen, E., Lännenpää, M., Julkunen-Tiitto, R. and Loreto, F. (2010) Volatile emissions and phenolic compound concentrations along a vertical profile of *Populus nigra* leaves exposed to realistic ozone concentrations. *Photosynthesis Research*, **104**, 61–74.

Findlay, S., Carreiro, M., Krischik, V. and Jones, C. G. (1996) Effects of damage to living plants on leaf litter quality. *Ecological Applications*, **6**, 269–275.

Finzi, A. C., Allen, A. S., DeLucia, E. H., Ellsworth, D. S. and Schlesinger, W. H. (2001) Forest litter production, chemistry, and decomposition following two years of free-air CO_2 enrichment. *Ecology*, **82**, 470–484.

Fraenkel, G. S. (1959) The *raison d'être* of secondary plant substances. *Science*, **129**, 1466–1470.

Francini, A., Nali, C., Picchi, V. and Lorenzini, G. (2007) Metabolic changes in white clover clones exposed to ozone. *Environmental and Experimental Botany*, **60**, 11–19.

Freiwald, V., Haikio, E., Julkunen-Tiitto, R., Holopainen, J. K. and Oksanen, E. (2008) Elevated ozone modifies the feeding behaviour of the common leaf weevil on hybrid aspen through shifts in developmental, chemical, and structural properties of leaves. *Entomologia Experimentalis et Applicata*, **128**, 66–72.

Gao, F., Zhu, S. R., Sun, Y. C. *et al.* (2008) Interactive effects of elevated CO_2 and cotton cultivar on tri-trophic interaction of *Gossypium hirsutum*, *Aphis gossyppii*, and *Propylaea japonica*. *Environmental Entomology*, **37**, 29–37.

Gebauer, R. L. E., Strain, B. R. and Reynolds, J. P. (1998) The effect of elevated CO_2 and N availability on tissue concentrations and whole plant pools of carbon-based secondary compounds in loblolly pine (*Pinus taeda*). *Oecologia*, **113**, 29–36.

Gleadow, R. M., Foley, W. J. and Woodrow, I. E. (1998) Enhanced CO_2 alters the relationship between photosynthesis and defence in cyanogenic *Eucalyptus cladocalyx* F. Muell. *Plant, Cell and Environment*, **21**, 12–22.

Goverde, M., Erhardt, A. and Stocklin, J. (2004) Genotype-specific response of a lycaenid herbivore to elevated carbon dioxide and phosphorus availability in calcareous grassland. *Oecologia*, **139**, 383–391.

Hahlbrock, K. and Scheel, D. (1989) Physiology and molecular-biology of phenylpropanoid metabolism. *Annual Review of Plant Physiology and Plant Molecular Biology*, **40**, 347–369.

Haikio, E., Makkonen, M., Julkunen-Tiitto, R. *et al.* (2009) Performance and secondary chemistry of two hybrid aspen (*Populus tremula L. × Populus tremuloides Michx.*) clones in long-term elevated ozone exposure. *Journal of Chemical Ecology*, **35**, 664–678.

Hall, M. C., Stiling, P., Moon, D. C., Drake, B. G. and Hunter, M. D. (2005) Effects of elevated CO_2 on foliar quality and herbivore damage in a scrub oak ecosystem. *Journal of Chemical Ecology*, **31**, 267–286.

Hall, M. C., Stiling, P., Moon, D. C., Drake, B. G. and Hunter, M. D. (2006) Elevated CO_2 increases the long-term decomposition rate of *Quercus myrtifolia* leaf litter. *Global Change Biology*, **12**, 568–577.

Hamilton, J. G., Zangerl, A. R., Berenbaum, M. R. *et al.* (2004) Insect herbivory in an intact forest understory under experimental CO_2 enrichment. *Oecologia*, **138**, 566–573.

Hartley, S. E., Jones, C. G., Couper, G. C. and Jones, T. H. (2000) Biosynthesis of plant phenolic compounds in elevated atmospheric CO_2. *Global Change Biology*, **6**, 497–506.

Hättenschwiler, S. and Bretscher, D. (2001) Isopod effects on decomposition of litter produced under elevated CO_2, N deposition and different soil types. *Global Change Biology*, **7**, 565–579.

Hättenschwiler, S. and Schafellner, C. (1999) Opposing effects of elevated CO_2 and N deposition on *Lymantria monacha* larvae feeding on spruce trees. *Oecologia*, **118**, 210–217.

Hättenschwiler, S. and Schafellner, C. (2004) Gypsy moth feeding in the canopy of a CO_2-enriched mature forest. *Global Change Biology*, **10**, 1899–1908.

Hättenschwiler, S., Bühler, S. and Körner, C. (1999) Quality, decomposition and isopod consumption of tree litter produced under elevated CO_2. *Oikos*, **85**, 271–281.

Heller, W., Rosemann, D., Osswald, W. *et al.* (1990) Biochemical response of Norway spruce (*Picea abies* (L.) Karsts) towards 14-month exposure to ozone and acid mist. 1. Effects on polyphenol and monoterpene metabolism. *Environmental Pollution*, **64**, 353–366.

Herms, D. A. and Mattson, W. J. (1992) The dilemma of plants: to grow or defend. *Quarterly Review of Biology*, **67**, 283–335.

Heyworth, C. J., Iason, G. R., Temperton, V., Jarvis, P. G. and Duncan, A. J. (1998) The effect of elevated CO_2 concentration and nutrient supply on carbon-based plant secondary metabolites in *Pinus sylvestris* L. *Oecologia*, **115**, 344–350.

Hillstrom, M. L. and Lindroth, R. L. (2008) Elevated atmospheric carbon dioxide and ozone alter forest insect abundance and community composition. *Insect Conservation and Diversity*, **1**, 233–241.

Himanen, S. J., Nissinen, A., Auriola, S. *et al.* (2008) Constitutive and herbivore-inducible glucosinolate concentrations in oilseed rape (*Brassica napus*) leaves are not affected by Bt Cry1Ac insertion but change under elevated atmospheric CO_2 and O_3. *Planta*, **227**, 427–437.

Holton, M. K., Lindroth, R. L. and Nordheim, E. V. (2003) Foliar quality influences tree-herbivore-parasitoid interactions: effects of elevated CO_2, O_3, and plant genotype. *Oecologia*, **137**, 233–244.

Hunt, M. G., Rasmussen, S., Newton, P. C. D., Parsons, A. J. and Newman, J. A. (2005) Near-term impacts of elevated CO_2, nitrogen and fungal endophyte-infection on *Lolium perenne* L. growth, chemical composition and alkaloid production. *Plant, Cell and Environment*, **28**, 1345–1354.

Jackson, D. M., Heagle, A. S. and Eckel, R. V. W. (1999) Ovipositional response of tobacco hornworm moths (Lepidoptera: Sphingidae) to tobacco plants grown under elevated levels of ozone. *Environmental Entomologist*, **28**, 566–571.

Jackson, D. M., Rufty, T. W., Heagle, A. S., Severson, R. F. and Eckel, R. V. W. (2000) Survival and development of tobacco hornworm larvae on tobacco plants grown under elevated levels of ozone. *Journal of Chemical Ecology*, **26**, 1–19.

Johnson, R. H. and Lincoln, D. E. (1990) Sagebrush and grasshopper responses to atmospheric carbon dioxide concentration. *Oecologia*, **84**, 103–110.

Jones, C. G. and Coleman, J. S. (1988) Plant stress and insect behavior: cottonwood, ozone and the feeding and oviposition preference of a beetle. *Oecologia*, **76**, 51–56.

Julkunen-Tiitto, R., Tahvanainen, J. and Silvola, J. (1993) Increased CO_2 and nutrient status changes affect phytomass and the production of plant defensive secondary chemicals in *Salix myrsinifolia* (Salisb.). *Oecologia*, **95**, 495–498.

Kainulainen, P., Holopainen, J. K., Hyttinen, H. and Oksanen, J. (1994) Effect of ozone on the biochemistry and aphid infestation of Scots pine. *Phytochemistry*, **35**, 39–42.

Kainulainen, P., Holopainen, J. K. and Holopainen, T. (1998) The influence of elevated CO_2 and O_3 concentrations on Scots pine needles: changes in starch and secondary metabolites over three exposure years. *Oecologia*, **114**, 455–460.

Kainulainen, P., Holopainen, T. and Holopainen, J. K. (2003) Decomposition of secondary compounds from needle litter of Scots pine grown under elevated CO_2 and O_3. *Global Change Biology*, **9**, 295–304.

Kangasjärvi, J., Talvinen, J., Utriainen, M. and Karjalainen, R. (1994) Plant defence systems induced by ozone. *Plant, Cell and Environment*, **17**, 783–794.

Kangasjärvi, J., Jaspers, P. and Kollist, H. (2005) Signalling and cell death in ozone-exposed plants. *Plant, Cell and Environment*, **28**, 1021–1036.

Kanoun, M., Goulas, M. J. P. and Biolley, J. P. (2001) Effect of a chronic and moderate ozone pollution on the phenolic pattern of bean leaves (*Phaseolus vulgaris* L. cv Nerina), relations with visible injury and biomass production. *Biochemical Systematics and Ecology*, **29**, 443–457.

Kanowski, J. (2001) Effects of elevated CO_2 on the foliar chemistry of seedlings of two rainforest trees from north-east Australia: implications for folivorous marsupials. *Austral Ecology*, **26**, 165–172.

Karl, T. R., Melillo, J. M. and Peterson, T. C. (eds.) (2009) *Global Climate Change Impacts in the United States*. New York: Cambridge University Press.

Karnosky, D. F., Pregitzer, K. S., Zak, D. R. *et al.* (2005) Scaling ozone responses of forest trees to the ecosystem level in a changing climate. *Plant, Cell and Environment*, **28**, 965–981.

Karowe, D. N., Seimens, D. H. and Mitchell-Olds, T. (1997) Species-specific response of glucosinolate content to elevated atmospheric CO_2. *Journal of Chemical Ecology*, **23**, 2569–2582.

Kasurinen, A., Riikonen, J., Oksanen, E., Vapaavuori, E. and Holopainen, T. (2006) Chemical composition and decomposition of silver birch leaf litter produced under elevated CO_2 and O_3. *Plant and Soil*, **282**, 261–280.

Kasurinen, A., Peltonen, P. A., Julkunen-Tiitto, R. *et al.* (2007) Effects of elevated CO_2 and O_3 on leaf litter phenolics and subsequent performance of litter-feeding soil macrofauna. *Plant and Soil*, **292**, 25–43.

Keen, N. T. and Taylor, O. C. (1975) Ozone injury in soybeans: isoflavonoid accumulation is related to necrosis. *Plant Physiology*, **55**, 731–733.

Kinney, K. K., Lindroth, R. L., Jung, S. M. and Nordheim, E. V. (1997) Effects of CO_2 and NO_3-availability on deciduous trees, phytochemistry and insect performance. *Ecology*, **78**, 215–230.

Knepp, R. G., Hamilton, J. G., Mohan, J. E. *et al.* (2005) Elevated CO_2 reduces leaf damage by insect herbivores in a forest community. *New Phytologist*, **167**, 207–218.

Koike, T., Tobita, H., Shibata, T. *et al.* (2006) Defense characteristics of seral deciduous broad-leaved tree seedlings grown under differing levels of CO_2 and nitrogen. *Population Ecology*, **48**, 23–29.

Kopper, B. J. and Lindroth, R. L. (2003a) Responses of trembling aspen (*Populus tremuloides*) phytochemistry and aspen blotch leafminer (*Phyllonorycter tremuloidiella*) performance to elevated level of CO_2 and O_3. *Agricultural and Forest Entomology*, **5**, 17–26.

Kopper, B. J. and Lindroth, R. L. (2003b) Effects of elevated carbon dioxide and ozone on the phytochemistry of aspen and performance of an herbivore. *Oecologia*, **134**, 95–103.

Koricheva, J., Larsson, S., Haukioja, E. and Keinänen, M. (1998) Regulation of woody plant secondary metabolism by resource availability: hypothesis testing by means of meta-analysis. *Oikos*, **83**, 212–226.

Kretzschmar, F. D., Aidar, M. P. M., Salgado, I. and Braga, M. R. (2009) Elevated CO_2 atmosphere enhances production of defense-related flavonoids in soybean elicited by NO and a fungal elicitor. *Environmental and Experimental Botany*, **65**, 319–329.

Kuokkanen, K., Julkunen-Tiitto, R., Keinänen, M., Niemelä, P. and Tahvanainen, J. (2001) The effect of elevated CO_2 and temperature on the secondary chemistry of *Betula pendula* seedlings. *Trees – Structure and Function*, **15**, 378–384.

Kuokkanen, K., Yan, S. C. and Niemela, P. (2003) Effects of elevated CO_2 and temperature on the leaf chemistry of birch *Betula pendula* (Roth) and the feeding behaviour of the weevil *Phyllobius maculicornis*. *Agricultural and Forest Entomology*, **5**, 209–217.

Laothawornkitkul, J., Taylor, J. E., Paul, N. D. and Hewitt, C. N. (2009) Biogenic volatile organic compounds in the Earth system. *New Phytologist*, **183**, 27–51.

Lavola, A. and Julkunen-Tiitto, R. (1994) The effect of elevated carbon dioxide and fertilization on primary and secondary metabolites in birch, *Betula pendula* (Roth). *Oecologia*, **99**, 315–321.

Lavola, A., Julkunen-Tiitto, R. and Pääkkönen, E. (1994) Does ozone stress change the primary or secondary metabolites of birch (*Betula pendula* Roth.)? *New Phytologist*, **126**, 637–642.

Lawler, I. R., Foley, W. J., Woodrow, I. E. and Cork, S. J. (1997) The effects of elevated CO_2 on the nutritional quality of *Eucalyptus* foliage and its interaction with soil nutrient and light availability. *Oecologia*, **109**, 59–68.

Lincoln, D. E. and Couvet, D. (1989) The effect of carbon supply on allocation to allelochemicals and caterpillar consumption of peppermint. *Oecologia*, **78**, 112–114.

Lindroth, R. L. (2010) Impacts of elevated atmospheric CO_2 and O_3 on forests: phytochemistry, trophic interactions, and ecosystem dynamics. *Journal of Chemical Ecology*, **36**, 2–21.

Lindroth, R. L., Kinney, K. K. and Platz, C. L. (1993) Responses of deciduous trees to elevated atmospheric CO_2: productivity, phytochemistry and insect performance. *Ecology*, **74**, 763–777.

Lindroth, R. L., Roth, S., Kruger, E. L., Volin, J. C. and Koss, P. A. (1997) CO_2-mediated changes in aspen chemistry: effects on gypsy moth performance and susceptibility to virus. *Global Change Biology*, **3**, 279–289.

Lindroth, R. L., Roth, S. and Nordheim, E. V. (2001) Genotypic variation in response of quaking aspen (*Populus tremuloides*) to atmospheric CO_2 enrichment. *Oecologia*, **126**, 371–379.

Liu, L. L., King, J. S., Booker, F. L. *et al.* (2009) Enhanced litter input rather than changes in litter chemistry drive soil carbon and nitrogen cycles under elevated CO_2: a microcosm study. *Global Change Biology*, **15**, 441–453.

Loreto, F. and Fares, S. (2007) Is ozone flux inside leaves only a damage indicator? Clues from volatile isoprenoid studies. *Plant Physiology*, **143**, 1096–1100.

Loreto, F., Pinelli, P., Manes, F. and Kollist, H. (2004) Impact of ozone on monoterpene emissions and evidence for an isoprene-like antioxidant action of monoterpenes emitted by *Quercus ilex* leaves. *Tree Physiology*, **24**, 361–367.

Loranger, G. I., Pregitzer, K. S. and King, J. S. (2004) Elevated CO_2 and O_3t concentrations differentially affect selected groups of the fauna in temperate forest soils. *Soil Biology and Biochemistry*, **36**, 1521–1524.

Luo, Y., Su, B., Currie, W. S. *et al.* (2004) Progressive nitrogen limitation of ecosystem responses to rising atmospheric carbon dioxide. *Bioscience*, **54**, 731–739.

Manninen, A. M., Holopainen, T., Lyytikäinen-Saarenmaa, P. and Holopainen, J. K. (2000) The role of low-level ozone exposure and mycorrhizas in chemical quality and insect herbivore performance on Scots pine seedlings. *Global Change Biology*, **6**, 111–121.

Matros, A., Amme, S., Kettig, B. *et al.* (2006) Growth at elevated CO_2 concentrations leads to modified profiles of secondary metabolites in tobacco cv. SamsunNN and to increased resistance against infection with potato virus Y. *Plant, Cell and Environment*, **29**, 126–137.

Mattson, W. J., Kuokkanen, K., Niemelä, P. *et al.* (2004) Elevated CO_2 alters birch resistance to Lagomorpha herbivores. *Global Change Biology*, **10**, 1402–1413.

McDonald, E. P., Agrell, J. and Lindroth, R. L. (1999) CO_2 and light effects on deciduous trees: growth, foliar chemistry, and insect performance. *Oecologia*, **119**, 389–399.

Meehan, T. D., Crossley, M. S. and Lindroth, R. L. (2010) Impacts of elevated CO_2 and O_3 on aspen leaf litter chemistry and earthworm and springtail productivity. *Soil Biology and Biochemistry*, doi:10.1016/j.soilbio.2010.03.019.

Mohan, J. E., Ziska, L. H., Schlesinger, W. H. *et al.* (2006) Biomass and toxicity responses of poison ivy (*Toxicodendron radicans*) to elevated atmospheric CO$_2$. *Proceedings of the National Academy of Sciences USA*, **103**, 9086–9089.

Mondor, E. B., Tremblay, M. N., Awmack, C. S. and Lindroth, R. L. (2004) Divergent pheromone-mediated insect behaviour under global atmospheric change. *Global Change Biology*, **10**, 1820–1824.

Muntifering, R. B., Chappelka, A. H., Lin, J. C., Karnosky, D. F. and Somers, G. L. (2006) Chemical composition and digestibility of *Trifolium* exposed to elevated ozone and carbon dioxide in a free-air (FACE) fumigation system. *Functional Ecology*, **20**, 269–275.

Newman, D. J., Cragg, G. M. and Snader, K. M. (2003) Natural products as sources of new drugs over the period 1981–2002. *Journal of Natural Products*, **66**, 1022–1037.

Norby, R. J., Cotrufo, M. F., Ineson, P., O'Neill, E. G. and Canadell, J. G. (2001) Elevated CO$_2$, litter chemistry, and decomposition: a synthesis. *Oecologia*, **127**, 153–165.

O'Neill, B. F., Zangerl, A. R., Dermody, O. *et al.* (2010) Impact of elevated levels of atmospheric CO$_2$ and herbivory on flavonoids of soybean (*Glycine max* Linnaeus). *Journal of Chemical Ecology*, **36**, 35–45.

Parsons, W. F. J., Lindroth, R. L. and Bockheim, J. G. (2004) Decomposition of *Betula papyrifera* leaf litter under the independent and interactive effects of elevated CO$_2$ and O$_3$. *Global Change Biology*, **10**, 1666–1677.

Parsons, W. F. J., Bockheim, J. G. and Lindroth, R. L. (2008) Independent, interactive, and species-specific responses of leaf litter decomposition to elevated CO$_2$ and O$_3$ in a northern hardwood forest. *Ecosystems*, **11**, 505–519.

Peltonen, P. A., Vapaavuori, E. and Julkunen-Tiitto, R. (2005) Accumulation of phenolic compounds in birch leaves is changed by elevated carbon dioxide and ozone. *Global Change Biology*, **11**, 1305–1324.

Peltonen, P. A., Julkunen-Tiitto, R., Vapaavuori, E. and Holopainen, J. K. (2006) Effects of elevated carbon dioxide and ozone on aphid oviposition preference and birch bud exudate phenolics. *Global Change Biology*, **12**, 1670–1679.

Peñuelas, J. and Estiarte, M. (1998) Can elevated CO$_2$ affect secondary metabolism and ecosystem function? *Trends in Ecology and Evolution*, **13**, 20–24.

Peñuelas, J., Estiarte, M., Kimball, B. A. *et al.* (1996) Variety of responses of plant phenolic concentration to CO$_2$ enrichment. *Journal of Experimental Botany*, **47**, 1463–1467.

Peñuelas, J., Estiarte, M. and Llusià, J. (1997) Carbon-based secondary compounds at elevated CO$_2$. *Photosynthetica*, **33**, 313–316.

Percy, K. E., Awmack, C. S., Lindroth, R. L. *et al.* (2002) Altered performance of forest pests under atmospheres enriched by CO$_2$ and O$_3$. *Nature*, **420**, 403–407.

Peters, H. A., Baur, B., Bazzaz, F. and Körner, C. (2000) Consumption rates and food preferences of slugs in a calcareous grassland under current and future CO$_2$ conditions. *Oecologia*, **125**, 72–81.

Petit, J. R., Jouzel, J., Raynaud, D. *et al.* (1999) Climate and atmospheric history of the past 420,000 years from the Vostok ice core, Antarctica. *Nature*, **399**, 429–436.

Pinto, D. M., Blande, J. D., Nykänen, R. *et al.* (2007) Ozone degrades common herbivore-induced plant volatiles: does this affect herbivore prey location by predators and parasitoids? *Journal of Chemical Ecology*, **33**, 683–694.

Pinto, D. M., Blande, J. D., Souza, S. R., Nerg, A. and Holopainen, J. K. (2010) Plant volatile organic compounds (VOCs) in ozone (O$_3$) polluted atmospheres: the ecological effects. *Journal of Chemical Ecology*, **36**, 22–34.

Räisänen, T., Ryyppö, A. and Kellomäki, S. (2008) Effects of elevated CO$_2$ and temperature on monoterpene emission of Scots pine (*Pinus sylvestris* L.). *Atmospheric Environment*, **42**, 4160–4171.

Reddy, G. V. P., Tossavainen, P., Nerg, A. M. and Holopainen, J. K. (2004) Elevated atmospheric CO_2 affects the chemical quality of *Brassica* plants and the growth rate of the specialist, *Plutella xylostella*, but not the generalist, *Spodoptera littoralis*. *Journal of Agricultural and Food Chemistry*, **52**, 4185–4191.

Reitz, S. R., Karowe, D. N., Diawara, M. M. and Trumble, J. T. (1997) Effects of elevated atmospheric carbon dioxide on the growth and linear furanocoumarin content of celery. *Journal of Agricultural and Food Chemistry*, **45**, 3642–3646.

Rossi, A. M., Stiling, P., Moon, D. C., Cattell, M. V. and Drake, B. G. (2004) Induced defensive response of myrtle oak to foliar insect herbivory in ambient and elevated CO_2. *Journal of Chemical Ecology*, **30**, 1143–1152.

Roth, S. K. and Lindroth, R. L. (1995) Elevated atmospheric CO_2: effects on phytochemistry, insect performance and insect-parasitoid interactions. *Global Change Biology*, **1**, 173–182.

Roth, S., McDonald, E. P. and Lindroth, R. L. (1997) Atmospheric CO_2 and soil water availability: consequences for tree–insect interactions. *Canadian Journal of Forest Research*, **27**, 1281–1290.

Roth, S., Lindroth, R. L., Volin, J. C. and Kruger, E. L. (1998) Enriched atmospheric CO_2 and defoliation: effects on tree chemistry and insect performance. *Global Change Biology*, **4**, 419–430.

Sager, E. P., Hutchinson, T. C. and Croley, T. R. (2005) Foliar phenolics in sugar maple (*Acer saccharum*) as a potential indicator of tropospheric ozone pollution. *Environmental Monitoring and Assessment*, **105**, 419–430.

Saleem, A., Loponen, J., Pihlaja, K. and Oksanen, E. (2001) Effects of long-term open-field ozone exposure on leaf phenolics of European silver birch (*Betula pendula* Roth). *Journal of Chemical Ecology*, **27**, 1049–1062.

Sallas, L., Kainulainen, P., Utriainen, J., Holopainen, T. and Holopainen, J. K. (2001) The influence of elevated O_3 and CO_2 concentrations on secondary metabolites of Scots pine (*Pinus sylvestris* L.) seedlings. *Global Change Biology*, **7**, 303–311.

Sandermann, H. Jr (1996) Ozone and plant health. *Annual Review of Phytopathology*, **34**, 347–366.

Saxe, H., Ellsworth, D. S. and Heath, J. (1998) Tree and forest functioning in an enriched CO_2 atmosphere. *New Phytologist*, **139**, 395–436.

Scherzer, A. J., Rebbeck, J. and Boerner, R. E. J. (1998) Foliar nitrogen dynamics and decomposition of yellow-poplar and eastern white pine during four seasons of exposure to elevated ozone and carbon dioxide. *Forest Ecology and Management*, **109**, 355–366.

Schonhof, I., Kläring, H. P., Krumbein, A. and Schreiner, M. (2007) Interaction between atmospheric CO_2 and glucosinolates in broccoli. *Journal of Chemical Ecology*, **33**, 105–114.

Shadkami, F., Helleur, R. J. and Cox, R. M. (2007) Profiling secondary metabolites of needles of ozone-fumigated white pine (*Pinus strobus*) clones by thermally assisted hydrolysis/methylation GC/MS. *Journal of Chemical Ecology*, **33**, 1467–1476.

Stiling, P. and Cornelissen, T. (2007) How does elevated carbon dioxide (CO_2) affect plant–herbivore interactions? A field experiment and meta-analysis of CO_2-mediated changes on plant chemistry and herbivore performance. *Global Change Biology*, **13**, 1823–1842.

Stiling, P., Rossi, A. M., Hungate, B. *et al.* (1999) Decreased leaf-miner abundance in elevated CO_2: reduced leaf quality and increased parasitoid attack. *Ecological Applications*, **9**, 240–244.

Stiling, P., Cattell, M., Moon, D. C. *et al.* (2002) Elevated atmospheric CO_2 lowers herbivore abundance, but increases leaf abscission rates. *Global Change Biology*, **8**, 658–667.

Stiling, P., Moon, D. C., Hunter, M. D. *et al.* (2003) Elevated CO_2 lowers relative and absolute herbivore density across all species of a scrub-oak forest. *Oecologia*, **134**, 82–87.

Stiling, P., Moon, D., Rossi, A., Hungate, B. A. and Drake, B. (2009) Seeing the forest for the trees: long-term exposure to elevated CO_2 increases some herbivore densities. *Global Change Biology*, **15**, 1895–1902.

Strain, B. R. and Bazzaz, F. A. (1983) Terrestrial plant communities. In E. R. Lemon (ed.) *The Response of Plants to Rising Levels of Atmospheric Carbon Dioxide*. Boulder, CO: Westview Press, 177–222.

Traw, M. B., Lindroth, R. L. and Bazzaz, F. A. (1996) Decline in gypsy moth (*Lymantria dispar*) performance in an elevated CO_2 atmosphere depends upon host plant species. *Oecologia*, **108**, 113–120.

Tylianakis, J. M., Didham, R. K., Bascompte, J. and Wardle, D. A. (2008) Global change and species interactions in terrestrial ecosystems. *Ecology Letters*, **11**, 1351–1363.

Valkama, E., Koricheva, J. and Oksanen, E. (2007) Effects of elevated O_3, alone and in combination with elevated CO_2, on tree leaf chemistry and insect herbivore performance: a meta-analysis. *Global Change Biology*, **13**, 184–201.

Vannette, R. L. and Hunter, M. D. (2011) Genetic variation in expression of defense phenotype may mediate evolutionary adaptation of *Asclepias syriaca* to elevated CO_2. *Global Change Biology*, **17**, 1277–1288.

Veteli, T. O., Kuokkanen, K., Julkunen-Tiitto, R., Roininen, H. and Tahvanainen, J. (2002) Effects of elevated CO_2 and temperature on plant growth and herbivore defensive chemistry. *Global Change Biology*, **8**, 1240–1252.

Vingarzan, R. (2004) A review of surface ozone background levels and trends. *Atmospheric Environment*, **38**, 3431–3442.

Vuorinen, T., Nerg, A. M., Ibrahim, M. A., Reddy, G. V. P. and Holopainen, J. K. (2004a) Emission of *Plutella xylostella*-induced compounds from cabbages grown at elevated CO_2 and orientation behavior of the natural enemies. *Plant Physiology*, **135**, 1984–1992.

Vuorinen, T., Nerg, A. M. and Holopainen, J. K. (2004b) Ozone exposure triggers the emission of herbivore-induced plant volatiles, but does not disturb tritrophic signalling. *Environmental Pollution*, **131**, 305–311.

Wennberg, P. O. and Dabdub, D. (2008) Rethinking ozone production. *Science*, **319**, 1624–1625.

Whittaker, J. B., Kristiansen, L. W., Mikkelsen, T. N. and Moore, R. (1989) Responses to ozone of insects feeding on a crop and weed species. *Environmental Pollution*, **62**, 89–101.

Williams, R. S., Lincoln, D. E. and Thomas, R. B. (1994) Loblolly pine grown under elevated CO_2 affects early instar pine sawfly performance. *Oecologia*, **98**, 64–71.

Williams, R. S., Thomas, R. B., Strain, B. R. and Lincoln, D. E. (1997) Effects of elevated CO_2, soil nutrient levels, and foliage age on the performance of two generations of *Neodiprion lecontei* (Hymenoptera: Diprionidae) feeding on loblolly pine. *Environmental Entomology*, **26**, 1312–1322.

Williams, R. S., Lincoln, D. E. and Norby, R. J. (1998) Leaf age effects of elevated CO_2-grown white oak leaves on spring-feeding lepidopterans. *Global Change Biology*, **4**, 235–246.

Williams, R. S., Norby, R. J. and Lincoln, D. E. (2000) Effects of elevated CO_2 and temperature-grown red and sugar maple on gypsy moth performance. *Global Change Biology*, **6**, 685–695.

Williams, R. S., Lincoln, D. E. and Norby, R. J. (2003) Development of gypsy moth larvae feeding on red maple saplings at elevated CO_2 and temperature. *Oecologia*, **137**, 114–122.

Wittig, V. E., Ainsworth, E. A., Naidu, S. L., Karnosky, D. F. and Long, S. P. (2009) Quantifying the impact of current and future tropospheric ozone on tree biomass, growth, physiology and biochemistry: a quantitative meta-analysis. *Global Change Biology*, **15**, 396–424.

Young, P. J., Arneth, A., Schurgers, G., Zeng, G. and Pyle, J. A. (2009) The CO_2 inhibition of terrestrial isoprene emission significantly affects future ozone projections. *Atmospheric Chemistry and Physics*, **9**, 2793–2803.

Yuan, J. S., Himanen, S. J., Holopainen, J. K., Chen, F. and Stewart, C. N. (2009) Smelling global climate change: mitigation of function for plant volatile organic compounds. *Trends in Ecology and Evolution*, **24**, 323–331.

Zavala, J. A., Casteel, C. L., Delucia, E. H. and Berenbaum, M. R. (2008) Anthropogenic increase in carbon dioxide compromises plant defense against invasive insects. *Proceedings of the National Academy of Sciences USA*, **105**, 5129–5133.

Ziska, L. H., Emche, S. D., Johnson, E. L. *et al.* (2005) Alterations in the production and concentration of selected alkaloids as a function of rising atmospheric carbon dioxide and air temperature: implications for ethno-pharmacology. *Global Change Biology*, **11**, 1798–1807.

Ziska, L., Panicker, S. and Wojno, H. (2008) Recent and projected increases in atmospheric carbon dioxide and the potential impacts on growth and alkaloid production in wild poppy (*Papaver setigerum* DC.). *Climatic Change*, **91**, 395–403.

Zvereva, E. L. and Kozlov, M. V. (2006) Consequences of simultaneous elevation of carbon dioxide and temperature for plant–herbivore interactions: a metaanalysis. *Global Change Biology*, **12**, 27–41.

CHAPTER EIGHT

The role of plant secondary metabolites in freshwater macrophyte–herbivore interactions: limited or unexplored chemical defences?

ELISABETH M. GROSS

Laboratoire Interactions Ecotoxicologie Biodiversité Ecosystémes, Université de Lorraine

ELISABETH S. BAKKER

Department of Aquatic Ecology, Netherlands Institute of Ecology

8.1   Introduction

Historically, herbivory on aquatic plants has been considered negligible. 'One could probably remove all the larger plants and substitute glass structures of the same form and surface texture without greatly affecting the immediate food relations', wrote Shelford (1918), cited in Hutchinson (1975) about grazing losses of submerged angiosperms. This misconception might have persisted for so long because grazing by zooplankton on phytoplankton has been the major focus in limnology for decades. Also, herbivore-related biomass losses of higher aquatic plants were estimated to be less than 10% of the total production (Wetzel, 1983). In the past two decades many studies have shown that multiple invertebrate and vertebrate herbivores feed on freshwater angiosperms and that herbivory on vascular plants is quantitatively equally important in terrestrial and freshwater habitats (Lodge, 1991; Newman, 1991; Cyr & Pace, 1993). Thus, we are now ready to critically consider the role of plant secondary metabolites (PSMs) in freshwater plant–herbivore interactions. Whereas the importance and tremendous variety of PSMs is well acknowledged in terrestrial plants and seaweeds, relatively little is known about the presence, levels, types and function of PSMs in freshwater plants (Lodge *et al.*, 1998; Sotka *et al.*, 2009). This is surprising because aquatic angiosperms and most of their insect herbivores are in fact secondarily aquatic, descendant from terrestrial ancestors (Newman, 1991). Thus, similarities in potential feeding deterrents and host-plant selection might be anticipated. Yet there may also be pronounced differences in plant–herbivore interactions in the aquatic environment. For example, water provides different physico-chemical conditions compared with air or soil, which should affect the dispersal of released compounds. Additionally, not all terrestrial

The Ecology of Plant Secondary Metabolites: From Genes to Global Processes, eds. Glenn R. Iason, Marcel Dicke and Susan E. Hartley. Published by Cambridge University Press. © British Ecological Society 2012.

plant families and growth forms have relatives underwater, and aquatic herbivores differ in species composition and diet selection from their terrestrial counterparts. These environmental, phylogenetic and ecological predispositions might have shaped the kinds of feeding deterrents that are present in freshwater systems.

Freshwater plant–herbivore interactions share some similarities with marine habitats, but also have distinct differences. In marine systems, as in freshwater, many generalist herbivores or omnivores such as fish and crustaceans are present. Some higher plants occur in the marine littoral, especially monocots belonging to the Alismatales, such as Hydrocharitaceae, Ruppiaceae and Zosteraceae. Yet the dominant primary producers in littoral marine habitats are various macroalgae belonging to the Phaeophyta, Rhodophyta or Chlorophyta. In freshwater systems, generally fewer macroalgae and more angiosperms are present in littoral zones. A comparison of plant–herbivore interactions across all biomes would be valuable, but is beyond the scope of this review. Marine and terrestrial plant–herbivore interactions have been compared by Hay (1991).

We review freshwater plant–herbivore interactions and the role that feeding deterrents and other bioactive PSMs may play in these interactions. We will conclude with suggestions for future research.

8.2   Plant–herbivore interactions in freshwater systems

There is no sharp boundary between land and water, but a gradual shift from a more terrestrial to a fully aquatic lifestyle. Adaptations to the aquatic habitat result in different growth forms along this transect from land into water, from emergent to floating to fully submerged macrophytes. Growth forms may vary depending on environmental conditions or ontogeny. Some plants exhibit an amphibious lifestyle, and occur either submerged or emergent. Commonly, higher aquatic plants together with some macroalgae (charophytes), water mosses and ferns are named 'macrophytes' in contrast to microphytic algae and cyanobacteria. Our review will be restricted to these macrophytes. The growth form of macrophytes has important consequences for aquatic plant–herbivore interactions, as macrophyte properties and herbivore access depend strongly on whether the plant or plant part is below or above the water surface. Compared with fully submerged plants, emergent plants need more structural tissue, thicker cell walls and a more complex cuticle to limit evapotranspiration and provide stability. This results in tougher plants with a higher C:N ratio in emergent plants or plant parts (Cloern et al., 2002; Demars & Edwards, 2008), which might reduce plant palatability for herbivores. Emergent and submerged macrophytes experience herbivory by different species: emergent macrophytes are accessible for terrestrial herbivores above the water surface, whereas their submerged parts and submerged macrophytes are eaten by aquatic herbivores. This applies especially to invertebrates and fish, as larger vertebrates (for example, beaver or coots) may be able to forage on both. Meta-analyses show that insect herbivory causes similar reductions in production in terrestrial and fresh-water plants (Newman, 1991; Lodge *et al.*, 1998). However, no distinction is made between emergent and submerged macrophytes and, therefore, it is difficult to compare the effects of terrestrial and aquatic invertebrates. A direct comparison of three amphibious plant species in the field showed that aquatic leaves exhibited more grazing damage by invertebrates than terrestrial leaves (Sand-Jensen & Jacobsen, 2002). The percentage grazing damage of *Nuphar lutea* was 2.9 times as large on submerged than on floating leaves (Cronin *et al.*, 1998) and the invasive *Alternanthera philoxeroides* exhibited a higher tolerance to herbivory in terrestrial than in aquatic habitats (Sun *et al.*, 2009). Therefore, current studies suggest that grazing pressure may be higher under water, but more plant species should be tested to establish whether this conclusion holds more generally.

Aquatic herbivores comprise vertebrates and invertebrates. Waterfowl and fish are the dominant vertebrate herbivores. Others are manatees (*Trichechus* sp.), beavers (*Castor* sp.), muskrats (*Ondatra zibethicus*), coypu (*Myocastor coypus*) and moose (*Alces alces*). Most of the vertebrates are facultatively herbivorous as they have an omnivorous diet. Only a very few rely almost entirely on aquatic plant tissue (e.g. grass carp (*Ctenopharyngodon idella*), red-crested pochard (*Netta rufina*), gadwall (*Anas strepera*), manatee and coypu). Others are mostly herbiv-orous but consume both aquatic and terrestrial plants (e.g. moose, beaver, mute swan (*Cygnus olor*), Greylag geese (*Anser anser*) and European wigeon (*Anas penelope*)). In Western Europe, mute swan and coot (*Fulica atra*) are probably quantitatively the most important vertebrate herbivores together with rudd (*Scardinius erythrophthalmus*), roach (*Rutilus rutilus*) and introduced grass carp. Important invertebrate herbivores are aquatic insects, crayfish, amphipods and snails. Aquatic Crustaceae are generally omnivorous, and the proportion of aquatic plants in their diet might depend on the food spectrum available (Alcorlo *et al.*, 2004; Berezina, 2007). Most aquatic snails graze on benthic algae, and only very few consume significant amounts of higher plant tissue, e.g. pond snails (*Lymnaea* sp.) and apple snails (*Pomacea* sp.) (Brönmark, 1989; Carlsson & Brönmark, 2006; Elger *et al.*, 2007). Only 20% of the aquatic insect species consume macrophytes, in contrast to terrestrial insects where 45% are estimated to consume vascular plants (Newman, 1991). More than 75% of aquatic insect herbivores are secondarily aquatic, mainly belonging to the Homoptera, Diptera, Coleoptera and Lepidoptera. These are also the most specialised herbivores, generally feeding only on very few plant taxa. Trichopterans, primarily aquatic insect herbivores, are more polyphagous (Newman, 1991; Cronin, 1998). Quantitatively important freshwater insect herbivores are members of two families of Coleoptera (Chrysomelidae, Curculionidae) and Lepidoptera (Pyralidae, Noctuidae), which also include high-impact terrestrial herbivores. Biological control agents are mostly from these families, and are successful in reducing certain macrophyte species owing to a high specialisation and strong effects on target plants (Wheeler *et al.*, 1998; Newman, 2004; Habeck & Balciunas, 2005; Wilson *et al.*, 2007).

The role of PSMs in freshwater plant–herbivore systems may depend on the amount of plant tissue in the diet and the range of food plants eaten by the herbivores. Whereas low levels of constitutive PSMs may be suitable to deter feeding of aquatic omnivores, macrophytes that experience grazing by specialist herbivores need higher levels of targeted PSMs (Hay & Fenical, 1988). Co-evolutionary relationships might be most pronounced between secondarily aquatic insects and their host plants. The classical terrestrial distinction in generalists and specialists falls short, given that only few aquatic herbivores rely solely on higher plants. We propose to use facultative and obligate herbivores to distinguish omnivores from true herbivores, and then to add the degree of specialisation, i.e. facultative generalist herbivore (e.g. amphipods) or obligate specialist herbivore (e.g. the weevil *Euhrychiopsis lecontei*). A similar terminology has recently been proposed for herbivorous mammals, pointing out the difficulty in assessing the diet breadth and relative contribution of differently defended food plants (Shipley *et al.*, 2009). Yet the consumption of plants by both herbivores and omnivores might be more pronounced in freshwater than in terrestrial systems, or at least warrant a detailed diet analysis of major herbivores to quantify the amount of plant tissue in the food.

8.3   Macrophyte palatability to herbivores

The impact of invertebrate and vertebrate herbivores on aquatic vascular plants varies considerably (Lodge *et al.*, 1998; Marklund *et al.*, 2002), but significant effects both on standing crop and on plant community composition have been frequently recorded (Painter & McCabe, 1988; Gross *et al.*, 2001; Hidding *et al.*, 2009). Herbivores have been shown to prefer certain plant species, both in no-choice and choice feeding trials in the laboratory as well as in field exclosure and enclosure studies. Herbivore preference or avoidance of certain plant species depends on plant structure, nutritional value and presence of chemical feeding deterrents (Cronin *et al.*, 2002). In feeding trials comparing palatability among plant species, herbivore consumption correlated positively with nitrogen (or protein) concentration (Elger & Lemoine, 2005), and negatively with plant dry-matter concentration (as a proxy for the amount of structural tissue) (Elger & Willby, 2003; Elger & Lemoine, 2005; Burlakova *et al.*, 2009) and phenolic concentration (Li *et al.*, 2004). However, frequently no relationships were found between any of these plant properties and consumption rates (Dorn *et al.*, 2001; Cronin *et al.*, 2002; Li *et al.*, 2005). This is most likely to be due to the hierarchical nature of the effects: plants that are high in structural or chemical defences and high in nitrogen are avoided, whereas plants with high nitrogen but low defences may be preferred (Lodge *et al.*, 1998). Structural and chemical defences prevent the feeding on plants with high nitrogen concentrations, as could be shown in various bioassays with artificial diets (Newman *et al.*, 1992; Bolser *et al.*, 1998; Cronin *et al.*, 2002). However, only a few studies have tried to separate structural, chemical and nutritional properties of macrophytes in feeding trials. Therefore, it remains unclear what the general prevalence is of PSMs in freshwater macrophytes and how important these are as feeding deterrents for aquatic herbivores.

The effectiveness of structural and chemical defences may further depend on the consumer. For large herbivores (e.g. swans) structural tissue may be much less of a feeding inhibitor than for small herbivores (e.g. snails). However, the presence of herbivorous waterfowl and fish caused a shift in dominance of submerged macrophytes in Lake Zwemlust, in the Netherlands: rigid hornworth (*Ceratophyllum demersum*) replaced the structurally undefended *Elodea nuttallii* (van Donk, 1998). The efficiency of structural defences can easily be shown by grinding up dried plant tissue and incorporating the powder in artificial diets (Bolser *et al.*, 1998; Cronin *et al.*, 1998). Some submerged macrophytes have distinct spines or teeth made out of silica, e.g. *C. demersum* or *Najas marina* ssp. *intermedia*, which will prevent at least some herbivores from feeding on them. Trichomes are present on young tissue of *Myriophyllum spicatum* (Godmaire & Nalewajko, 1990), but their role in herbivore defence has never been investigated. In general, submerged macrophytes might be less structurally defended than floating-leaved or emergent plants, because they do not contain much lignin and have slender, thin leaves. However, the examples cited above show that this is not universally true. We need a detailed analysis of the different types of structural defences in the different growth forms of freshwater plants, and how they deter invertebrate or vertebrate herbivores.

The low prevalence of structural defence might be counteracted by higher chemical defences in submerged macrophytes. At present, we do not have enough information to answer this question. Maybe submerged macrophytes reinvaded the aquatic habitat because they tried to escape their specialised herbivores. But under water, other, more generalist or omnivorous herbivores are attacking freshwater plants. Also, most submerged freshwater plants belong to monocots or more basal dicots, and thus may have phylogenetic constraints on more evolved PSMs (see below). We will only be able to answer the question of the prevalence of chemical defences when we have more information on herbivore-deterrent PSMs in all different growth forms of freshwater plants.

Omnivorous crustaceans might incorporate only small amounts of fresh plant tissue in their diet. Among the herbivores, obligate herbivores, especially specialists with a tight co-evolutionary feeding relationship to their food plant, are likely to have a different tolerance for chemical defences and accept plants that are difficult food for facultative generalist herbivores or omnivores. Weevils specialised on *Myriophyllum spicatum* are insensitive to tannins (Marko *et al.*, 2005), which are effective against facultative generalist herbivorous fish (Pipalova, 2002). In aquatic systems, chemical defences are most effective against larger, generalist herbivores (Hay *et al.*, 1987; Parker *et al.*, 2007). As large herbivores usually have the strongest impact on macrophyte production (Lodge *et al.*, 1998) and most of the freshwater herbivores are omnivores and at least facultative generalists, this would suggest that secondary metabolites should be commonly found in freshwater macrophytes.

8.4   Effects of secondary metabolites on aquatic herbivores

Little is known about the effect of secondary metabolites on aquatic herbivores. Differences in phenolic contents are frequently considered to affect feeding rates when multiple plant species are offered in feeding trials (Lodge, 1991; Cronin *et al.*, 2002). However, this is only a correlation and no proof for the effect of phenolic compounds as feeding deterrents. The pulmonate snail *Radix swinhoei* rejected high tannin-containing *Myriophyllum spicatum* compared with several other submerged macrophytes in choice and no-choice bioassays (Li *et al.*, 2004). Tannins may be particularly effective against facultative and obligate generalist herbivores. But some obligate specialist insect herbivores, especially lepidopteran larvae, have evolved specific adaptations to deal with enhanced tannin concentrations. Higher hydrolysable tannin contents in *M. spicatum* compared with the tannin-free *Potamogeton perfoliatus* led to a slower growth of larvae of the aquatic moth *Acentria ephemerella* (Choi *et al.*, 2002) and affected gut bacteria (Walenciak *et al.*, 2002). However, specific physico-chemical conditions in the gut enable *Acentria* larvae to tolerate tannin-rich diets (Gross *et al.*, 2008), similar to terrestrial tannin-tolerant Lepidoptera such as *Lymantria dispar* (Appel & Maines, 1995).

Tannins and other PSMs can even be beneficial for herbivores, in combatting pathogen attacks or other stressors, as has been shown in terrestrial herbivores (Hunter & Schultz, 1993; Forbey & Foley, 2009; Lisonbee *et al.*, 2009). For example, the freshwater amphipod *Hyalella azteca* selectively consumes roots of *Berula erecta*, which are rich in coumarins, and this herbivore is less preferred by predators if it includes more *Berula* roots in its diet (Rowell & Blinn, 2003). However, the potential benefits of PSMs have not been readily investigated in aquatic systems (Sotka *et al.*, 2009).

Host-plant preferences by herbivores cannot be deduced solely from consumption or growth assays, as feeding may be influenced by life stage and previous feeding experience. Larvae of the aquatic Lepidoptera *Munroessa gyralis* fed on several aquatic macrophyte families and are thus obligate generalist herbivores, but pupation occurred exclusively on *Nymphaea*, indicating that different life stages need specific host plants (Dorn *et al.*, 2001). Weevil larvae (*Galerucella nymphaeae*) preferentially consumed either *Nuphar*, *Polygonum* or *Brasenia*, depending on the dominant macrophytes in their home lakes (Cronin *et al.*, 1999). Moreover, preference for either two species within the Nymphaeaceae or Polygonaceae possibly resulted in the formation of morphologically distinct host races of the weevil (Pappers *et al.*, 2002).

Chemical cues can also influence host-plant preference. For example, chemical cues attracted specialised aquatic insect herbivores (*Hyporhygma quadripunctatum*, Diptera, *Bagous americanus*, Curculionidae and the two chrysomelid beetles *Donacia cincticornis* and *Galerucella nymphaeae*) to their preferred water lily host plants, whereas plant toughness or structural defences were the main feeding determinants for a facultative generalist herbivore, caddisfly larvae *Limnephilus infernalis* (Cronin *et al.*, 1998). The weevil *Euhrychiopsis lecontei* is attracted to high concentrations of uracil, glycerol and an unidentified compound exuded by fast-growing *M. spicatum* (Marko *et al.*, 2005). Mixtures of polyoles, pyrimidines and nucleosides exuded by aquatic plant roots act synergistically in attracting other aquatic insects, specifically mosquito larvae (Serandour *et al.*, 2008). The pulmonate snail *Biomphalaria glabrata* is attracted by decaying *Lemna paucicostata* in which the end-products of microbial decomposition, short-chain carboxylic acids, may play an important role (Sterry *et al.*, 1983). The ubiquitous role of chemical cues in terrestrial systems (Arimura *et al.*, 2009; Gershenzon *et al.*, Chapter 4; van Dam, Chapter 10; Schuman & Baldwin, Chapter 15; Dicke *et al.*, Chapter 16) and the important role chemicals play in freshwater food webs in general (Burks & Lodge, 2002) suggests that chemical cues should be widely active in plant–herbivore interactions and deserve further attention.

8.5 Identification of herbivore deterrents in aquatic plants

The most basic testing of whether macrophytes possess chemical substances that affect feeding rates has been done by making hydrophilic or lipophilic extracts, applying those to standard food items and testing whether they increase or decrease consumption rates compared with controls. In this way 39 macrophyte species have been tested, of which 49% contained herbivore deterrents, whereas two species increased feeding in some herbivores (Bolser & Hay, 1998; Bolser *et al.*, 1998; Cronin, 1998; Kubanek *et al.*, 2001; Cronin *et al.*, 2002; Prusak *et al.*, 2005; Parker *et al.*, 2006; Erhard *et al.*, 2007; Miller & Provenza, 2007). Feeding was reduced in 54% of the emergent and 36% of the submerged macrophytes, whereas only one floating macrophyte species was tested, and this deterred feeding. Some care should be taken to interpret these numbers as several of the emergent macrophytes are more terrestrial plants (e.g. *Eupatorium capillifolium* and *Galium tinctorium*). Plant species for these tests were collected from field sites based on availability, and selection was not biased towards plants suspected to contain herbivore deterrents. Thus, it can be concluded that macrophytes commonly contain secondary metabolites that act as feeding deterrents. Further purification of active compounds requires a bioassay-directed fractionation of crude extracts. Most studies use methods established in marine systems to detect consumption rates of agar-based diets impregnated with plant extracts, fractionated extracts or pure compounds. So far, this is feasible only with certain fish, amphipods, isopods or crayfish, but very difficult with more specialised herbivores such as lepidopteran larvae (Erhard *et al.*, 2007).

It is difficult to assess whether freshwater plants, especially submerged macrophytes, contain more hydrophilic deterrents. The few studies comparing hydrophilic or lipophilic extracts rather indicate a stronger deterrency of lipophilic extracts (Bolser *et al.*, 1998; Prusak *et al.*, 2005). Even among phenolic compounds, we find a wide range of different structures from strongly water-soluble to highly lipophilic ones. We can only answer this question by identifying more deterrents in freshwater plants.

As yet, only very few feeding deterrents have been structurally identified from freshwater plants (Bolser *et al.*, 1998). These include lignans in *Saururus cernuus*, phenylpropanoids in *Micranthemum umbrosum*, habenariol in *Habenaria repens*, glucosinolates in *Nasturtium officinale* and flavonoids in *Elodea nuttallii*. Lignans in emergent parts of the lizard's tail plant *Saururus cernuus* are active against crayfish (Kubanek *et al.*, 2000, 2001). The major feeding deterrent of the shoreline-dwelling orchid *Habenaria repens* (Wilson *et al.*, 1999) is an unusual ester, habenariol. Glucosides of this ester (habenarioside) occur in another *Habenaria* species (Cota, 2008). Given that the glycosylation pattern of secondary metabolites can influence the feeding behaviour of herbivores (Harborne, 1988), we cannot rule out that habenariol glucosides are also deterrents or precursors. Interestingly, both the lipophilic and hydrophilic extract of *Habenaria repens* inhibited crayfish feeding (Bolser *et al.*, 1998).

Micranthemum umbrosum, a worldwide distributed aquarium and aquaculture dicotylous submerged macrophyte, contains several feeding deterrents, phenylpropanoid derivatives, against carp and crayfish (Parker *et al.*, 2006). Besides elemicin, three lignoids were isolated by bioassay-directed fractionation, and all four compounds deterred feeding by at least one of the following consumers, North American crayfish *Procambarus spiculifer* and *P. acutus*, and grass carp (*Ctenopharyngodon idella*).

When the active substance is known, it can be manipulated to experimentally test its importance as feeding deterrent. Synthetic analogues revealed that both the allyl and methoxy moieties of elemecin contributed to feeding deterrence, and that the disruption of the lactone moiety of the podophyllin-derivative reduced its deterrence (Lane & Kubanek, 2006). Phenylethyl iso-thiocyanate set free from glucosinolates in fresh *Nasturtium officinale* leaves impaired feeding by gammarids (Newman *et al.*, 1992, 1996).

8.6   Types of PSMs in freshwater macrophytes

Many of the currently known active herbivore deterrents in macrophytes are phenolic compounds. Phenolic compounds are widely distributed in plants and are frequently investigated in aquatic plant–herbivore interactions. In general, submerged macrophytes have much lower phenolic content com-pared with emergent or floating leaved macrophytes (Smolders *et al.*, 2000) and with terrestrial vegetation. The major reasons for this difference might be the low or missing lignification and less exposure to ultraviolet light of submerged macrophytes. However, submerged Haloragaceae (*Myriophyllum* spp, *Proserpinaca palustris*) possess very high concentrations of hydrolysable tannins (8–20% of dry mass) (Choi *et al.*, 2002; Hempel *et al.*, 2009). These concentrations are about one order of magnitude higher than in other sub-merged macrophytes, and may be related to their dicotyledon phylogeny, while most other temperate submerged species are monocots.

Several alkaloids have been proposed as chemical defences in submerged macrophytes (Ostrofsky & Zettler, 1986), but no bioassays have been per-formed. Although terpenoids and sterols occur in freshwater macrophytes, their role in feeding preferences has so far not been established, even though these compounds affect plant–herbivore interactions in terrestrial and marine systems (Sotka *et al.*, 2009).

Whether there are phylogenetic constraints on the occurrence of PSMs in freshwater plants can at present only be speculated. The occurrence of aquatic plants in the plant kingdom is not uniformly distributed. Some orders are exclusively aquatic while others have several aquatic families or at least genera (Cook, 1999). Many freshwater angiosperms are at very basal positions in the current angiosperm phylogeny (Nymphaeales, Magnoliids), many are monocots (among the submerged macrophytes predominantly from the Alismatales), some are between monocots and dicots (Ceratophyllales, Ranunculales) and only a few species are true eudicots (Bremer *et al.*, 2009). More advanced PSMs such as pyrrolizidine alkaloids or cardiac glycosides are only found in higher evolved dicots (Harborne, 1988, p. 189), and there are only few, rather terrestrial, plants found in wetlands that belong to these families. Recent views on the relationship between phylogeny and phyto-chemistry question the artificial separation into monocots and dicots (Larsson, 2007). We strongly recommend the evaluation of biosynthetic capacities for PSMs in aquatic angiosperms to elucidate whether there are limits based on phylogenetic relationships.

8.7   Variation in levels of PSMs in macrophytes

Given the scarce knowledge on PSMs in freshwater macrophytes, it is not surprising that very little is known about their inter- and intra-specific variability, and factors accounting for this. Seasonal and within-plant variation of tannins were observed in *M. spicatum*, with higher nitrogen and phosphorus content in the autumn and in apical meristems (Hempel *et al.*, 2009). Increased light availability increased the concentration of phenolic compounds in *Vallisneria natans* (Li *et al.*, 2005), *Potamogeton amplifolius* and *Nuphar advena* (Cronin and Lodge, 2003), whereas increased nutrient availability also increased phenolic concentration in *P. amplifolius*. A differential regulation of the pool of phenolic compounds and the major allelochemical tellimagrandin II was based on light intensity and nitrogen availability in *M. spicatum* (Gross, 2003). A more detailed analysis with individual compounds is needed to reassess factors influencing intra-specific variability, e.g. as proposed in the carbon–nutrient balance hypothesis and its application in field studies (Bryant *et al.*, 1983; Bauer *et al.*, 2009).

Induced defences have been recorded in aquatic plant–herbivore interactions, but our current knowledge is simply limited by the number of investigations. Induced defences against generalist crayfish feeding on *Nuphar luteum macrophyllum* were observed, but the nature of the induced chemicals was not identified (Bolser & Hay, 1998). Similarly, artificially damaged leaves of *Potamogeton coloratus* reduced palatability for the caddis larvae *Triaenodes bicolor* (Jeffries, 1990). Out of 21 wetland species, only *Eupatorium capillifolium*, which contains pyrrolizidine alkaloids (Conner *et al.*, 2000), was less palatable to a generalist crayfish after the plant was mechanically damaged, indicating that constitutive defences were more common than induced defences in this set of plant species (Prusak *et al.*, 2005). Inducible chemical defence was suggested for *M. spicatum* but not for *Elodea canadensis* (Lemoine *et al.*, 2009). Our own observations support possible herbivore-specific inducible defences in *M. spicatum* (Onion, 2004; Rid *et al.*, 2008; E. M. Gross, unpublished data). Inducible chemical defences might be present in freshwater macrophytes, but are even less explored than constitutive defences.

8.8   Conclusions and future directions

When writing this review, we recognised that a new compilation of all the information on freshwater plant–herbivore interactions is much needed, since the major reviews by Lodge (1991) and Newman (1991) are 20 years old, and more recent reviews (Newman, 2004; Sotka *et al.*, 2009) have addressed only certain aspects. The past two decades brought a significant change in the perception of freshwater plant–herbivore interactions, and we now have a good understanding of the types of herbivores and some perception of their preferences for certain host plants. We now also have information on PSMs in freshwater plants, but most compounds were identified because of their role as allelopathic, antimicrobial or ethnobotanical/pharmacological agents, and their role in plant–herbivore interactions remains unclear.

Future work will benefit from strong interdisciplinary research. We need to know more about the effects of generalist and specialist herbivores, and the importance of plants in the diet of omnivores. More studies should be performed with true aquatic plant species, i.e. those fully adapted to life under water. The present studies are often biased towards emergent and partly terrestrial plants, and thus tell only half of the story.

More efforts should be made to work with obligate herbivores, especially specialists, since those may have developed specific metabolisation and/or detoxification mechanisms to the PSMs present in their host plants. We might even gain more information on possible feeding cues in freshwater plants. We recommend the development of suitable feeding bioassays to test fractionated extracts or purified compounds.

The secondary aquatic nature of both macrophytes and certain herbivorous insects offers great potential for (co-)evolutionary studies. Knowing when herbivorous secondary aquatic insects reinvaded the water will help to solve the question of why secondary aquatic plants moved back into the water. Molecular clock analyses (Calonje et al., 2009) will allow a phylogenetic reconstruction at the species or genus level. When applied to pairs of freshwater plants and associated herbivores, this method could help in elucidating the timing of new aquatic plant–herbivore associations derived from terrestrial ancestors. Maybe this analysis will provide new insights into the host-plant range of certain herbivores, as has been done for terrestrial plant–herbivore associations (Becerra et al., 2009). Differences in the palatability of plants to different herbivores, or the same herbivore at different locations or seasons, might be caused by variation in the levels of PSMs. Thus, more efforts are needed to identify where, when and how much PSMs are produced in host plants. We should aim to test more PSMs for their effect on herbivores feeding on aquatic plants, and also to identify new active compounds. Lane and Kubanek (2006) offer the first studies to identify the structural features responsible for feeding deterrency of known PSMs. Similar studies could be used to test and modify previously identified PSMs from other studies in plant–herbivore interactions.

Plant preference is guided by different plant traits, and this might differ depending on the herbivore. We should identify hierarchies of plant traits (i.e. structure, nutrient content, PSMs) for the different groups of herbivores (from facultative generalist herbivores to obligate specialist herbivores). PSMs may act differently on herbivores depending on their ontogeny or life cycle. It is also important to elucidate chemical cues that attract herbivores to certain plants. Aquatic plants, and submerged macrophytes, produce volatile organic compounds (VOCs), the typical green leaf volatiles found in terrestrial plants (E. M. Gross, unpublished data). Aquatic herbivores might also induce changes in VOCs that potentially attract parasitoids or predators, and establish aquatic tri-trophic allelochemical interactions. Possibly there are differences compared with terrestrial habitats because of different volatilisations, which might in turn affect the function and evolution of the respective compounds.

In conclusion, freshwater plant–herbivore interactions occur in an exciting habitat, where water and land merge. The peculiarities of this system might allow us to identify the early evolution of angiosperm–herbivore interactions, since many freshwater plants have a very basal phylogeny. Exchange with marine and terrestrial ecologists working in this field, as well as cooperation with natural product chemists and pharmacologically interested researchers, can improve our knowledge of PSMs and their role in the freshwater ecosystem. We conclude that chemical defences in freshwater macrophytes are not limited but largely unexplored.

References

Alcorlo, P., Geiger, W. and Otero, M. (2004) Feeding preferences and food selection of the red swamp crayfish, *Procambarus clarkii*, in habitats differing in food item diversity. *Crustaceana*, **77**, 435–453.

Appel, H. M. and Maines, L. W. (1995) The influence of host-plant on gut conditions of gypsy moth (*Lymantria dispar*) caterpillars. *Journal of Insect Physiology*, **41**, 241–246.

Arimura, G.-I., Matsui, K. and Takabayashi, J. (2009) Chemical and molecular ecology of herbivore-induced plant volatiles: proximate factors and their ultimate functions. *Plant and Cell Physiology*, **50**, 911–923.

Bauer, N., Blaschke, U., Beutler, E. *et al.* (2009) Seasonal and interannual dynamics of polyphenols in *Myriophyllum verticillatum* and their allelopathic activity on *Anabaena variabilis*. *Aquatic Botany*, **91**, 110–116.

Becerra, J. X., Noge, K. and Venable, D. L. (2009) Macroevolutionary chemical escalation in an ancient plant–herbivore arms race. *Proceedings of the National Academy of Sciences USA*, **106**, 18062–18066.

Berezina, N. (2007) Food spectra and consumption rates of four amphipod species from the North-West of Russia. *Fundamental and Applied Limnology*, **168**, 317–326.

Bolser, R. C. and Hay, M. E. (1998) A field test of inducible resistance to specialist and generalist herbivores using the water lily *Nuphar luteum*. *Oecologia*, **116**, 143–153.

Bolser, R. C., Hay, M. E., Lindquist, N., Fenical, W. and Wilson, D. (1998) Chemical defenses of freshwater macrophytes against crayfish herbivory. *Journal of Chemical Ecology*, **24**, 1639–1658.

Bremer, B., Bremer, K., Chase, M. W. *et al.* (2009) An update of the Angiosperm Phylogeny Group classification for the orders and families of flowering plants: APG III. *Botanical Journal of the Linnean Society*, **161**, 105–121.

Brönmark, C. (1989) Interactions between epiphytes, macrophytes and fresh-water snails – a review. *Journal of Molluscan Studies*, **55**, 299–311.

Bryant, J. P., Chapin Iii, F. S. and Klein, D. R. (1983) Carbon/nutrient balance of boreal plants in relation to vertebrate herbivory. *Oikos*, **40**, 357–368.

Burks, R. L. and Lodge, D. M. (2002) Cued in: advances and opportunities in freshwater chemical ecology. *Journal of Chemical Ecology*, **28**, 1901–1917.

Burlakova, L. E., Karatayev, A. Y., Padilla, D. K., Cartwright, L. D. and Hollas, D. (2009) Wetland restoration and invasive species: apple snail (*Pomacea insularum*) feeding on native and invasive aquatic plants. *Restoration Ecology*, **17**, 433–440.

Calonje, M., Martin-Bravo, S., Dobes, C. *et al.* (2009) Non-coding nuclear DNA markers in phylogenetic reconstruction. *Plant Systematics and Evolution*, **282**, 257–280.

Carlsson, N. O. L. and Brönmark, C. (2006) Size-dependent effects of an invasive herbivorous snail *(Pomacea canaliculata)* on macrophytes and periphyton in Asian wetlands. *Freshwater Biology*, **51**, 695–704.

Choi, C., Bareiss, C., Walenciak, O. and Gross, E. M. (2002) Impact of polyphenols on the growth of the aquatic herbivore *Acentria ephemerella* (Lepidoptera: Pyralidae). *Journal of Chemical Ecology*, **28**, 2223–2235.

Cloern, J. E., Canuel, E. A. and Harris, D. (2002) Stable carbon and nitrogen isotope composition of aquatic and terrestrial plants of the San Francisco Bay estuarine system. *Limnology and Oceanography*, **47**, 713–729.

Conner, W. E., Boada, R., Schroeder, F. C. *et al.* (2000) Chemical defense: bestowal of a nuptial alkaloidal garment by a male moth on its mate. *Proceedings of the National Academy of Sciences USA*, **97**, 14406–14411.

Cook, C. D. K. (1999) The number and kinds of embryo-bearing plants which have become aquatic: a survey. *Perspectives in Plant Ecology, Evolution and Systematics*, **2**, 79–102.

Cota, B. B., Magalhaes, A., Pimenta, A. M. C. *et al.* (2008) Chemical constituents of *Habenaria petalodes* Lindl. (Orchidaceae). *Journal of the Brazilian Chemical Society*, **19**, 1098–1104.

Cronin, G. (1998) Influence of macrophyte structure, nutritive value, and chemistry on the feeding choices of a generalist crayfish. In E. Jeppesen, M. Sondergaard and K. Christoffersen (eds.) *The Structuring Role of Macrophytes in Lakes*. New York: Springer-Verlag.

Cronin, G. and Lodge, D. M. (2003) Effects of light and nutrient availability on the growth, allocation, carbon/nitrogen balance, phenolic chemistry, and resistance to herbivory of two freshwater macrophytes. *Oecologia*, **137**, 32–41.

Cronin, G., Wissing, K. D. and Lodge, D. M. (1998) Comparative feeding selectivity of herbivorous insects on water lilies: aquatic vs. semi-terrestrial insects and submersed vs. floating leaves. *Freshwater Biology*, **39**, 243–257.

Cronin, G., Schlacher, T., Lodge, D. M. and Siska, E. L. (1999) Intraspecific variation in feeding preference and performance of *Galerucella nymphaeae* (Chrysomelidae: Coleoptera) on aquatic macrophytes. *Journal of the North American Benthological Society*, **18**, 391–405.

Cronin, G., Lodge, D. M., Hay, M. E. *et al.* (2002) Crayfish feeding preferences for freshwater macrophytes: the influence of plant structure and chemistry. *Journal of Crustacean Biology*, **22**, 708–718.

Cyr, H. and Pace, M. L. (1993) Magnitude and patterns of herbivory in aquatic and terrestrial ecosystems. *Nature*, **361**, 148–150.

Demars, B. O. L. and Edwards, A. C. (2008) Tissue nutrient concentrations in aquatic macrophytes: comparison across biophysical zones, surface water habitats and plant life forms. *Chemistry and Ecology*, **24**, 413–422.

Dorn, N. J., Cronin, G. and Lodge, D. M. (2001) Feeding preferences and performance of an aquatic lepidopteran on macrophytes: plant hosts as food and habitat. *Oecologia*, **128**, 406–415.

Elger, A. and Willby, N. J. (2003) Leaf dry matter content as an integrative expression of plant palatability: the case of freshwater macrophytes. *Functional Ecology*, **17**, 58–65.

Elger, A. and Lemoine, D. (2005) Determinants of macrophyte palatability to the pond snail *Lymnaea stagnalis*. *Freshwater Biology*, **50**, 86–95.

Elger, A., De Boer, T. and Hanley, M. E. (2007) Invertebrate herbivory during the regeneration phase: experiments with a freshwater angiosperm. *Journal of Ecology*, **95**, 106–114.

Erhard, D., Pohnert, G. and Gross, E. M. (2007) Chemical defense in *Elodea nuttallii* reduces feeding and growth of aquatic herbivorous Lepidoptera. *Journal of Chemical Ecology*, **33**, 1646–1661.

Forbey, J. S. and Foley, W. J. (2009) PharmEcology: a pharmacological approach to understanding plant–herbivore interactions. An introduction to the symposium. *Integrative and Comparative Biology*, **49**, 267–273.

Godmaire, H. and Nalewajko, C. (1990) Structure and development of secretory trichomes on *Myriophyllum spicatum* L. *Aquatic Botany*, **37**, 99–121.

Gross, E. M. (2003) Differential response of tellimagrandin II and total bioactive hydrolysable tannins in an aquatic angiosperm to changes in light and nitrogen. *Oikos*, **103**, 497–504.

Gross, E. M., Johnson, R. L. and Hairston, N. G. (2001) Experimental evidence for changes in submersed macrophyte species composition caused by the herbivore *Acentria ephemerella* (Lepidoptera). *Oecologia*, **127**, 105–114.

Gross, E. M., Brune, A. and Walenciak, O. (2008) Gut pH, redox conditions and oxygen levels in an aquatic caterpillar: potential effects on the fate of ingested tannins. *Journal of Insect Physiology*, **54**, 462–471.

Habeck, D. H. and Balciunas, J. K. (2005) Larvae of Nymphulinae (Lepidoptera : Pyralidae) associated with *Hydrilla verticillata* (Hydrocharitaceae) in North Queensland. *Australian Journal of Entomology*, **44**, 354–363.

Harborne, J. B. (1988) *Introduction to Ecological Biochemistry*, 3rd edn. London: Academic Press.

Hay, M. E. (1991) Marine terrestrial contrasts in the ecology of plant-chemical defenses against herbivores. *Trends in Ecology and Evolution*, **6**, 362–365.

Hay, M. E., Duffy, J. E., Pfister, C. A. and Fenical, W. (1987) Chemical defense against different marine herbivores: are amphipods insect equivalents? *Ecology*, **68**, 1567–1580.

Hay, M. E. and Fenical, W. (1988) Marine plant–herbivore interactions – the ecology of chemical defense. *Annual Review of Ecology and Systematics*, **19**, 111–145.

Hempel, M., Grossart, H. P. and Gross, E. M. (2009) Community composition of bacterial biofilms on two submerged macrophytes and an artificial substrate in a pre-alpine lake. *Aquatic Microbial Ecology*, **58**, 79–94.

Hidding, B., Nolet, B. A., De Boer, T., De Vries, P. P. and Klaassen, M. (2009) Compensatory growth in an aquatic plant mediates exploitative competition between seasonally tied herbivores. *Ecology*, **90**, 1891–1899.

Hunter, M. D. and Schultz, J. C. (1993) Induced plant defenses breached – phytochemical induction protects an herbivore from disease. *Oecologia*, **94**, 195–203.

Hutchinson, G. E. (1975) *A Treatise on Limnology* Volume III. *Limnological Botany*. New York: John Wiley and Sons.

Jeffries, M. (1990) Evidence of induced plant defenses in a pondweed. *Freshwater Biology*, **23**, 265–270.

Kubanek, J., Fenical, W., Hay, M. E., Brown, P. J. and Lindquist, N. (2000) Two antifeedant lignans from the freshwater macrophyte *Saururus cernuus*. *Phytochemistry*, **54**, 281–287.

Kubanek, J., Hay, M. E., Brown, P. J., Lindquist, N. and Fenical, W. (2001) Lignoid chemical defenses in the freshwater macrophyte *Saururus cernuus*. *Chemoecology*, **11**, 1–8.

Lane, A. L. and Kubanek, J. (2006) Structure–activity relationship of chemical defenses from the freshwater plant *Micranthemum umbrosum*. *Phytochemistry*, **67**, 1224–1231.

Larsson, S. (2007) The 'new' chemosystematics: phylogeny and phytochemistry. *Phytochemistry*, **68**, 2904–2908.

Lemoine, D. G., Barrat-Segretain, M. H. and Roy, A. (2009) Morphological and chemical changes induced by herbivory in three common aquatic macrophytes. *International Review of Hydrobiology*, **94**, 282–289.

Li, Y. K., Yu, D. and Yan, X. (2004) Are polyphenolics valuable in anti-herbivory strategies of submersed freshwater macrophytes? *Archiv für Hydrobiologie*, **161**, 391–402.

Li, Y. K., Yu, D., Xu, X. W. and Xie, Y. G. (2005) Light intensity increases the susceptibility of *Vallisneria natans* to snail herbivory. *Aquatic Botany*, **81**, 265–275.

Lisonbee, L. D., Villalba, J. J., Provenza, F. D. and Hall, J. O. (2009) Tannins and self-medication: implications for sustainable parasite control in herbivores. *Behavioural Processes*, **82**, 184–189.

Lodge, D. M. (1991) Herbivory on fresh-water macrophytes. *Aquatic Botany*, **41**, 195–224.

Lodge, D. M., Cronin, G., van Donk, E. and Froelich, A. J. (1998) Impact of herbivory on plant standing crop: comparison among biomes, between vascular and nonvascular plants, and among freshwater herbivore taxa. In E. Jeppesen, M. Sondergaard, M. Sondergaard and K. Christoffersen (eds.) *The Structuring Role of Submerged Macrophytes in Lakes*. New York: Springer.

Marklund, O., Sandsten, H., Hansson, L. A. and Blindow, I. (2002) Effects of waterfowl and fish on submerged vegetation and macroinvertebrates. *Freshwater Biology*, **47**, 2049–2059.

Marko, M. D., Newman, R. M. and Gleason, F. K. (2005) Chemically mediated host-plant selection by the milfoil weevil: a freshwater insect–plant interaction. *Journal of Chemical Ecology*, **31**, 2857–2876.

Miller, S. A. and Provenza, F. D. (2007) Mechanisms of resistance of freshwater macrophytes to herbivory by invasive juvenile common carp. *Freshwater Biology*, **52**, 39–49.

Newman, R. M. (1991) Herbivory and detritivory on fresh-water macrophytes by invertebrates: a review. *Journal of the North American Benthological Society*, **10**, 89–114.

Newman, R. M. (2004) Invited review – biological control of Eurasian watermilfoil by aquatic insects: basic insights from an applied problem. *Archiv für Hydrobiologie*, **159**, 145–184.

Newman, R. M., Hanscom, Z. and Kerfoot, W. C. (1992) The watercress glucosinolate-myrosinase system: a feeding deterrent to caddisflies, snails and amphipods. *Oecologia*, **92**, 1–7.

Newman, R. M., Kerfoot, W. C. and Hanscom, Z. (1996) Watercress allelochemical defends high-nitrogen foliage against consumption: effects on freshwater invertebrate herbivores. *Ecology*, **77**, 2312–2323.

Onion, A. (2004) Herbivore resistance in invasive and native *Myriophyllum spicatum* and *Myriophyllum heterophyllum*. MSc thesis, Department of Ecology and Evolutionary Biology, Cornell University: Ithaca, NY.

Ostrofsky, M. L. and Zettler, E. R. (1986) Chemical defences in aquatic plants. *Journal of Ecology*, **74**, 279–287.

Painter, D. S. and Mccabe, K. J. (1988) Investigation into the disappearance of Eurasian watermilfoil from the Kawartha Lakes, Canada. *Journal of Aquatic Plant Management*, **26**, 3–12.

Pappers, S. M., Van Der Velde, G. and Ouborg, N. J. (2002) Host preference and larval performance suggest host race formation in *Galerucella nymphaeae*. *Oecologia*, **130**, 433–440.

Parker, J. D., Collins, D. O., Kubanek, J. *et al.* (2006) Chemical defenses promote persistence of the aquatic plant *Micranthemum umbrosum. Journal of Chemical Ecology*, **32**, 815–833.

Parker, J. D., Caudill, C. C. and Hay, M. E. (2007) Beaver herbivory on aquatic plants. *Oecologia*, **151**, 616–625.

Pipalova, I. (2002) Initial impact of low stocking density of grass carp on aquatic macrophytes. *Aquatic Botany*, **73**, 9–18.

Prusak, A. C., O'Neal, J. and Kubanek, J. (2005) Prevalence of chemical defenses among freshwater plants. *Journal of Chemical Ecology*, **31**, 1145–1160.

Rid, S., Hesselschwerdt, J. and Gross, E. M. (2008) Induziert *Lymnaea stagnalis* Verteidigungsmechanismen in *Myriophyllum spicatum*? Konstanz: Jahrestagung der DGL/ Deutsche Gesellschaft für Limnologie.

Rowell, K. and Blinn, D. W. (2003) Herbivory on a chemically defended plant as a predation deterrent in *Hyalella azteca. Freshwater Biology*, **48**, 247–254.

Sand-Jensen, K. and Jacobsen, D. (2002) Herbivory and growth in terrestrial and aquatic populations of amphibious stream plants. *Freshwater Biology*, **47**, 1475–1487.

Serandour, J., Reynaud, S., Willison, J. *et al.* (2008) Ubiquitous water-soluble molecules in aquatic plant exudates determine specific insect attraction. *PLOS One*, **3**, e3350.

Shipley, L. A., Forbey, J. S. and Moore, B. D. (2009) Revisiting the dietary niche: when is a mammalian herbivore a specialist? *Integrative and Comparative Biology*, **49**, 274–290.

Smolders, A. J. P., Vergeer, L. H. T., van der Velde, G. and Roelofs, J. G. M. (2000) Phenolic contents of submerged, emergent and floating leaves of aquatic and semi-aquatic macrophyte species: why do they differ? *Oikos*, **91**, 307–310.

Sotka, E. E., Forbey, J., Horn, M. *et al.* (2009) The emerging role of pharmacology in understanding consumer–prey interactions in marine and freshwater systems. *Integrative and Comparative Biology*, **49**, 291–313.

Sterry, P. R., Thomas, J. D. and Patience, R. L. (1983) Behavioural response of *Biomphalaria glabrata* (Say) to chemical factors from aquatic macrophytes including decaying *Lemna paucicostata* (Hegelm ex Engelm). *Freshwater Biology*, **13**, 465–476.

Sun, Y., Ding, J. Q. and Rena, M. X. (2009) Effects of simulated herbivory and resource availability on the invasive plant, *Alternanthera philoxeroides* in different habitats. *Biological Control*, **48**, 287–293.

van Donk, E. (1998) Switches between clear and turbid water states in a biomanipulated lake (1986–1996): the role of herbivory on macrophytes. In E. Jeppesen, M. Sondergaard, M. Sondergaard and K. Christoffersen (eds.) *The Structuring Role of Submerged Macrophytes in Lakes*. New York: Springer.

Walenciak, O., Zwisler, W. and Gross, E. M. (2002) Influence of *Myriophyllum spicatum*-derived tannins on gut microbiota of its herbivore *Acentria ephemerella. Journal of Chemical Ecology*, **28**, 2045–2056.

Wetzel, R. G. (1983) *Limnology*. Fort Worth, TX: Saunders College Publishing.

Wheeler, G. S., Van, T. K. and Center, T. D. (1998) Herbivore adaptations to a low-nutrient food: weed biological control specialist *Spodoptera pectinicornis* (Lepidoptera: Noctuidae) fed the floating aquatic plant *Pistia stratiotes. Environmental Entomology*, **27**, 993–1000.

Wilson, D. M., Fenical, W., Hay, M., Lindquist, N. and Bolser, R. (1999) Habenariol, a freshwater feeding deterrent from the aquatic orchid *Habenaria repens* (Orchidaceae). *Phytochemistry*, **50**, 1333–1336.

Wilson, J. R. U., Ajuonu, O., Center, T. D. *et al.* (2007) The decline of water hyacinth on Lake Victoria was due to biological control by *Neochetina* spp. *Aquatic Botany*, **87**, 90–93.

CHAPTER NINE

The soil microbial community and plant foliar defences against insects

ALAN C. GANGE

School of Biological Sciences, Royal Holloway, University of London

RENÉ ESCHEN

CABI

VIVIANE SCHROEDER

School of Biological Sciences, Royal Holloway, University of London

9.1 Soil microbial communities

No plant in nature grows in a soil devoid of microorganisms. Plant roots are surrounded by a rich microbial community, which reaches greatest levels of abundance and diversity in the zone immediately surrounding the root, a micro-habitat known as the rhizosphere. It has been claimed that the rhizosphere is where most biodiversity on Earth exists (Hinsinger *et al.*, 2009) and it is certainly one of the most dynamic and important ecosystems, through effects on plant growth and thus crop production and the structure and function of natural communities (Barrios, 2007).

The microbial community associated with plant roots contains a diverse array of bacteria, protozoa and fungi, some of which can be antagonistic to plant growth (pathogens), while others may appear to be benign or to have a range of beneficial effects. These latter effects include improved nutrient uptake by roots, chiefly through fixation and cycling of nitrogen, and mineralisation and uptake of phosphorus. Furthermore, soil microbes may increase plant growth by the synthesis of phytohormones (Costacurta & Vanderleyden, 1995), antagonism of deleterious soil bacteria and fungi by antibiotic production or depriving them of iron (Kloepper *et al.*, 1980), alleviation of salt and drought stress (Evelin *et al.*, 2009), enhancement of photosynthesis, and increasing resistance to foliar pathogens and insect predators (van der Ent *et al.*, 2009). The fact that root-associated microorganisms can alter the resistance of foliar tissues to insect herbivores is a relatively recent discovery, and the aim of this review is to document these interactions and to explore their mechanisms.

Pathogenic microbes that attack roots may have devastating effects on their plant hosts. This fact alone may mean that experiments involving root pathogens and foliar-feeding insects are hard to accomplish, and may be why very few exist in the literature. Equally important are groups of

The Ecology of Plant Secondary Metabolites: From Genes to Global Processes, eds. Glenn R. Iason, Marcel Dicke and Susan E. Hartley. Published by Cambridge University Press. © British Ecological Society 2012.

beneficial microbes in the rhizosphere, which include non-pathogenic plant growth-promoting rhizobacteria (PGPR) (Lugtenberg & Kamilova, 2009) and mycorrhizal fungi (Smith & Read, 2008). From the ecological point of view, PGPR seem to be less well studied, with the majority of the literature encompassing 'model' plants such as *Arabidopsis* or those of horticultural or agronomic interest (van Loon, 2007). However, our knowledge of the molecular and biochemical mechanisms by which these bacteria interact with plants is very good, and there is enormous potential for the ecological consequences of these interactions to be studied and understood (Pieterse & Dicke, 2007). Meanwhile there are a number of studies of the effects of mycorrhizal fungi on resistance of foliar tissues to insects (Koricheva *et al.*, 2009), but the mechanisms are much less clearly understood. We will discuss each of these different groups of microbes in turn and end with a consideration of the future directions which research needs to follow.

9.2   Root pathogenic fungi

Plant pathogenic fungi can cause a number of chemical changes in their hosts that have consequences for insects that subsequently feed upon infected tissues (Stout *et al.*, 2006). The vast majority of these studies involve pathogens and herbivores that attack the foliar tissues of plants. Studies of the interactions between root pathogens and foliar-feeding insects are extremely rare. These are mostly restricted to long-lived host species such as trees.

One good example of this spatially separated interaction is provided by Carter-Wientjes *et al.* (2004). These authors reared larvae of the soybean looper moth (*Pseudoplusia includens*) on plants with or without root infection by the charcoal rot fungus, *Macrophomina phaseolina*. The study was conducted over 2 years, and in the first year larvae consumed more foliage from, and were larger on, infected plants. However, results were inconsistent, as no effect was seen in the second year of study.

The most extensive series of root pathogen and foliar-feeding insect studies involve fungal and insect pests of coniferous trees in North America and Canada (Hertert *et al.*, 1975; James & Goheen, 1981; Gara *et al.*, 1984). Inconsistency of insect response to infected trees is a feature of these studies also, but in the majority of cases, pathogens appear to predispose trees to attack by bark or foliar-feeding beetles. The strength of the interaction appears to depend on the identity of the fungi and the insects; more virulent pathogens such as species of *Armillaria* tend to have greater effects than less virulent species (Lewis & Lindgren, 2002).

The mechanism by which diseased trees are more attractive to beetles appears to involve plant secondary metabolites (PSMs). Species of *Armillaria* (honey fungus) infecting roots can affect the phenolic and phenyl propanoid content of needles, as well as volatile emissions from foliage (Madziaraborusiewicz &

Strzelecka, 1977; Nebeker *et al.*, 1995). These changes can then render foliar tissues more susceptible to attack. However, the mechanism may be considerably more complicated than a simple change in PSMs. Bonello *et al.* (2003) studied the effects of the fungus *Heterobasidion annosum* (a widespread root pathogen of coniferous trees throughout the northern hemisphere) on foliar chemistry of *Pinus ponderosa*. They found that infected trees had higher levels of ferulic acid, but reduced levels of lignification. The reduction in lignin facilitated invasion by other fungi, with which bark beetles are associated.

These examples show that in coniferous forests, the interactions between root pathogens and foliar insects are of ecological and economic importance. However, it is unknown whether root pathogens affect foliar PSMs and insect herbivores in less long-lived plants, such as herbs and grasses. It is highly likely that they do, as the reverse interaction (effect of feeding on pathogen infection) has been documented in a few cases (Leath & Byers, 1977; Moellenbeck *et al.*, 1992). As with trees, these effects are often inconsistent too, and likely to depend upon the identity of the insect(s), fungus and plant under consideration (Stout *et al.*, 2006).

9.3   Plant growth-promoting rhizobacteria (PGPR)
9.3.1   Effects on plant growth

PGPR are non-pathogenic bacteria that have direct or indirect effects on plant growth. Direct effects may come about by antagonistic effects on plant pathogens (Compant *et al.*, 2005). Soils that contain these bacteria are known as 'suppressive', while those in which a pathogen causes disease are known as 'conducive'. It has been known for some time that mixing a small amount of suppressive soil with a larger amount of a conducive soil will render the latter suppressive (Schroth & Hancock, 1982). To date, there are no studies of the effects of suppressive soils on foliar-feeding insects, although Piskiewicz *et al.* (2009) showed that soil bacteria and fungi control populations of the herbivorous nematode, *Tylenchorhynchus ventralis* in sand dune systems.

In the absence of pathogens, however, PGPR may also improve the growth of plants by acting as biofertilisers or phytostimulators, or alleviating stress. Free-living nitrogen fixers such as *Azospirillum* can increase plant yield by enhancing nitrogen availability, while other species may solubilise phosphate, thereby increasing growth (Lugtenberg & Kamilova, 2009). Indeed, many species of rhizobacteria are constituents of over 150 formulations of biofertilisers and biopesticides (Schenk *et al.*, 2008). However, a feature of the literature is that the effects of PGPR on plant growth vary between species of bacteria and strains within species. Furthermore, the effects of each PGPR can differ depending on the identity of the plant species. As an illustration of such effects, Çakmakçi *et al.* (2007) inoculated wheat (*Triticum aestivum*) and spinach (*Spinacia oleracea*) with a number of different PGPR (*Bacillus, Paenibacillus* and *Pseudomonas*) species. In wheat, all PGPR strains significantly increased fresh weight of plants compared to the uninoculated controls. But this was not the case in spinach, where *Bacillus megaterium*, *Bacillus* spp. M13 and *B. cereus* failed to increase plant growth. Differences in the effect of particular strains on different host plants were also apparent. For example, *B. megaterium* increased the yield of wheat by 48%, but in spinach, this was only 9%. Meanwhile, *Paenibacillus polymyxa* RC05 was equally effective in both species.

These authors also measured changes in the antioxidant enzyme activities of leaves in the different PGPR treatments and found the same pattern of variation between bacteria within a plant and between plants with the same bacterium. For example, *B. megaterium* increased glutathione reductase concentrations in wheat by 200%, but reduced this in spinach by 16%. Clearly, from this and other studies (Lugtenberg & Kamilova, 2009), PGPR have the ability to alter both plant growth and foliar chemistry, but the extent of these effects depends very much on the identity of both bacterium and plant. This will translate into variable effects on PSMs, and thus insects that feed upon these plants.

9.3.2   PGPR and constitutive defence levels

Plants produce a colossal array of PSMs, which have activity against a range of organisms, as illustrated by the breadth of papers within this volume. Constitutive defences are those present in foliar tissues on a continuous basis, separating them from induced defences, which are only produced when a plant is attacked by a predator or pathogen.

Some authors have examined the effects of PGPR on PSMs, in the absence of pathogens and insects. The majority of studies have measured phenolic content of leaves when roots were inoculated with different PGPR. Lavania *et al.* (2006) inoculated *Piper betle* (betel vine) with *Serratia marcescens* NBI1213 and found that this increased levels of a number of phenolics in leaves, including gallic, chlorogenic, caffeic and ferulic acids. An induced response was also seen, as plants accumulated the greatest levels of these chemicals when inoculated with both *S. marcescens* and the root rot pathogen, *Phytophthora nicotianae*. Similar increases in gallic, ferulic and also tannic and cinnamic acids were found within the leaves of *Pisum sativum* (pea) when roots were inoculated with two other PGPR, *Pseudomonas fluorescens* (strain Pf4) and *P. aeruginosa*, by Singh *et al.* (2002). Meanwhile, del Amor *et al.* (2008) found that *Capsicum annuum* (sweet pepper) inoculated with a mixture of *Azospirillum brasilense* and *Pantoea dispersa* had higher levels of total phenolics in green fruits, but as these ripened, the difference between treatments became non-significant. PGPR can also increase the concentration of alkaloids in foliar tissues, as Jaleel *et al.* (2007) found that ajmalicine levels were higher in plants of *Catharanthus roseus* (Madagascar periwinkle) when inoculated with *P. fluorescens*.

The exact mechanism for the change in phenolics caused by PGPR is unclear but could be due to activation of the phenylpropanoid pathway (van Peer *et al.* 1991; Mabrouk *et al.*, 2007). In the latter paper, inoculation of *P. sativum* with *Rhizobium leguminosarum* caused increases in polyphenolox-idase (PPO) and phenylalanine ammonia lyase (PAL) genes, both suggestive of an important role for the phenylpropanoid/isoflavanoid pathways. The end result was a significant accumulation of phenolics in inoculated plants. Meanwhile, Kandan *et al.* (2002) also suggested that *P. fluorescens* stimulated the phenylpropanoid pathway, leading to an increase in phenolics in tomato (*Solanum lycopersicum*). A review of the biochemical mechanisms and genes involved in phenylpropanoid metabolism is provided by Vogt (2010).

9.3.3 PGPR and induced defence levels

Fifty years ago, Ross (1961) showed that if a pathogen attacked a plant, then non-infected tissues were subsequently more resistant to attack. This phenomenon has been termed 'systemic acquired resistance' (SAR) (Durrant & Dong, 2004) and is effective against a broad range of pathogenic microbes (Walters & Heil, 2007). SAR is associated with increased levels of salicylic acid (SA) and the coordinated expression of a set of genes encoding pathogenesis-related (PR) proteins (van Loon, 1997).

Thirty years after the work by Ross, Alström (1991), van Peer *et al.* (1991) and Wei *et al.* (1991) independently demonstrated for the first time that PGPR could also reduce pathogen infection on foliar tissues. Given that the PGPR and the attacking organism were spatially separated, it was clear that this effect must be due to changes within the plant, and the phenomenon was termed 'induced systemic resistance' (ISR) (Kloepper *et al.*, 1992). Although ISR appears to be functionally similar to SAR, in that it acts unspecifically against pathogens, its underlying mechanisms are different, ISR being regulated mainly by the plant hormones jasmonic acid (JA) and ethylene (van der Ent *et al.*, 2009) and SAR being regulated by the plant hormone salicylic acid (SA) (Pieterse & Dicke, 2007). Furthermore, unlike SAR, ISR resulting from PGPR activity seems to require less transcriptome reprogramming (Verhagen *et al.*, 2004), although accumulation of the regulatory protein NPR1 still occurs (Pieterse *et al.*, 1998). These induced defence responses are not mutually exclusive, as van Wees *et al.* (2000) showed that simultaneous activation of SAR and ISR is possible, creating an additive level of resistance against pathogens. There has been a large amount of work on the biochemical and molecular genetic processes involved in PGPR mediation of ISR, and there are many comprehensive reviews of the subject, with the most recent ones being by de Vleesschauwer and Höfte (2009) and van der Ent *et al.* (2009). The former paper presents a large data table, which summarises the studies in which PGPR affect ISR and where the pathway has been elucidated. A summary of the bacterial and plant species that are involved in these studies is presented in

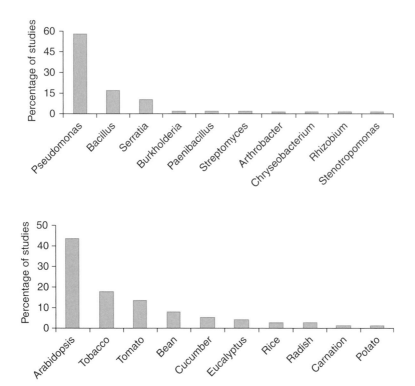

Figure 9.1 Percentage of studies involving PGPR-mediated host-plant ISR and in which the biochemical pathway has been elucidated. Upper graph depicts bacterial species, lower graph depicts plant species. Data taken from de Vleesschauwer and Höfte (2009).

Figure 9.1. These data show that 10 bacterial and 10 plant species have been extensively studied, and in each case there is one dominant bacterial genus or plant species. For PGPR, this is *Pseudomonas* (mostly *fluorescens* or *putida*) while for plants it is *Arabidopsis thaliana*. Thus, it can be seen that our knowledge of PGPR effects on ISR is limited to a narrow range of both microbial and herbaceous plant species. Indeed, as a recent review shows, PGPR effects on ISR in woody plants have never been studied, even though induced defence reactions are very well recorded in trees (Eyles *et al.*, 2010).

There are frequent reports of plants that express ISR induced by PGPR showing an accelerated defence response to attack by a pathogen or insect herbivore (Verhagen *et al.*, 2004; van Wees *et al.*, 2008). This phenomenon is known as 'priming' or 'getting ready for battle', and a review of the topic is provided by Conrath (2009). Priming is characterised by an enhanced defence response that is faster and stronger when previously unattacked tissues are attacked and has been shown to be an efficient form of defence, in which the benefits outweigh the costs (van Hulten *et al.*, 2006).

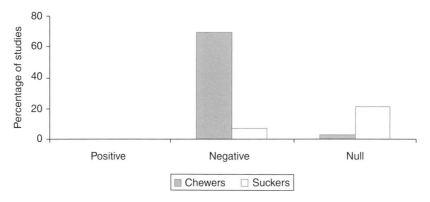

Figure 9.2 Summary of the effects of PGPR-mediated changes in PSMs on insect performance. Data expressed as the proportion of all 29 experiments that show positive, negative or null effects on chewing or sucking insects. Data summarised from Bong and Sikorowski (1991), Yao *et al.* (1996a, b), Zehnder *et al.* (1997a, b, 2001), Qingwen *et al.* (1998), Ramamoorthy *et al.* (2008), Commare *et al.* (2002), de Vos *et al.* (2007), Hanafi *et al.* (2007), Saravanakumar *et al.* (2007, 2008), Herman *et al.* (2008), van Oosten *et al.* (2008), Boutard-Hunt *et al.* (2009), Senthilraja *et al.* (2010) and Valenzuela-Soto *et al.* (2010).

To date, 18 papers, describing a total of 29 separate experiments, have investigated the effects of PGPR-ISR on the performance of foliar-feeding insects. The literature is, therefore, not sufficiently robust for a meta-analysis to be conducted, but a summary of the effects seen with chewing and sucking insects is presented in Figure 9.2. First, one should note that the literature is heavily biased towards just a few plant and insect species and that 28 of the 29 experiments have involved insects with a broad host range. The plants involved are similar to those in Figure 9.1, and the most commonly used are *Cucumus sativus* (cucumber), *Oryza sativa* (rice), *Solanum lycopersicum* (tomato) and *Capsicum annuum* (sweet pepper). Perhaps the most striking feature of Figure 9.2 is that in no study has a positive effect of PGPR inoculation of a plant ever been found on the performance of foliar-feeding insects. This is in contrast to the effects of SAR on insects, where positive, null and negative effects have been recorded (Stout *et al.*, 2006). With ISR, the majority of studies show negative effects, and it is clear that chewing insects seem to be adversely affected most of the time. In contrast, a greater proportion of studies with sucking insects has found no effect, and negative effects are rarer. Again, a note of caution should be sounded, in that six out of the seven null results involve the generalist sucking insect *Myzus persicae*.

In several cases, the biochemical and molecular genetic pathways have been determined for the PGPR effects on insects. Zehnder *et al.* (1997a,b) found that PGPR-induced resistance of cucumber against cucumber beetles was due to reductions in foliar levels of cucurbitacin, a powerful beetle feeding stimulant. Valenzuela-Soto *et al.* (2010) found that the resistance effect in tomato was due to a combination of JA-dependent and independent responses, the latter involving accumulation of PSMs and activation of PR (pathogenesis-related) proteins. Meanwhile, Saravanakumar *et al.* (2008) recorded higher levels of polyphenoloxidase and lipoxygenase in foliage of rice, when inoculated with PGPR. Perhaps the most interesting of these studies is that of van Oosten *et al.* (2008) with *Arabidopsis*; it is the only one to have used an insect with a relatively narrow diet breadth, and differences were found in the performance of this species (*Pieris rapae*) and that of a more generalist feeder, *Spodoptera exigua*. Although based on just one study, this fact is especially interesting when one considers the responses of specialist and generalist insect herbivores to mycorrhizal colonisation of their host plants (see Section 9.4.2, below). *Spodoptera exigua*, as a generalist insect, seems to be much more responsive to the JA-dependent defences induced by PGPR, while *P. rapae* is unaffected (van Oosten *et al.*, 2008).

Two other interesting aspects of these studies deserve comment. The first is that the effect of PGPR on foliar feeders seems to be just as noticeable in field conditions as it is in controlled situations (Saravanakumar *et al.*, 2007, 2008). Given that a diverse array of bacteria exists in field soils, it might be expected that the inoculated species would be lost or outcompeted in their effects on plants and thus herbivores (van Veen *et al.*, 1997). The fact that this does not happen is most encouraging from the point of view of field applications in agriculture and also the understanding of the role of these bacteria in ecological communities of plants and insects. It also means that there is the potential for commercial products containing these bacteria to be produced and to be applied to field crops, resulting in increases in yields.

The second interesting fact is that populations of natural enemies (spiders and hymenopteran parasitoids) of herbivores have been shown to be higher on plants treated with PGPR (Commare *et al.*, 2002; Saravanakumar *et al.*, 2008), although this is not always the case (van Oosten *et al.*, 2008). As JA is important in the production of volatiles that attract higher trophic levels (Schuman & Baldwin, Chapter 15; Dicke *et al.*, Chapter 16), it might be thought that the involvement of the JA pathway in ISR by PGPR might make plants more attractive to natural enemies. However, given that there are often antagonistic interactions between direct and indirect defences, volatile production could be unaffected. Clearly, this is an area that would merit further study, particularly in view of the possible applications to pest control.

The conclusion from these studies is that PGPR seem to be able to reduce the performance of chewing insects, through ISR, while having little effect on sucking insects. However, the range of plants, insects and microbes is extremely narrow, and this would be a most rewarding area for ecological study.

9.4   Plant growth-promoting fungi (PGPF)

In contrast to bacteria, studies with non-pathogenic species of fungi and their growth-promoting effects on plants are much less common, and, to our knowledge, not a single published example of the effects of free-living PGPF in the soil and foliar-feeding insects exists. This is despite the fact that a number of fungal species have influences on ISR similar to PGPR. Examples include species of *Trichoderma*, and non-pathogenic species or strains of *Fusarium*, *Penicillium* and *Pythium*. A review of the effects that these fungi have on plants, including effects on ISR, priming and PSMs, is provided by Trillas and Segarra (2009). The latter paper combines those fungi entirely within a root, and mycorrhizas (where the majority of the fungal hyphae are exterior to the root), as 'endophytes', but here we make a distinction between the types as their effects on plant growth and understanding of interactions with insects differ.

9.4.1   Endophyte fungi

Endophyte fungi exist entirely within plant tissues and some of those that inhabit roots appear to have the ability to increase the growth of plants. A good example is *Piriformospora indica*, which has a global distribution and the capability of associating with a wide variety (over 140) of host plants (Oelmuller *et al.*, 2009). This fungus is capable of alleviating drought stress and accelerating growth rate in a variety of plants, leading to higher yields in plants with the fungus (Achatz *et al.*, 2010; Sun *et al.*, 2010).

Piriformospora indica also seems to be able to induce chemical changes in foliage that may have effects on insect herbivores. This species elicits ISR in *Arabidopsis* (Stein *et al.*, 2008) and has antagonistic effects on a range of foliar pathogens and root-feeding nematodes (Shoresh *et al.*, 2010). However, the mechanism of action within plants is unclear, as Waller *et al.* (2005) found that *P. indica* could cause ISR, without any reliance on JA, SA or ethylene, and instead seemed to increase antioxidant levels in leaves.

Another endophyte species that occurs in roots is *Acremonium strictum*, and this fungus can increase plant biomass of marram grass (*Ammophila arenaria*) through antagonistic effects on root-feeding nematodes (Hol *et al.*, 2007). It also has activity against a variety of plant pathogens (Choi *et al.*, 2009). However, of more interest to this review is that this species has been shown to reduce the performance of both sucking and chewing insects, attacking the foliar parts of tomato and broad bean (*Vicia faba*) plants (Vidal, 1996; Jaber & Vidal, 2010). Indeed, the latter paper shows that the endophyte effects are passed on to the F1 generation of insects, and the authors suggest that this is due to the accumulation of PSMs (though these were not measured) by insects feeding on infected plants. Clearly, PGPF have the potential to alter PSM levels in foliage of a range of plants, and this would also merit further study.

9.4.2   Arbuscular mycorrhizal fungi

Arbuscular mycorrhizal fungi (AMF) associate with the roots of about 80% of the world's vascular plants, wherein they form a mutualistic relationship, donating phosphate and other nutrients to the plant in return for a supply of organic carbon. In contrast to PGPR, a considerable amount of information exists on the interactions between these fungi and foliar-feeding insects, with relatively little knowledge of the mechanisms involved (Gange, 2007).

The first clear demonstration that AMF can affect the performance of foliar-feeding insects, mediated through PSMs, was provided by Gange and West (1994). In that study, plants of *Plantago lanceolata* grown in a field with reduced mycorrhizal colonisation suffered elevated levels of attack by foliar-chewing insects. When foliage of these and mycorrhizal plants were fed to larvae of the generalist chewing insect, *Arctia caja*, larval growth on mycorrhizal plants was half of that on non-mycorrhizal plants. Levels of the iridoid glycosides aucubin and catalpol were significantly higher in mycorrhizal plants, suggestive that the fungi can elevate PSMs in foliar tissues, rendering these more resistant to insect attack.

Since this paper, a number of other studies have examined insect–mycorrhizal interactions, and a recent meta-analysis has summarised the effects found (Koricheva *et al.*, 2009). These authors found that overall, generalist chewing insects seem to be negatively affected by AMF, while specialist chewing and sucking insects are positively affected. Thus, these results mirror to a large extent those found with PGPR (though we must exercise caution in such a comparison, given the lack of experimental data) (Section 9.3.3, above). They also provide strong inferential evidence that AMF effects on foliar feeders are mediated by changes in PSMs, if one assumes that generalist insects are negatively affected by these chemicals but specialists respond positively (Schoonhoven *et al.*, 2005).

We say 'strong inferential evidence' because in the majority of other studies no information on PSMs is provided, and it is usually unclear whether the authors investigated these but did not find changes, or did not seek them at all (Vannette & Hunter, 2009). Indeed, when differences in PSMs are found, it is difficult to determine whether these are changes in constitutive levels or whether induced resistance has occurred. Two recent papers provide exciting evidence that both types of defence changes may happen in response to colonisation by AMF (Bennett *et al.* 2009; Kempel *et al.*, 2010). In the former study, *P. lanceolata* was grown in association with a number of different AMF. Some fungal species increased the levels of iridoid glycosides in the absence of herbivory (constitutive defence), while others did not. Meanwhile, when plants were attacked by larvae of the butterfly *Junonia coenia*, some AMF species prevented any induction of glycosides, while others seemed to enhance induction. These results are important, because they show how the identity of AMF species is critical in understanding effects on PSMs and also help to explain the inconsistent results that have been found with these fungi and PSMs in *P. lanceolata* (Fontana *et al.*, 2009). Furthermore, they may explain the apparent ecological specificity of AMF, in which insect responses depend upon the identity of the fungi in the root system (Gange *et al.*, 2005). Indeed, Nishida *et al.* (2010), working with herbivorous mites, have shown that in *Lotus japonicus* plants, the induction of elevated levels in leaf phenolics (and thus mite performance) depends upon the identity of the AMF in the root system.

In the second study, Kempel *et al.* (2010) studied how the mycorrhizal fungus *Glomus intraradices* affects induction of defences in plants, caused by insect attack, in six herbaceous plant species. In three of these (*Festuca rubra*, *Artemisia vulgaris* and *Senecio jacobaea*), induction of defences (as measured by herbivore performance) only occurred in mycorrhizal plants. In the other three plant species (*Deschampsia flexuosa*, *Poa pratensis* and *Plantago lanceolata*), no effect of AMF on defence induction was found.

AMF therefore seem to affect the induction of plant defences more frequently than altering constitutive levels. The mechanism by which this occurs is also unclear. It may involve changes in defence-related gene expression in the absence of an attacking organism (Liu *et al.*, 2007) or through priming of plant tissues, in a similar manner to that elicited by PGPR (Pozo & Azcón-Aguilar, 2007; Trillas & Segarra, 2009). Furthermore, AMF colonisation appears to be related to repression of SA pathways and elevated JA production, leading to defence induction (Pozo & Azcón-Aguilar, 2007). The latter paper is currently the most comprehensive review of mycorrhizal-induced resistance to insects in plants.

AMF can also play a role in extended multi-trophic interactions of the plants they colonise. For example, they can affect the levels of parasitism of herbivorous insects (Gange *et al.*, 2003; Hempel *et al.*, 2009) and it is thought that this may be due to the fungi affecting volatile organic compound (VOC) emissions from foliage (Guerrieri *et al.*, 2004). Indeed, Fontana *et al.* (2009) show that emission of volatiles was significantly higher in plants of *P. lanceolata* when colonised by *G. intraradices*. Levels of one compound, (Z)-3-hexenyl acetate, were similar in unattacked mycorrhizal plants and herbivore-damaged plants, suggesting that the AM fungus can alter the PSM content of tissues to attract natural enemies such as parasitoids. However, this effect also appears to be species- or context-specific, as Leitner *et al.* (2010) have shown that VOC (mainly sesquiterpenoids) emission in *Medicago truncatula* can be influenced by species of AMF, but that the effect depends upon the genotype of the plant.

9.5 Assembling the microbial community

We began this review by stating that plants do not grow in sterile soil and that the soil microbial community with which they are associated is abundant and diverse. This is certainly true of all field situations but unfortunately does not apply to every controlled study performed in the laboratory. It is well known that PGPR and AMF act and interact to affect plant growth and the PSM content of foliage (Artursson *et al.*, 2006). However, throughout the literature, many studies of mycorrhizal fungi and plants have been performed with 'control' plants grown in a 'sterile' soil, and a comparison made with others inoculated with a mycorrhizal fungus. Koide and Li (1989) proposed that all AMF experiments should involve a control treatment that contains a non-mycorrhizal community. However, even recent papers on AMF and foliar PSMs still have controls containing sterile soil only (e.g. Kempel *et al.*, 2010; Leitner *et al.*, 2010). Thus, while these papers may provide most interesting results, interpretation of the experiments is complicated by the lack of an entire microbial community.

A good illustration of the interactions that can occur between PGPR and AMF, and their effects on PSMs, is given by Selvaraj *et al.* (2008). In this experiment, plants of *Begonia malabarica* were inoculated with the AMF *G. mosseae*, the PGPR *Bacillus coagulans* and the PGPF *Trichoderma viride*. In this study, there was an appropriate sterile control, to enable statistical comparison of the full factorial design. Several important conclusions can be drawn from this study. First, AMF, PGPR and PGPF all demonstrated an ability to increase total phenolics in leaf tissue. Similar results were also found with flavonoids and alkaloids. Second, the greatest effect on PSMs was seen in the complete microbial community treatment, with phenolics being significantly higher than in any of the two-factor treatments. Finally, if one compares the control with the AMF treatment only, the difference was greater than that between the PGPR and PGPR+AMF or between AMF and PGPR+AMF treatments, indicating that comparison with sterile controls overestimates the effects of the microbes.

9.6 Conclusions and future directions

We have tried to show that the various members of the soil microbial community, be they detrimental (pathogens) or beneficial to plant growth, can have effects on PSMs that may translate into effects on foliar-feeding insects. However, these effects are still little explored and present many areas for future study and exploitation. Understanding how foliar pest insects respond to hosts with diseased roots could have important consequences in crop or forestry systems, for example. Some groups are better studied, and plant growth-promoting rhizobacteria have been shown to induce systemic resistance (ISR) in plants, mediated by the plant hormones jasmonic acid and ethylene. In particular, they induce a state in which plant defence is more responsive when foliage is attacked, a phenomenon known as priming. There is an extensive literature on the molecular mechanisms behind ISR and priming, with less on the ecological aspects. Arbuscular mycorrhizal fungi (AMF) can increase both constitutive and induced levels of PSMs in foliage, thereby rendering tissues more resistant to some insect herbivores. In contrast to the rhizobacteria, most studies are ecological and there is less understanding of the molecular mechanisms. There is limited evidence that specialist insects either do not respond or show positive reactions to PGPR or AMF colonisation of plants, while generalist insects show negative reactions. These differences suggest a pivotal role for PSMs in the mediation of interactions between soil microbes and foliar-feeding insects. However, the vast majority of experiments study just one group of microbes, and there is a great need for studies that involve more complex microbial communities.

Perhaps the most difficult but interesting question for ecologists to address is: why do soil microbes elevate the levels of PSMs in foliage, rendering these tissues more resistant to insect herbivores? Furthermore, as we have tried to illustrate, the vast majority of our knowledge of the effects of soil fungi and bacteria come from studies involving plants of horticultural interest or where the genetics of the plant are well known (e.g. *Arabidopsis*). There is indeed a great deal of merit in using such systems, where the knowledge gained may be applied in crop production or the mechanism more easily understood (Adesemoye & Kloepper, 2009). However, our understanding of these organisms in the structuring and function of natural plant communities is negligible and again restricted to a few 'model' plants, such as *P. lanceolata*. PSMs in plants are known to change seasonally (Koricheva & Barton, Chapter 3) while soil microbial communities are known to change both within and across seasons (Maharning *et al.*, 2009). It is unknown whether the foliar PSM changes are linked to those in the soil microbial community, and this would be a fascinating area for further study.

One of the most influential papers in plant ecology is that of Herms and Mattson (1992) in which it is argued that as plants are resource-limited, it is often impossible for resources to be allocated to maximise both growth and defence. The soil microbial community appears to enable plants to overcome resource limitation to an extent, since many of the studies reviewed here report increases in foliar PSMs and plant biomass. As the microbial community is heterogeneous in space and time, the effects on PSMs are likely to be unpredictable for a foliar-feeding insect, leading to variation in performance from one plant to another. The extent to which this variation affects the structure of natural communities of plants and insects is unknown and would be an interesting area for future study.

Perhaps it is simply a question of foliar-feeding insects responding over evolutionary time to the PSM content of foliage, which has been influenced by the soil microbial community. Indeed, Gange *et al.* (2002) found a relationship in the UK flora between the degree of mycorrhizal affinity of plants and the proportion of insects associated with those plants that are specialists. Plant families that contain many mycorrhizal species have insect herbivore loads that are dominated by specialists, while families with few mycorrhizal species are attacked mostly by generalists. It was suggested that the changes in PSM content of foliage caused by the mycorrhiza have led to the evolution of insect specialism. Whether such effects on insect diet breadth are also caused by PGPR or PGPF is entirely unknown.

One possible mechanism for interactions between soil microbes and above-ground attackers and defences is through the secretion of root exudates. It is well known that the quality and quantity of exudates structures the microbial community (Narula *et al.*, 2009). It may be postulated that if herbivores reduce the amount of exudation or alter its composition, this could be to the detriment of the microbial community, and thus it would be in the interests of this community to increase the resistance of foliar tissues to herbivores, through the most cost-efficient system available, namely ISR (Paterson, 2003). However, recent evidence suggests that such a carbon-limitation hypothesis is unlikely to apply to the microbial community, since defoliation appears to have little effect either on rhizobacteria (Vestergard *et al.*, 2008) or AM fungi (Barto & Rillig, 2010). Clearly, the field of soil microbial communities and their effects on PSMs is ready for substantial cross-disciplinary research, to unravel the mechanisms of these fascinating interactions (Pieterse & Dicke, 2007).

Acknowledgements

We are grateful to the Natural Environment Research Council, the Biotechnology and Biosciences Research Council and Symbio Ltd for funding our work on soil microbes, plant growth, insects and PSM.

References

Achatz, B., von Ruden, S., Andrade, D. *et al.* (2010) Root colonization by *Piriformospora indica* enhances grain yield in barley under diverse nutrient regimes by accelerating plant development. *Plant and Soil*, **333**, 59–70.

Adesemoye, A. O. and Kloepper, J. W. (2009) Plant–microbe interactions in enhanced fertilizer-use efficiency. *Applied Microbiology and Biotechnology*, **85**, 1–12.

Alström, S. (1991) Induction of disease resistance in common bean susceptible to halo blight bacterial pathogen after seed bacterization with rhizosphere pseudomonads. *Journal of General and Applied Microbiology*, **37**, 495–501.

Artursson, V., Finlay, R. D. and Jansson, J. K. (2006) Interactions between arbuscular mycorrhizal fungi and bacteria and their potential for stimulating plant growth. *Environmental Microbiology*, **8**, 1–10.

Barrios, E. (2007) Soil biota, ecosystem services and land productivity. *Ecological Economics*, **64**, 269–285.

Barto, E. K. and Rillig, M. C. (2010) Does herbivory really suppress mycorrhiza? A meta-analysis. *Journal of Ecology*, **98**, 745–753.

Bennett, A. E., Bever, J. D. and Bowers, M. D. (2009) Arbuscular mycorrhizal fungal species suppress inducible plant responses and alter defensive strategies following herbivory. *Oecologia*, **160**, 711–719.

Bonello, P., Storer, A. J., Gordon, T. R. and Wood, D. L. (2003) Systemic effects of *Heterobasidion annosum* on ferulic acid glucoside and lignin of presymptomatic ponderosa pine phloem, and potential effects on bark-beetle-associated fungi. *Journal of Chemical Ecology*, **29**, 1167–1182.

Bong, C. F. J. and Sikorowski, P. P. (1991) Effects of cytoplasmic polyhedrosis virus and bacterial contamination on growth and development of the corn earworm, *Helicoverpa zea* (Lepidoptera: Noctuidae). *Journal of Invertebrate Pathology*, **57**, 406–412.

Boutard-Hunt, C., Smart, C. D., Thaler, J. and Nault, B. A. (2009) Impact of plant growth-promoting rhizobacteria and natural enemies on *Myzus persicae* (Hemiptera: Aphididae) infestations in pepper. *Journal of Economic Entomology*, **102**, 2183–2191.

Çakmakçi, R., Erat, M., Erdogan, U. and Dönmez, M. E. (2007) The influence of plant growth-promoting rhizobacteria on growth and enzyme activities in wheat and spinach plants. *Journal of Plant Nutrition and Soil Science*, **170**, 288–295.

Carter-Wientjes, C. H., Russin, J. S., Boethel, D. J., Griffin, J. L. and McGawley, E. C. (2004) Feeding and maturation by soybean looper (Lepidoptera: Noctuidae) larvae on soybean affected by weed, fungus, and nematode pests. *Journal of Economic Entomology*, **97**, 14–20.

Choi, G. J., Kim, J. C., Jang, K. S. *et al.* (2009) Biocontrol activity of *Acremonium strictum* BCP against *Botrytis* diseases. *Plant Pathology Journal*, **25**, 165–171.

Commare, R. R., Nandakumara, R., Kandana, A. *et al.* (2002) *Pseudomonas fluorescens* based bio-formulation for the management of sheath blight disease and leaffolder insect in rice. *Crop Protection*, **21**, 671–677.

Compant, S., Duffy, B., Nowak, J., Clément, C. and Barka, E. A. (2005) Use of plant growth-promoting bacteria for biocontrol of plant diseases: principles, mechanisms of action, and future prospects. *Applied and Environmental Microbiology*, **71**, 4951–4959.

Conrath, U. (2009) Priming of induced plant defense responses. *Advances in Botanical Research*, **51**, 361–395.

Costacurta, A. and Vanderleyden, J. (1995) Synthesis of phytohormones by plant associated bacteria. *Critical Reviews in Microbiology*, **21**, 1–18.

de Vleesschauwer, D. and Höfte, M. (2009) Rhizobacteria-induced systemic resistance. *Advances in Botanical Research*, **51**, 223–281.

de Vos, M., van Oosten, V. R., Jander, G., Dicke, M. and Pieterse, C. M. J. (2007) Plants under attack: multiple interactions with insects and microbes. *Plant Signaling and Behavior*, **2**, 527–529.

del Amor, F. M., Serrano-Martínez, A., Fortea, M. I., Legua, P. and Núñez-Delicado, E. (2008) The effect of plant-associative bacteria (*Azospirillum* and *Pantoea*) on the fruit quality of sweet pepper under limited nitrogen supply. *Scientia Horticulturae*, **117**, 191–196.

Durrant, W. E. and Dong, X. (2004) Systemic acquired resistance. *Annual Review of Phytopathology*, **42**, 185–209.

Evelin, H., Kapoor, R. and Giri, B. (2009) Arbuscular mycorrhizal fungi in alleviation of salt stress: a review. *Annals of Botany*, **104**, 1263–1280.

Eyles, E., Bonello, P., Ganley, R. and Mohammed, C. (2010) Induced resistance to pests and pathogens in trees. *New Phytologist*, **185**, 893–908.

Fontana, A., Reichelt, M., Hempel, S., Gershenzon, J. and Unsicker, S. B. (2009) The effects of arbuscular mycorrhizal fungi on direct and indirect defense metabolites of *Plantago lanceolata* L. *Journal of Chemical Ecology*, **35**, 833–843.

Gange, A. C. (2007) Insect–mycorrhizal interactions: patterns, processes and consequences. In T. Ohgushi, T. Craig and P. W. Price (eds.) *Indirect Interaction Webs: Nontrophic Linkages through Induced Plant Traits.* Cambridge University Press, 124–144.

Gange, A. C. and West, H. M. (1994) Interactions between arbuscular mycorrhizal fungi and foliar-feeding insects in *Plantago lanceolata* L. *New Phytologist*, **128**, 79–87.

Gange, A. C., Stagg, P. G. and Ward, L. K. (2002) Arbuscular mycorrhizal fungi affect phytophagous insect specialism. *Ecology Letters*, **5**, 11–15.

Gange. A. C., Brown, V. K. and Aplin, D. M. (2003) Multitrophic links between arbuscular mycorrhizal fungi and insect parasitoids. *Ecology Letters*, **6**, 1051–1055.

Gange, A. C., Brown, V. K. and Aplin, D. M. (2005) Ecological specificity of arbuscular mycorrhizae: evidence from foliar- and seed-feeding insects. *Ecology*, **86**, 603–611.

Gara, R. I., Geiszler, D. R. and Littke, W. R. (1984) Primary attraction of the mountain pine beetle to lodgepole pine in Oregon. *Annals of the Entomological Society of America*, **77**, 333–334.

Guerrieri E., Lingua G., Digilio, M. C., Massa, N. and Berta, G. (2004) Do interactions between plant roots and the rhizosphere affect parasitoid behaviour? *Ecological Entomology*, **29**, 753–756.

Hanafi, A., Traoré, M., Schnitzler, W. H. and Woitke, M. (2007) Induced resistance of tomato to whiteflies and *Pythium* with the PGPR *Bacillus subtilis* in a soilless crop grown under greenhouse conditions. *Acta Horticulturae*, **747**, 315–322.

Hempel, S., Stein, C., Unsicker, S. B. *et al.* (2009) Specific bottom-up effects of arbuscular mycorrhizal fungi across a plant–herbivore–parasitoid system. *Oecologia*, **160**, 267–277.

Herman, M. A. B., Nault, B. A. and Smart, C. D. (2008) Effects of plant growth-promoting rhizobacteria on bell pepper production and green peach aphid infestations in New York. *Crop Protection*, **27**, 996–1002.

Herms, D. A. and Mattson, W. J. (1992) The dilemma of plants – to grow or defend. *Quarterly Review of Biology*, **67**, 283–335.

Hertert, H. D., Miller, D. L. and Partridge, A. D. (1975) Interactions of bark beetles (Coleoptera:Scolytidae) and root rot pathogens in northern Idaho. *Canadian Entomologist*, **107**, 899–904.

Hinsinger, P., Bengough, A. G., Vetterlein, D. and Young, I. M. (2009) Rhizosphere: biophysics, biogeochemistry and ecological relevance. *Plant and Soil*, **321**, 117–152.

Hol, W. H. G., de la Pena, E., Moens, M. and Cook, R. (2007) Interaction between a fungal endophyte and root herbivores of *Ammophila arenaria*. *Basic and Applied Ecology*, **8**, 500–509.

Jaber, L. R. and Vidal, S. (2010) Fungal endophyte negative effects on herbivory are enhanced on intact plants and maintained in a subsequent generation. *Ecological Entomology*, **35**, 25–36.

Jaleel, C. A., Manivannan, P., Sankar, B. *et al.* (2007) *Pseudomonas fluorescens* enhances biomass yield and ajmalicine production in *Catharanthus roseus* under water deficit stress. *Colloids and Surfaces B: Biointerfaces*, **60**, 7–11.

James, R. L. and Goheen, D. J. (1981) Conifer mortality associated with root disease and insects in Colorado. *Plant Disease*, **65**, 506–507.

Kandan, A., Commare, R. R., Nandakumar, R. *et al.* (2002) Induction of phenylpropanoid metabolism by *Pseudomonas fluorescens* against tomato spotted wilt virus in tomato. *Folia Microbiologica*, **47**, 121–129.

Kempel, A. Schmidt, A. K. Brandl, R. and Schädler, M. (2010) Support from the underground: induced plant resistance depends on arbuscular mycorrhizal fungi. *Functional Ecology*, **24**, 293–300.

Kloepper, J. W., Leong, J., Teintze, M. and Schroth, M. N. (1980) Enhanced plant growth by siderophores produced by plant growth-promoting rhizobacteria. *Nature*, **286**, 885–886.

Kloepper, J. W., Tuzun, S. and Kuć, J. A. (1992) Proposed definitions related to induced disease resistance. *Biocontrol Science and Technology*, **2**, 349–351.

Koide, R. T. and Li, M. (1989) Appropriate controls for vesicular arbuscular mycorrhizal research. *New Phytologist*, **111**, 35–44.

Koricheva, J., Gange, A. C. and Jones, T. (2009) Effects of mycorrhizal fungi on insect herbivores: a meta-analysis. *Ecology*, **90**, 2088–2097.

Lavania, M., Chauhan, P. S., Chauhan, S. V. S., Singh, H. B. and Nautiyal, C. S. (2006) Induction of plant defense enzymes and phenolics by treatment with plant growth-promoting rhizobacteria *Serratia marcescens* NBRI1213. *Current Microbiology*, **52**, 363–368.

Leath, K. T. and Byers, R. A. (1977) Interaction of *Fusarium* root rot with pea aphid and potato leafhopper feeding on forage legumes. *Phytopathology*, **67**, 226–229.

Leitner, M., Kaiser, R., Hause, B., Boland, W. and Mithöfer, A. (2010) Does mycorrhization influence herbivore-induced volatile emission in *Medicago truncatula*? *Mycorrhiza*, **20**, 89–101.

Lewis, K. J. and Lindgren, B. S. (2002) Relationship between spruce beetle and tomentosus root disease: two natural disturbance agents of spruce. *Canadian Journal of Forest Research*, **32**, 31–37.

Liu, J., Maldonado-Mendoza, I., Lopez-Meyer, M. *et al.* (2007) Arbuscular mycorrhizal symbiosis is accompanied by local and systemic alterations in gene expression and an increase in disease resistance in the shoots. *Plant Journal*, **50**, 529–544.

Lugtenberg, B. J. J. and Kamilova, F. (2009) Plant growth-promoting rhizobacteria. *Annual Review of Microbiology*, **63**, 541–556.

Mabrouk, Y., Simier, P., Delavault, P. *et al.* (2007) Molecular and biochemical mechanisms of defence induced in pea by *Rhizobium leguminosarum* against *Orobanche crenata*. *Weed Research*, **47**, 452–460.

Madziaraborusiewicz, K. and Strzelecka, H. (1977) Conditions of spruce (*Picea excelsa*) infestation by engraver beetle (*Ips typographus*) in mountains of Poland I. Chemical composition of volatile oils from healthy trees and those infested with honey fungus (*Armillaria mellea*). *Journal of Applied Entomology*, **83**, 409–415.

Maharning, A. R., Mills, A. A. S. and Adl, S. M. (2009) Soil community changes during secondary succession to naturalized grasslands. *Applied Soil Ecology*, **41**, 137–147.

Moellenbeck, D. J., Quisenberry, S. S. and Colyer, P. D. (1992) *Fusarium* crown-rot development in alfalfa stressed by threecornered alfalfa hopper (Homoptera: Membracidae) feeding. *Journal of Economic Entomology*, **58**, 1442–1449.

Narula, N., Kothe, E. and Behl, R. K. (2009) Role of root exudates in plant-microbe interactions. *Journal of Applied Botany and Food Quality*, **82**, 122–130.

Nebeker, T. E., Schmitz, R. F., Tisdale, R. A. and Hobson, K. R. (1995) Chemical and nutritional status of dwarf mistletoe, *Armillaria* root-rot, and *Comandra* blister rust infected trees which may influence tree susceptibility to bark beetle attack. *Canadian Journal of Botany*, **73**, 360–369.

Nishida, T., Katayama, N., Izumi, N. and Ohgushi, T. (2010) Arbuscular mycorrhizal fungi species-specifically affect induced plant responses to a spider mite. *Population Ecology*, **52**, 507–515.

Oelmuller, R., Sherameti, I., Tripathi, S. and Varma, A. (2009) *Piriformospora indica*, a cultivable root endophyte with multiple biotechnological applications. *Symbiosis*, **49**, 1–17.

Paterson, E. (2003) Importance of rhizodeposition in the coupling of plant and microbial productivity. *European Journal of Soil Science*, **54**, 741–750.

Pieterse, C. M. J. and Dicke, M. (2007) Plant interactions with microbes and insects: from molecular mechanisms to ecology. *Trends in Plant Science*, **12**, 564–569.

Pieterse, C. M. J., van Wees, S. C. M., van Pelt, J. A. et al. (1998) A novel signalling pathway controlling induced systemic resistance in *Arabidopsis*. *Plant Cell*, **10**, 1571–1580.

Piskiewicz, A. M., Duyts, H. and van der Putten, W. H. (2009) Soil microorganisms in coastal foredunes control the ectoparasitic root-feeding nematode *Tylenchorhynchus ventralis* by local interactions. *Functional Ecology*, **23**, 621–626.

Pozo, M. J. and Azcón-Aguilar, C. (2007) Unraveling mycorrhiza-induced resistance. *Current Opinion in Plant Biology*, **10**, 393–398.

Qingwen, Z., Ping, L., Gang, W. and Qingnian, C. (1998) On the biochemical mechanism of induced resistance of cotton to cotton bollworm by cutting off young seedling at plumular axis. *Acta Phytophylactica Sinica*, **25**, 209–212.

Ramamoorthy, V., Viswanathan, R., Raguchander, T., Prakasam, V. and Samiyappan, R. (2008) Induction of systemic resistance by plant growth promoting rhizobacteria in crop plants against pests and diseases. *Crop Protection*, **20**, 1–20.

Ross, A. F. (1961) Systemic acquired resistance induced by localized virus infections in plants. *Virology*, **14**, 340–358.

Saravanakumar, D., Muthumeena, K., Lavanya, N. et al. (2007) *Pseudomonas*-induced defence molecules in rice plants against leaffolder (*Cnaphalocrocis medinalis*) pest. *Pest Management Science*, **63**, 714–721.

Saravanakumar, D., Lavanya, N., Muthumeena, B. et al. (2008) *Pseudomonas fluorescens* enhances resistance and natural enemy population in rice plants against leaffolder pest. *Journal of Applied Entomology*, **132**, 469–479.

Schenk, P. M., McGrath, K. C., Lorito, M. and Pieterse, C. M. J. (2008) Plant–microbe and plant–insect interactions meet common grounds. *New Phytologist*, **179**, 251–256.

Schoonhoven, L. M., van Loon, J. J. A. and Dicke, M. (2005) *Insect–Plant Biology*. Oxford: Oxford University Press.

Schroth, M. N. and Hancock, J. G. (1982) Disease-suppressive soil and root-colonizing bacteria. *Science*, **216**, 1376–1381.

Selvaraj, T., Rajeshkumar, S., Nisha, M. C., Wondimu, L. and Tesso, M. (2008) Effect of *Glomus mosseae* and plant growth promoting rhizomicroorganisms (PGPR's) on growth, nutrients and content of secondary metabolites in *Begonia malabarica* Lam. *Maejo International Journal of Science and Technology*, **2**, 516–525.

Senthilraja, G., Anand, T., Durairaj, C. et al. (2010) A new microbial consortia containing entomopathogenic fungus, *Beauveria bassiana* and plant growth promoting rhizobacteria, *Pseudomonas fluorescens* for simultaneous management of leafminers and collar rot disease in groundnut. *Biocontrol Science and Technology*, **20**, 449–464.

Shoresh, M., Harman, G. E. and Mastouri, F. (2010) Induced systemic resistance and plant responses to fungal biocontrol agents. *Annual Review of Phytopathology*, **48**, 21–23.

Singh, U. P., Sarma, B. K., Singh, D. P. and Bahadur, A. (2002) Plant growth-promoting rhizobacteria-mediated induction of phenolics in pea (*Pisum sativum*) after infection with *Erysiphe pisi*. *Current Microbiology*, **44**, 396–400.

Smith, S. E. and Read, D. J. (2008) *Mycorrhizal Symbiosis*, 3rd edn. London: Academic Press.

Stein, E., Molitor, A., Kogel, K.-H. and Waller, F. (2008) Systemic resistance in *Arabidopsis* conferred by the mycorrhizal fungus *Piriformospora indica* requires jasmonic acid signaling and the cytoplasmic function of NPR1. *Plant and Cell Physiology*, **49**, 1747–1751.

Stout, M. J., Thaler, J. S. and Thomma, B. P. H. J. (2006) Plant-mediated interactions between pathogenic microorganisms and herbivorous arthropods. *Annual Review of Entomology*, **51**, 663–689.

Sun, C. A., Johnson, J., Cai, D. G. et al. (2010) *Piriformospora indica* confers drought tolerance in Chinese cabbage leaves by stimulating antioxidant enzymes, the expression of drought-related genes and the plastid-localized CAS protein. *Journal of Plant Physiology*, **167**, 1009–1017.

Trillas, M. I. and Segarra, G. (2009) Interactions between nonpathogenic fungi and plants. *Advances in Botanical Research*, **51**, 321–359.

Valenzuela-Soto, J. H., Estrada-Hernández, M. G., Ibarra-Laclette, E. and Délano-Frier, J. P. (2010) Inoculation of tomato plants (*Solanum lycopersicum*) with growth-promoting *Bacillus subtilis* retards whitefly *Bemisia tabaci* development. *Planta*, **231**, 397–410.

van der Ent, S., Van Wees, S. C. M. and Pieterse, C. M. J. (2009) Jasmonate signalling in plant interactions with resistance-inducing beneficial microbes. *Phytochemistry*, **70**, 1581–1588.

van Hulten, M., Pelser, M., van Loon, L. C., Pieterse, C. M. J. and Ton, J. (2006) Costs and benefits of priming for defense in *Arabidopsis*. *Proceedings of the National Academy of Sciences USA*, **103**, 5602–5607.

van Loon, L. C. (1997) Induced resistance and the role of pathogenesis related proteins. *European Journal of Plant Pathology*, **103**, 753–765.

van Loon, L. C. (2007) Plant responses to plant-growth promoting rhizobacteria. *European Journal of Plant Pathology*, **119**, 243–254.

van Oosten, V. R., Bodenhausen, N., Reymond, P. *et al.* (2008) Differential effectiveness of microbially induced resistance against herbivorous insects in *Arabidopsis*. *Molecular Plant-Microbe Interactions*, **21**, 919–930.

van Peer, R., Niemann, G. J. and Schippers, B. (1991) Induced resistance and phytoalexin accumulation in biological control of Fusarium wilt of carnation by *Pseudomonas* sp. strain WCS417r. *Phytopathology*, **81**, 728–734.

van Veen, J. A., van Overbeek, L. S. and van Elsas, J. D. (1997) Fate and activity of microorganisms introduced into soil. *Microbiology and Molecular Biology Reviews*, **61**, 121–135.

van Wees, S. C. M., de Swart, E. A. M., van Pelt, J. A., van Loon, L. C. and Pieterse, C. M. J. (2000) Enhancement of induced disease resistance by simultaneous activation of salicylate- and jasmonate-dependent defense pathways in *Arabidopsis thaliana*. *Proceedings of the National Academy of Sciences USA*, **97**, 8711–8716.

van Wees, S. C. M., van der Ent, S. and Pieterse, C. M. J. (2008) Plant immune responses triggered by beneficial microbes. *Current Opinion in Plant Biology*, **11**, 443–448.

Vannette, R. L. and Hunter, M. D. (2009) Mycorrhizal fungi as mediators of defence against insect pests in agricultural systems. *Agricultural and Forest Entomology*, **11**, 351–358.

Verhagen, B. W. M., Glazebrook, J., Zhu, T. *et al.* (2004) The transcriptome of rhizobacteria-induced systemic resistance in *Arabidopsis*. *Molecular Plant-Microbe Interactions*, **17**, 895–908.

Vestergard, M. Henry, F. Rangel-Castro, J. I. *et al.* (2008) Rhizosphere bacterial community composition responds to arbuscular mycorrhiza, but not to reductions in microbial activity induced by foliar cutting. *FEMS Microbiology Ecology*, **64**, 78–89.

Vidal, S. (1996) Changes in suitability of tomato for whiteflies mediated by a non-pathogenic endophytic fungus. *Entomologia Experimentalis et Applicata*, **80**, 272–274.

Vogt, T. (2010) Phenylpropanoid biosynthesis. *Molecular Plant*, **3**, 2–20.

Waller, F., Achatz, B., Baltruschat, H. *et al.* (2005) The endophytic fungus *Piriformospora indica* reprograms barley to salt-stress tolerance, disease resistance, and higher yield. *Proceedings of the National Academy of Sciences USA*, **102**, 13386–13391.

Walters, D. and Heil, M. (2007) Costs and trade-offs associated with induced resistance. *Physiological and Molecular Plant Pathology*, **71**, 3–17.

Wei, G., Kloepper, J. W. and Tuzun, S. (1991) Induction of systemic resistance of cucumber to *Colletotrichum orbiculare* by select strains of plant growth promoting rhizobacteria. *Phytopathology*, **81**, 1508–1512.

Yao, C., Wei, G., Zehnder, G. W., Shelby, R. A. and Kloepper, J. W. (1996a) Induced systemic resistance against bacterial wilt of cucumber by select plant growth-promoting rhizobacteria. *Phytopathology*, **84**, 1082.

Yao, C., Zehnder, G. W., Bauske, E. and Kloepper, J. W. (1996b) Relationship between cucumber beetle (Coleoptera: Chrysomelidae) density and incidence of bacterial wilt. *Journal of Economic Entomology*, **89**, 510–514.

Zehnder, G. W., Kloepper, J. W., Tuzun, S. *et al.* (1997a) Insect feeding on cucumber mediated by rhizobacteria-induced plant resistance. *Entomologia Experimentalis et Applicata*, **83**, 81–85.

Zehnder, G. W., Kloepper J. W., Yao, C. and Wei, G. (1997b) Induction of systemic resistance against cucumber beetles (Coleoptera: Chrysomelidae) by plant growth-promoting rhizobacteria. *Journal of Economic Entomology*, **90**, 391–396.

Zehnder, G. W., Murphy, J. F., Sikora, E. J. and Kloepper, J. W. (2001) Application of rhizobacteria for induced resistance. *European Journal of Plant Pathology*, **107**, 39–50.

CHAPTER TEN

Phytochemicals as mediators of aboveground–belowground interactions in plants

NICOLE M. VAN DAM

Department of Ecogenomics, Institute for Water and Wetland Research,
Radboud University

10.1   Introduction

Since Fraenkel (1959) put secondary plant compounds back on the research agenda for their role as mediators of plant–herbivore interactions, thousands of papers have been published on this topic. In the 150 years before Fraenkel's paper, plant secondary metabolites (PSMs) were mainly analysed for medicinal purposes or for their application as textile dyes or spices (Hartmann, 2007). In those times, secondary metabolites were merely considered to be the waste products of the plant's primary metabolism, despite the publication of an early monograph in the late nineteenth century on the role of plant metabolites as protection against snail damage (Stahl, 1888). In the past 50 years of research, we have finally come to acknowledge the fact that there may be ecological and evolutionary reasons for plants having become true 'chemical factories' (Gershenzon *et al.*, Chapter 4). It is estimated that plants may produce over 200 000 different compounds, the majority of which are classified as secondary metabolites (Pichersky & Gang, 2000). Despite their name, secondary plant compounds have been found to fulfil essential functions in plant defence against herbivores and in attracting beneficial organisms, such as pollinators and symbionts (Harborne, 1993). Even within one single class of biosynthetically related compounds such as the terpenes, there may be an overwhelming variety of 30 000 structures (Hartmann, 2007). Based on the level of sequence homology in the genes coding for the biosynthetic enzymes producing this dazzling diversity of phytochemicals, it has been postulated that the diversity we observe today may have evolved from several rounds of single gene and whole genome duplications in the past (Maere *et al.*, 2005). After these duplication events, the duplicated genes may have subfunctionalised and assumed slightly different biosynthetic functions, thereby increasing the diversity of plant secondary compounds (Mitchell-Olds & Clauss, 2002). Interestingly, it has been found that genes involved in the

The Ecology of Plant Secondary Metabolites: From Genes to Global Processes, eds. Glenn R. Iason, Marcel Dicke and Susan E. Hartley. Published by Cambridge University Press. © British Ecological Society 2012.

synthesis of secondary metabolites and responses to biotic stresses are more often retained than genes involved in hormonal signalling and primary metabolism (Maere *et al.*, 2005). The differences in retention rates between genes in these different classes lend additional support to the hypothesis that the diversity of secondary metabolites represents an important adaptive trait which helps plants to survive in a world full of herbivores (Jones & Firn, 1991; Maere *et al.*, 2005). Small variations in chemical structures may indeed have profound effects on herbivore performance, as was shown in *Barbarea vulgaris* plants. This species was found to be polymorphic for glucosinolate type. The majority of the plants in the natural populations that were surveyed in northwestern Europe outside the Netherlands contained mainly glucobarbarin (S-2-hydroxy-phenylethylglucosinolate). In about half the Dutch populations that were sampled, however, 2–20% of the plants contained mainly gluconasturtiin (2-phenylethylglucosinolate; van Leur *et al.*, 2006). This minor difference of only one hydroxyl group in the glucosinolate structure significantly affected the performance of both shoot and root herbivores. Plants with glucobarbarin were highly resistant to larvae of the generalist lepidopteran herbivore *Mamestra brassicae*, whereas larvae of the cabbage root fly *Delia radicum* performed better on the chemotypes with glucobarbarin (van Leur *et al.*, 2008a, b). Controlled pollination experiments showed that glucosinolate chemotype is a qualitatively heritable trait, implying that selection for either chemotype may occur in natural populations depending on the relative abundance of various insects in the community. As *B. vulgaris* is not fully sequenced yet, and these types of glucosinolates are not produced by the model plant species *Arabidopsis thaliana*, it is as yet unknown whether this polymorphism might have arisen from a gene duplication event or results from a loss-of-function mutation.

Interestingly, the role of secondary metabolites as resistance factors has mainly been studied for aboveground plant parts and their associated insect communities. With the possible exception of agricultural pests such as the larvae of various root flies feeding on cabbage, carrot and onion (Städler & Schoni, 1990; Hopkins *et al.*, 1997; Johnson & Gregory, 2006), and root-lesion nematodes *Pratylenchus* spp.; Potter *et al.*, 1999), little attention has been paid to the role of secondary metabolites as defences against belowground feeding herbivores. In some optimal defence theories it has been formally postulated that roots would have lower defence levels than shoots, because the probability of attack is lower belowground (Zangerl & Rutledge, 1996). Roots contain, however, just as rich a variety of PSMs as shoots do. Depending on the type of secondary metabolite that is analysed and the ontogenetic stage at which root and shoot levels are compared, the level of secondary plant compounds in the root may be even higher than in the shoot (Kaplan *et al.*, 2008; van Dam *et al.*, 2009). As for aboveground herbivores, it has been found that root herbivores use specific chemicals emitted or exuded by the roots as cues to locate their host plant (Johnson & Gregory, 2006). This and the fact that belowground herbivores can do as much, or sometimes even more, damage to wild plants as aboveground feeding herbivores (Maron, 1998; Gerber *et al.*, 2007) indicates that PSMs fulfil similar roles belowground as they do aboveground.

10.2   Induction of phytochemicals

Some 25 years after Fraenkel's paper, the first reports on inducible plant resistance started to appear in the scientific literature. It was reported that the levels of PSMs were not always constant, or constitutively expressed, but might increase after herbivores or pathogens had started to attack the plant. Such changes in secondary compound levels after herbivore feeding were called 'induced responses'. It must be realised though that defence compounds are seldom produced exclusively when plants are under attack: most plants produce low levels of these compounds constitutively as well. The term 'induced resistance' only applies to those cases where the induced response did indeed result in reduced herbivory after the first feeding event (Karban & Myers, 1989). The term 'induced defences' was coined to indicate those cases in which the induced resistance indeed increased plant fitness because of the induction (Karban & Myers, 1989). In most cases, the effect on plant fitness has not been explicitly assessed. It was not clear in all cases that the induced responses increased plant resistance; sometimes induction resulted in an increased susceptibility to the same or other species of herbivores. This phenomenon was called 'induced susceptibility' (Karban & Baldwin, 1997). Induced susceptibility does not always result from lower levels of phytochemicals after induction. It may also result from increased attractiveness to specialist insects that use species-specific induced defence compounds as cues to locate their host plant (Hopkins *et al.*, 2009).

Induced responses – independent of the effect they may have on plant fitness or the herbivore – have been reported in many different plant species, ranging from short-lived small herbs such as *Arabidopsis thaliana* to large perennial plants such as pine, birch and oak (Karban & Myers, 1989; Agrawal, 1999). Equally diverse are the types of phytochemicals – alkaloids, glucosinolates, cardenolides, terpenes, phenolics – that may be induced and the herbivores and pathogens that may induce these changes (aphids, beetles, caterpillars, fungi, bacteria). Induced responses may directly affect herbivores on the plant by increasing the levels of toxins or digestibility reducers in the damaged plant tissue (Karban & Baldwin, 1997). Additionally, plants may produce volatile organic compounds (VOCs) or extrafloral nectar to attract natural enemies that will attack the herbivore. The latter form of induced responses are called 'indirect defences' (Dicke & Sabelis, 1988; Schuman & Baldwin, Chapter 15; Dicke *et al.*, Chapter 16).

In an ecological sense, there may be different reasons why induced responses are preferred over constitutive ones. First, induced responses may reduce the cost of constantly producing secondary metabolites conferring resistance (Agrawal, 1999; Heil & Baldwin, 2002; Strauss *et al.*, 2002). Even though the metabolic costs of producing secondary metabolites may not be substantial (for a calculation see Gershenzon, 1994), it has been shown that under constraining conditions that more resemble the situation in natural plant communities, fitness costs may arise. *Nicotiana attenuata* plants lacking protease inhibitors conferring resistance to herbivores, for example, are superior competitors to conspecifics that do produce these defences (Glawe *et al.*, 2003). Moreover, when these wild tobacco plants are artificially triggered to increase their defence levels when in competition for limited resources, they produce fewer seeds than their uninduced neighbours (van Dam & Baldwin, 1998). This is not the case for every plant species, however, as *Brassica rapa* plants were found to be able to 'grow and defend' (Siemens *et al.*, 2002).

Other 'ecological costs' (Strauss *et al.*, 2002) may arise when high constitutive levels of chemical defences prohibit plants from attracting beneficial organisms, such as pollinators and symbiotic microorganisms. Moreover, specific defence compounds such as alkaloids and glucosinolates may also serve as cues for specialist herbivores to localise their host plant (Hopkins *et al.*, 2009). The chances of being recognised by adapted herbivores may be reduced by producing species-specific defence compounds only when needed.

10.3   Signalling of induced responses

The largest benefit of employing induced responses – compared with constitutive defences – is probably that the plant can tailor its response to the herbivore that is feeding. Herbivore feeding triggers the induction of plant hormones, such as jasmonic acid (JA), ethylene (ET) and salicylic acid (SA). The extent to which these signalling hormones are produced depends on the herbivore species (Diezel *et al.*, 2009). Interactions between the hormones determine which genes are transcribed and – eventually – which defences are produced (Kahl *et al.*, 2000; Diezel *et al.*, 2009). Molecular studies analysing transcription profiles of plants infested with different types of herbivores have indeed shown that plants exhibit different gene expression profiles after aphid rather than caterpillar or thrips feeding (de Vos *et al.*, 2005). Based on the feeding type, but especially on compounds in the oral secretions of the herbivores, plants can 'sense' which of the many herbivores is feeding on their leaves (Vanholme *et al.*, 2004; Diezel *et al.*, 2009). This may enable plants to induce those responses that are the most effective. For example, when a specialist is feeding on the plant it may be more effective for the plant to attract the herbivore's natural enemies by increasing volatile emissions than to increase concentrations of phytochemicals to which the herbivore is adapted (Kahl *et al.*, 2000).

Similarly to constitutive phytochemicals, our knowledge of induced root responses, their regulatory mechanisms and their specificity towards different root herbivores is limited. In a recent review, Rasmann and Agrawal (2008) listed several examples of plants showing locally induced chemical responses, mainly by insect root herbivores. Terpenoids in the roots of cotton plants were found to increase after feeding by click beetle larvae (*Agriotes lineatus*; Bezemer *et al.*, 2004). Feeding by cabbage root fly larvae (*Delia* spp.) increased the levels of glucosinolates, lignin and volatiles in cabbage roots (Hopkins *et al.*, 1995; Neveu *et al.*, 2002). Similarly, it was found that feeding by larvae of the invasive pest *Diabrotica virgifera virgifera* on maize roots induced a specific volatile, (*E*)-β-caryophyllene (Rasmann *et al.*, 2005). Both in cabbage and in maize, as well as in *Thuja* roots infested with vine weevil larvae (*Otiorhynchus sulcatus*), the volatiles that were induced attracted natural enemies of the root herbivores (van Tol *et al.*, 2001; Neveu *et al.* 2002; Rasmann *et al.*, 2005; Hiltpold *et al.*, 2009). These examples show that also belowground, plants may employ indirect defences, which may be used in crop protection programmes (Degenhardt *et al.*, 2009).

Interestingly enough, it is seldom assessed whether root-induced compounds classified as direct defences enhance resistance to other belowground herbivores of the same or other species. In cases where these behavioural responses have been recorded, they usually concern oviposition decisions of the aboveground adult stages of the same herbivores (Städler & Schoni, 1990; Johnson & Gregory, 2006). The paucity of data may be largely because it is difficult to assess feeding behaviour of soil-dwelling herbivores. For insect herbivores, novel non-invasive technologies such as X-ray tomography may help to analyse insect herbivore choice behaviour in soils (Johnson *et al.*, 2007). When combined with chemical and molecular techniques, these visualisation techniques can show whether herbivore feeding behaviour in the soil correlates with the magnitude and direction of induced root responses.

Our knowledge of the molecular mechanisms governing induced root responses mainly comes from the nematode literature. These omnipresent root feeders produce various proteins in their salivary glands which are injected into root cells upon infection, either to establish a feeding cell or to digest the cell content (Vanholme *et al.*, 2004; Davis *et al.*, 2008). Similar to the oral secretions of leaf-feeding herbivores, the salivary compounds of nematodes trigger signalling cascades causing specific gene transcription events. The phytohormones SA, ET and indole acetic acid (IAA) have been found to be involved in nematode-induced responses (Branch *et al.*, 2004; Curtis, 2007;

Fuller *et al.*, 2007), indicating that nematode-induced responses may overlap or interact with responses induced by chewing herbivores.

Little attention has been paid to whether the nematode-induced gene transcription leads to the induction of phytochemicals in the root. Root knot nematodes can significantly increase the local levels of nicotine and several phenolic compounds known to serve as defences in tobacco roots (Kaplan *et al.*, 2008). Similarly, root knots of *Meloidogyne hapla* on *Brassica nigra* contained concentrations of glucosinolates five times as high as in the surrounding roots or in roots of uninfested plants (Figure 10.1).

As the induction by root knot nematodes is quite localised, it is not clear whether the increased glucosinolate levels reduce the infestation rates of other nematodes feeding on the roots. It was found that *Brassica napus* plants with high nasturtiin levels were more resistant to infestation by the root lesion nematode *Pratylenchus penetrans* (Potter *et al.*, 1999). As nematodes are true masters of plant manipulation, it is also possible that the high levels of glucosinolates in the plant cells forming the root knot provide a layer of protection to *M. hapla*, for example against nematicidal microorganisms or carnivorous nematodes dwelling in the soil. This and many

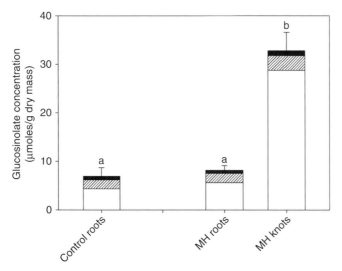

Figure 10.1 Glucosinolate levels in roots infested with the root knot nematode *Meloidogyne hapla* (MH; $n = 5$ pools of 4 plants) and control roots ($n = 6$ pools of 4 plants). The root knots of MH plants were dissected from the other roots, and extracted separately. In control roots, sections of similar root sizes were analysed for comparison. Different letters indicate significant differences in total glucosinolate levels between samples (*t*-test, $p < 0.01$). White = sinigrin; hatched = gluconasturtiin; black = total indole glucosinolates (glucobrassicin, 4-hydroxyglucobrassicin and neoglucobrassicin).

other potential ecological effects of induced root responses still await experimental testing.

10.5   Systemic induced responses cause aboveground–belowground interactions between herbivores

Locally induced responses in most cases also cause systemic responses in undamaged plant parts. Even though the JA peak after herbivore induction may rapidly spread throughout the leaf (Schittko *et al.*, 2000) the signalling hormones themselves are not the exclusive signals triggering systemic responses. Small peptides, such as systemin (Constabel *et al.*, 1998), jasmonic acid conjugates such as JA-Ile (Staswick & Tiryaki, 2004) and even sugars such as trehalose may be involved in systemic signalling (Bae *et al.*, 2005). Systemic induction patterns of plant defences are often found to follow existing source–sink relations and vascular architecture of the plant (Jones *et al.*, 1993; Orians *et al.*, 2000; Gomez & Stuefer, 2006). This may constrain systemic responses to organs that are strong sinks or orthostichous to the location of induction.

   Ecologists were the first to assess that root herbivores affect the performance of shoot-feeding insects, long before chemical ecologists realised that systemically induced responses may play a role. Both greenhouse and field manipulations of root herbivore densities showed that populations of shoot herbivores are affected by the presence of root herbivores (Moran & Whitham, 1990; Masters & Brown, 1992; Masters, 1995). Root herbivores were even found to affect higher trophic levels and pollinators, which in turn affected plant reproductive output (Masters *et al.*, 2001; Poveda *et al.*, 2005; Soler *et al.*, 2005). Most of the early studies on aboveground–belowground interactions reported positive effects of root herbivores on their aboveground counterparts. Based on this, it was postulated that root damage primarily caused drought stress responses in the aboveground parts by reducing water uptake (Masters & Brown, 1992). This led to increased soluble nutrient levels in the shoots, thereby benefiting shoot herbivore performance (Masters *et al.*, 1993; Blossey & Hunt-Joshi, 2003). Shoot herbivory, on the other hand, was hypothesised to reduce the photosynthetic capacity of the plant by removing leaf area. This would reduce the allocation of resources to the roots, which in turn would negatively affect root herbivore performance (Masters & Brown, 1992). It was also recognised that herbivory, or the application of JA as a mimic thereof, may trigger the reallocation of primary compounds to the undamaged organ (Henkes *et al.*, 2008). Reallocation of resources after herbivory may be part of the plant's strategy to tolerate root or shoot herbivory by re-growing the lost tissue from stored reserves (Nuñez-Farfan *et al.*, 2007). Reallocation of primary compounds may also affect the herbivores that are feeding on the undamaged compartment, as nutritional quality of the tissue may change considerably.

This paradigm rapidly shifted as soon as PSMs were included in the debate on which mechanisms may mediate interactions between aboveground and belowground herbivores (van der Putten *et al.*, 2001; van Dam *et al.*, 2003). Indeed, it was found that plants that were naturally or artificially induced at their roots showed increased phytochemical levels in their shoots, and vice versa. These studies also found that root induction had a negative effect on shoot-feeding herbivores, which contrasted with the earlier studies (Bezemer *et al.*, 2004; van Dam *et al.*, 2005; van Dam & Raaijmakers, 2006; Kaplan *et al.*, 2008). Partly, these contrasting effects may be due to differences in the feeding types of the herbivores that were used to assess the aboveground effects. The earlier ecological studies mainly analysed the effects on aboveground aphids, whereas the chemical-ecological studies primarily assessed effects on lepidopteran larvae. Caterpillars consume leaf mass, whereas aphids mainly ingest phloem sap via their stylet. Even though phloem sap may contain similar phytochemicals to the leaf, the concentration of these compounds may be very different (Cole, 1997). Additionally, it may matter which belowground herbivore is used to induce the roots. Phytophagous nematodes, for example, may affect aboveground defence levels differently from chewing herbivores (van Dam *et al.*, 2005; Wurst & van der Putten, 2007). Most likely, nematode damage and larval feeding trigger different signalling cascades, resulting in different systemic responses. Several studies have shown that the effect of systemic induced responses may go both ways between the roots and the shoots (Feng *et al.*, 2010; see also van Dam & Heil, 2011 for a review). However, currently there is much more evidence for the effect of root herbivores affecting shoot herbivores than the reverse, for similar reasons as mentioned above in the section about root defence allocation; it is much more difficult to assess root herbivore performance than that of shoot herbivores.

The negative effect of root herbivores on shoot herbivores, and vice versa, is not always due to the direct induction of phytochemicals in the shoots. Just like rhizosphere bacteria, root herbivores may prime shoots to respond faster upon aboveground damage (van Dam *et al.*, 2005; Kaplan *et al.*, 2008; Pozo *et al.*, 2008; Pineda *et al.*, 2010; Gange *et al.*, Chapter 9). Whether phytochemicals are directly induced or indirectly affect the strength and speed of aboveground induction may depend on the class to which the phytochemical belongs. It was postulated that the variety of responses triggered by root herbivores may provide plants with even more options to optimise their defence responses towards aboveground herbivores (Kaplan *et al.*, 2008).

Phytochemicals were also involved in the effects that root herbivores have on higher trophic levels. These effects may either be caused by changes in the nutritional quality of the aboveground herbivore or by changes in induced volatile profiles that serve as cues for predators and parasitoids. *Pieris brassicae* larvae raised on root-induced *Brassica* plants were indeed less suitable hosts for *Cotesia* parasitoids than those reared on control plants (Soler *et al.*, 2005; Qiu *et al.*, 2009). Interestingly, root herbivory also altered the aboveground emissions of volatiles, based on which parasitic wasps of the shoot herbivore avoided these plants with less suitable caterpillar hosts (Soler *et al.*, 2007). Metabolomic analyses of plants and caterpillars showed that specific, induced phytochemicals may be transferred to the caterpillar, which in turn may affect the performance of the third trophic level (Jansen *et al.*, 2009). Root herbivory was also found to alter the volatile pattern of plants. In both maize and cabbage plants, root herbivores caused specific changes in plant volatile profiles or reduced volatile emissions overall (Rasmann & Turlings, 2007; Soler *et al.*, 2007). These changes in the volatile profiles affected the choice of aboveground parasitoids, usually steering them away from the root-infested plants. This implies that root herbivores may both directly and indirectly affect the efficacy of aboveground indirect defences, which raises the question of how plants can optimise aboveground-induced indirect defences to their own advantage in natural environments where roots are commonly infested by herbivores. Assessing the molecular mechanisms underlying cross-talk between aboveground and belowground responses in plants will help ecologists to better understand how plants integrate multiple signals in their natural environment (van Dam & Heil, 2011).

10.6   Conclusions and future directions

Systemic induction of phytochemicals clearly plays a role in mediating inter-actions between aboveground and belowground organisms. Recent studies have shown that small ubiquitous soil organisms, such as nematodes, may play a larger role in determining aboveground defence levels than previously thought. It is also an open question whether insect root herbivores trigger similar signalling cascades in the roots to those triggered by shoot-feeding insects in the leaves. Another question is whether similar signalling compounds trigger systemic responses when the original damage is done in the roots. Untargeted genomic and metabolomic approaches are invaluable to reveal the differences in signalling events and phytochemical responses between root and shoot induction. In the history of phytochemistry, new discoveries of the ecological roles of phytochemicals were often driven by technological innovations (Hartmann, 2007). Interestingly, genomic approaches may also show whether drought responses are involved in sys-temic responses to root herbivores, as postulated by Masters and Brown in 1992, after all. Recently it was shown that abscisic acid signalling, which mediates drought responses in plants, is also involved in aboveground–

belowground interactions (Pozo *et al.*, 2008; Erb *et al.*, 2009). Above all, more research efforts should be dedicated to experiments that address whether the phytochemicals induced by the many root and shoot organisms that surround plants in their natural environments can result in an optimal defence response. Only these kinds of ecologically inspired experiments will help us to better understand the '*raison d'être*' (Fraenkel, 1959) of systemic induced phytochemicals mediating interactions between shoot and root herbivores.

Acknowledgements
The author thanks Ciska Raaijmakers and Henk Duyts at NIOO-KNAW for help with the nematode identification and chemical analyses of nematode-infested roots. She also thanks Glenn Iason, Sue Hartley and Marcel Dicke for inviting her to present her work at the BES symposium in Sussex, UK, and Marcel Dicke, Glenn Iason and an anonymous reviewer for providing insightful comments on an earlier version of this chapter.

References

Agrawal, A. A. (1999) Induced plant defense: evolution of induction and adaptive phenotypic plasticity. In A. A. Agrawal, S. Tuzun and E. Bent (eds.) *Induced Plant Defenses Against Pathogens and Herbivores*. St Paul, MN: APS Press, 251–268.

Bae, H. H., Herman, E., Bailey, B., Bae, H. J. and Sicher, R. (2005) Exogenous trehalose alters *Arabidopsis* transcripts involved in cell wall modification, abiotic stress, nitrogen metabolism, and plant defense. *Physiologia Plantarum*, **125**, 114–126.

Bezemer, T. M., Wagenaar, R., van Dam, N. M., van der Putten, W. H. and Wäckers, F. L. (2004) Above- and below-ground terpenoid aldehyde induction in cotton, *Gossypium herbaceum*, following root and leaf injury. *Journal of Chemical Ecology*, **30**, 53–67.

Blossey, B. and Hunt-Joshi, T. R. (2003) Belowground herbivory by insects: influence on plants and aboveground herbivores. *Annual Review of Entomology*, **48**, 521–547.

Branch, C., Hwang, C. F., Navarre, D. A. and Williamson, V. M. (2004) Salicylic acid is part of the Mi-1-mediated defense response to root-knot nematode in tomato. *Molecular Plant–Microbe Interactions*, **17**, 351–356.

Cole, R. A. (1997) The relative importance of glucosinolates and amino acids to the development of two aphid pests *Brevicoryne brassicae* and *Myzus persicae* on wild and cultivated brassica species. *Entomologia Experimentalis et Applicata*, **85**, 121–133.

Constabel, C. P., Yip, L. and Ryan, C. A. (1998) Prosystemin from potato, black nightshade, and bell pepper: primary structure and biological activity of predicted system in polypeptides. *Plant Molecular Biology*, **36**, 55–62.

Curtis, R. H. C. (2007) Do phytohormones influence nematode invasion and feeding site establishment? *Nematology*, **9**, 155–160.

Davis, E. L., Hussey, R. S., Mitchum, M. G. and Baum, T. J. (2008) Parasitism proteins in nematode–plant interactions. *Current Opinion in Plant Biology*, **11**, 360–366.

Degenhardt, J., Hiltpold, I., Kollner, T. G. *et al.* (2009) Restoring a maize root signal that attracts insect-killing nematodes to control a major pest. *Proceedings of the National Academy of Sciences USA*, **106**, 13213–13218.

de Vos, M., Van Oosten, V. R., Van Poecke, R. M. P. *et al.* (2005) Signal signature and transcriptome changes of *Arabidopsis* during pathogen and insect attack. *Molecular Plant–Microbe Interactions*, **18**, 923–937.

Dicke, M. and Sabelis, M. W. (1988) How plants obtain predatory mites as bodyguards. *Netherlands Journal of Zoology*, **38**, 148–165.

Diezel, C., von Dahl, C. C., Gaquerel, E. and Baldwin, I. T. (2009) Different lepidopteran elicitors account for cross-talk in herbivory-induced phytohormone signaling. *Plant Physiology*, **150**, 1576–1586.

Erb, M., Flors, V., Karlen, D. *et al.* (2009) Signal signature of aboveground-induced resistance upon belowground herbivory in maize. *Plant Journal*, **59**, 292–302.

Feng, Y. J., Wang, J. W. and Jin, Q. (2010) Asian corn borer (*Ostrinia furnacalis*) damage induced systemic response in chemical defence in Bt corn (*Zea mays* L.). *Allelopathy Journal*, **26**, 101–112.

Fraenkel, G. S. (1959) The *raison d'être* of secondary plant substances. *Science*, **129**, 1466–1470.

Fuller, V. L., Lilley, C. J., Atkinson, H. J. and Urwin, P. E. (2007) Differential gene expression in *Arabidopsis* following infection by plant-parasitic nematodes *Meloidogyne incognita* and *Heterodera schachtii*. *Molecular Plant Pathology*, **8**, 595–609.

Gerber, E., Hinz, H. L. and Blossey, B. (2007) Interaction of specialist root and shoot herbivores of *Alliaria petiolata* and their impact on plant performance and reproduction. *Ecological Entomology*, **32**, 357–365.

Gershenzon, J. (1994) Metabolic costs of terpenoid accumulation in higher plants. *Journal of Chemical Ecology*, **20**, 1281–1328.

Glawe, G. A., Zavala, J. A., Kessler, A., van Dam, N. M. and Baldwin, I. T. (2003) Ecological costs and benefits correlated with trypsin protease inhibitor production in *Nicotiana attenuata*. *Ecology*, **84**, 79–90.

Gomez, S. and Stuefer, J. F. (2006) Members only: induced systemic resistance to herbivory in a clonal plant network. *Oecologia*, **147**, 461–468.

Harborne, J. B. (1993) *Introduction to Ecological Biochemistry*, 4th edn. London: Elsevier Academic Press.

Hartmann, T. (2007). From waste products to ecochemicals: fifty years research of plant secondary metabolism. *Phytochemistry*, **68**, 2831–2846.

Heil, M. and Baldwin, I. T. (2002) Fitness costs of induced resistance: emerging experimental support for a slippery concept. *Trends in Plant Science*, **7**, 61–67.

Henkes, G. J., Thorpe, M. R., Minchin, P. E. H., Schurr, U. and Rös, U. S. R. (2008) Jasmonic acid treatment to part of the root system is consistent with simulated leaf herbivory, diverting recently assimilated carbon towards untreated roots within an hour. *Plant, Cell and Environment*, **31**, 1229–1236.

Hiltpold, I., Toepfer, S., Kuhlmann, U., Turlings, T. C. J. (2010) How maize root volatiles affect the efficacy of entomopathogenic nematodes in controlling the western corn rootworm. *Chemoecology*, **20**, 155–162.

Hopkins, R. J., Birch, A. N. E., Griffiths, D. W., Morrison, I. M. and McKinlay, R. G. (1995) Changes in the dry-matter, sugar, plant fiber and lignin contents of swede, rape and kale roots in response to turnip root fly (*Delia floralis*) larval damage. *Journal of the Science of Food and Agriculture*, **69**, 321–328.

Hopkins, R. J., Birch, A. N. E., Griffiths, D. W. *et al.* (1997) Leaf surface compounds and oviposition preference of turnip root fly *Delia floralis*: the role of glucosinolate and nonglucosinolate compounds. *Journal of Chemical Ecology*, **23**, 629–643.

Hopkins, R. J., van Dam, N. M. and van Loon, J. J. A. (2009) Role of glucosinolates in insect–plant relationships and multitrophic interactions. *Annual Review of Entomology*, **54**, 57–83.

Jansen, J. J., Allwood, J. W., Marsden-Edwards, E. *et al.* (2009) Metabolomic analysis of the interaction between plants and herbivores. *Metabolomics*, **5**, 150–161.

Johnson, S. N. and Gregory, P. J. (2006) Chemically-mediated host-plant location and selection by root-feeding insects. *Physiological Entomology*, **31**, 1–13.

Johnson, S. N., Crawford, J. W., Gregory, P. J. *et al.* (2007) Non-invasive techniques for investigating and modelling root-feeding insects in managed and natural systems. *Agricultural and Forest Entomology*, **9**, 39–46.

Jones, C. G. and Firn, R. D. (1991) On the evolution of plant secondary chemical diversity. *Philosophical Transactions of the Royal Society of London B: Biological Sciences*, **333**, 273–280.

Jones, C. G., Hopper, R. F., Coleman, J. S. and Krischik, V. A. (1993) Control of systemically induced herbivore resistance by plant vascular architecture. *Oecologia*, **93**, 452–456.

Kahl, J., Siemens, D. H., Aerts, R. J. *et al.* (2000) Herbivore-induced ethylene suppresses a direct defense but not a putative indirect defense against an adapted herbivore. *Planta*, **210**, 336–342.

Kaplan, I., Halitschke, R., Kessler, A., Sardanelli, S. and Denno, R. F. (2008) Constitutive and induced defenses to herbivory in above- and belowground plant tissues. *Ecology*, **89**, 392–406.

Karban, R. and Baldwin, I. T. (1997) *Induced Responses to Herbivory*. Chicago, IL: University of Chicago Press.

Karban, R. and Myers, J. H. (1989) Induced plant responses to herbivory. *Annual Review of Ecology and Systematics*, **20**, 331–348.

Maere, S., de Bodt, S., Raes, J. *et al.* (2005) Modeling gene and genome duplications in eukaryotes. *Proceedings of the National Academy of Sciences USA*, **102**, 5454–5459.

Maron, J. L. (1998) Insect herbivory above- and belowground: individual and joint effects on plant fitness. *Ecology*, **79**, 1281–1293.

Masters, G. J. (1995) The impact of root herbivory on aphid performance – field and laboratory evidence. *Acta Oecologica International Journal of Ecology*, **16**, 135–142.

Masters, G. J. and Brown, V. K. (1992) Plant-mediated interactions between two spatially separated insects. *Functional Ecology*, **6**, 175–179.

Masters, G. J., Brown, V. K. and Gange, A. C. (1993) Plant mediated interactions between aboveground and belowground insect herbivores. *Oikos*, **66**, 148–151.

Masters, G. J., Jones, T. H. and Rogers, M. (2001) Host-plant mediated effects of root herbivory on insect seed predators and their parasitoids. *Oecologia*, **127**, 246–250.

Mitchell-Olds, T. and Clauss, M. J. (2002) Plant evolutionary genomics. *Current Opinion in Plant Biology*, **5**, 74–79.

Moran, N. A. and Whitham, T. G. (1990) Interspecific competition between root-feeding and leaf-galling aphids mediated by host-plant resistance. *Ecology*, **71**, 1050–1058.

Neveu, N., Grandgirard, J., Nenon, J. P. and Cortesero, A. M. (2002) Systemic release of herbivore-induced plant volatiles by turnips infested by concealed root-feeding larvae *Delia radicum* L. *Journal of Chemical Ecology*, **28**, 1717–1732.

Nuñez-Farfan, J., Fornoni, J. and Valverde, P. L. (2007) The evolution of resistance and tolerance to herbivores. *Annual Review of Ecology, Evolution and Systematics*, **38**, 541–566.

Orians, C. M., Pomerleau, J. and Ricco, R. (2000) Vascular architecture generates fine scale variation in systemic induction of proteinase inhibitors in tomato. *Journal of Chemical Ecology*, **26**, 471–485.

Pichersky, E. and Gang, D. R. (2000) Genetics and biochemistry of secondary metabolites in plants: an evolutionary perspective. *Trends in Plant Science*, **5**, 439–445.

Pineda A., Zheng, S.-J., van Loon, J. J. A., Pieterse, C. M. J. and Dicke, M. (2010) Helping plants to deal with insects: the role of beneficial soil-borne microbes. *Trends in Plant Science*, **15**, 507–514.

Potter, M. J., Vanstone, V. A., Davies, K. A., Kirkegaard, J. A. and Rathjen, A. J. (1999) Reduced susceptibility of *Brassica napus* to *Pratylenchus neglectus* in plants with elevated root levels of 2-phenylethyl glucosinolate. *Journal of Nematology*, **31**, 291–298.

Poveda, K., Steffan-Dewenter, I., Scheu, S. and Tscharntke, T. (2005) Effects of decomposers and herbivores on plant performance and aboveground plant-insect interactions. *Oikos*, **108**, 503–510.

Pozo, M. J., van der Ent, S., van Loon, L. C. and Pieterse, C. M. J. (2008) Transcription factor MYC2 is involved in priming for enhanced defense during rhizobacteria-induced systemic resistance in *Arabidopsis thaliana*. *New Phytologist*, **180**, 511–523.

Qiu, B. L., Harvey, J. A., Raaijmakers, C. E., Vet, L. E. M. and van Dam, N. M. (2009) Nonlinear effects of plant root and shoot jasmonic acid application on the performance of *Pieris brassicae* and its parasitoid *Cotesia glomerata*. *Functional Ecology*, **23**, 496–505.

Rasmann, S. and Agrawal, A. A. (2008) In defense of roots: a research agenda for studying plant resistance to belowground herbivory. *Plant Physiology*, **146**, 875–880.

Rasmann, S. and Turling, T. C. J. (2008) Simultaneous feeding by aboveground and belowground herbivores attenuates plant-mediated attraction of their respective natural enemies. *Ecology Letters*, **10**, 926–936.

Rasmann, S., Kollner, T. G. Degenhardt, J. *et al.* (2005) Recruitment of entomopathogenic nematodes by insect-damaged maize roots. *Nature*, **434**, 732–737.

Schittko, U., Preston, C. A. and Baldwin, I. T. (2000) Eating the evidence? *Manduca sexta* larvae can not disrupt specific jasmonate induction in *Nicotiana attenuata* by rapid consumption. *Planta*, **210**, 343–346.

Siemens, D. H., Garner, S. H., Mitchell-Olds, T. and Callaway, R. M. (2002) Cost of defense in the context of plant competition: *Brassica rapa* may grow and defend. *Ecology*, **83**, 505–517.

Soler, R., Bezemer, T. M., van der Putten, W. H., Vet, L. E. M. and Harvey, J. A. (2005) Root herbivore effects on above-ground herbivore, parasitoid and hyperparasitoid performance via changes in plant quality. *Journal of Animal Ecology*, **74**, 1121–1130.

Soler, R., Harvey, J. A., Kamp, A. F. D. *et al.* (2007) Root herbivores influence the behaviour of an aboveground parasitoid through changes in plant-volatile signals. *Oikos*, **116**, 367–376.

Städler, E. and Schoni, R. (1990) Oviposition behavior of the cabbage root fly, *Delia radicum* (L), influenced by host plant-extracts. *Journal of Insect Behavior*, **3**, 195–209.

Stahl, E. (1888) Eine biologische Studie über die Schutzmittel der Pflanzen gegen Schneckenfrass. *Jenaer Zeitung fuer Naturwissenschaften*, **15**, 557–684.

Staswick, P. E. and Tiryaki, I. (2004) The oxylipin signal jasmonic acid is activated by an enzyme that conjugates it to isoleucine in Arabidopsis. *Plant Cell*, **16**, 2117–2127.

Strauss, S. Y., Rudgers, J. A., Lau, J. A. and Irwin, R. A. (2002) Direct and ecological costs of resistance to herbivory. *Trends in Ecology and Evolution*, **17**, 278–285.

van Dam, N. M. and Baldwin, I. T. (1998) Cost of jasmonate-induced responses in plants competing for limited resources. *Ecology Letters*, **1**, 30–33.

van Dam, N. M. and Heil, M. (2011) Multitrophic interactions below-ground and aboveground: en route to the next level. *Journal of Ecology*, **99**, 77–88.

van Dam, N. M. and Raaijmakers, C. E. (2006) Local and systemic induced responses to cabbage root fly larvae (*Delia radicum*) in *Brassica nigra* and *B. oleracea*. *Chemoecology*, **16**, 17–24.

van Dam, N. M., Harvey, J. A., Wackers, F. L. *et al.* (2003) Interactions between aboveground and belowground induced responses against phytophages. *Basic and Applied Ecology*, **4**, 63–77.

van Dam, N. M., Raaijmakers, C. E. and van der Putten, W. H. (2005) Root herbivory reduces growth and survival of the shoot feeding specialist *Pieris rapae* on *Brassica nigra*. *Entomologia Experimentalis et Applicata*, **115**, 161–170.

van Dam, N. M., Tytgat, T. O. G. and
    Kirkegaard, J. A. (2009) Root and shoot
    glucosinolates: a comparison of their
    diversity, function and interactions in
    natural and managed ecosystems.
    Phytochemistry Reviews, **8**,
    171–186.
van der Putten, W. H., Vet, L. E. M., Harvey, J. A.
    and Wäckers, F. L. (2001) Linking above- and
    belowground multitrophic interactions of
    plants, herbivores, pathogens and their
    antagonists. *Trends in Ecology and Evolution*,
    16, 547–554.
van Leur, H., Raaijmakers, C. E. and van
    Dam, N. M. (2006) A heritable glucosinolate
    polymorphism within natural populations
    of *Barbarea vulgaris*. *Phytochemistry*, **67**,
    1214–1223.
van Leur, H., Raaijmakers, C. E. and van
    Dam, N. M. (2008a) Reciprocal interactions
    between the cabbage root fly (*Delia radicum*)
    and two glucosinolate phenotypes of
    Barbarea vulgaris. *Entomologia Experimentalis et
    Applicata*, **128**, 312–322.

van Leur, H., Vet, L. E. M., van der Putten, W. H.
    and van Dam, N. M. (2008b) *Barbarea vulgaris*
    glucosinolate phenotypes differentially
    affect performance and preference of
    two different species of lepidopteran
    herbivores. *Journal of Chemical Ecology*, **34**,
    121–131.
van Tol, R. W. H. M., van der Sommen, A. T. C.,
    Boff, M. I. C. *et al.* (2001) Plants protect their
    roots by alerting the enemies of grubs.
    Ecology Letters, **4**, 292–294.
Vanholme, B., De Meutter, J., Tytgat, T. *et al.* (2004)
    Secretions of plant-parasitic nematodes: a
    molecular update. *Gene*, **332**, 13–27.
Wurst, S. and van der Putten, W. H. (2007) Root
    herbivore identity matters in plant-
    mediated interactions between root and
    shoot herbivores. *Basic and Applied Ecology*, **8**,
    491–499.
Zangerl, A. R. and Rutledge, C. E. (1996) The
    probability of attack and patterns of
    constitutive and induced defense: a test of
    optimal defense theory. *American Naturalist*,
    147, 599–608.

CHAPTER ELEVEN

Plant secondary metabolites and the interactions between plants and other organisms: the potential of a metabolomic approach

SUE E. HARTLEY

Department of Biology, University of York

RENÉ ESCHEN

CABI

JULIA M. HORWOOD, LYNNE ROBINSON and ELIZABETH M. HILL

School of Life Sciences, University of Sussex

11.1   Introduction

The central role of plant secondary metabolites (PSMs) in mediating the ecological interactions between plants and other organisms is both well known and well studied, particularly in the case of the defensive responses of plants against attack by herbivores or pathogens (Dangl & Jones, 2001; Kessler & Baldwin, 2002). Furthermore, because plants face many simultaneous threats (Maleck & Dietrich, 1999; Paul *et al.*, 2000), the chemical changes within plants in response to one attacking organism can influence the behaviour and performance of many others (Thaler *et al.*, 2002; Biere *et al.*, 2004). Thus, chemically mediated plant-based interactions have significant consequences for individual species, ecological communities and ecosystem function, so gaining an in-depth understanding of the chemical basis of these interactions is vital for ecologists (van der Putten, 2003; Dicke, 2006; Schuman & Baldwin, Chapter 15; Dicke *et al.*, Chapter 16).

Recent advances in ecological genomics have demonstrated the complexity of plant responses to biotic challenges at the molecular level: hundreds of genes are now known to be up- or down-regulated in response to herbivore or pathogen attack (Zheng & Dicke, 2008). This has been of great benefit in understanding the molecular basis of plant defence, but microarray data alone cannot unravel the complexity and variability in plant responses, many of which are specific to particular types of natural enemy, and/or vary according to environmental conditions (Kant & Baldwin, 2007). Genomic analysis needs to be supported by manipulative experiments which assess all the metabolic responses

The Ecology of Plant Secondary Metabolites: From Genes to Global Processes, eds. Glenn R. Iason, Marcel Dicke and Susan E. Hartley. Published by Cambridge University Press. © British Ecological Society 2012.

of plants to environmental challenges as well as the molecular ones – so-called metabolomic approaches. Metabolomics is the systematic analysis of the set of metabolites synthesised by an organism at a particular 'snapshot' in time and can be described as providing the link between genotypes and phenotypes (Fiehn, 2002; Macel *et al.*, 2010). Assuming that this set of metabolites reflects the interactions the plant is having with its abiotic and biotic environment, chemical ecologists can use this technique to study the mechanisms underpinning these interactions (Bundy *et al.*, 2009), including those between plants and other organisms such as herbivores and pathogens (Allwood *et al.*, 2008).

A metabolomic approach can provide many advantages over other methods in gaining a better understanding of the chemically mediated interactions between plants and other organisms (Prince & Pohnert, 2010; Verpoorte *et al.*, 2010). Simultaneous identification and quantification of all the metabolites within a plant, whilst ambitious, produces a complete and unbiased 'profile' of the physiological behaviour of the plant in response to attack, whether by herbivores, pathogens or both. This is a significant advance over the narrow analysis of small groups of specific compounds because, for example, it can detect a suite of simultaneous metabolic changes and so allow key mechanisms to be elucidated, even when they are complex and involve multiple chemical signals. It also allows for the relatively rapid assessment of similarities or contrasts between plant responses to different kinds of damage (Jansen *et al.*, 2009). A metabolomic approach is not only an effective method to identify the induced changes underpinning plant defensive responses to attack; it is also a powerful way to understand patterns in plant investment in constitutive defences, i.e. secondary metabolites present prior to attack. For example, using a metabolomic analysis can assess how plants allocate PSMs to different plant tissues, whether different kinds of PSM are restricted to different parts of the plant, and to what extent there are trade-offs in allocation to one type of PSM versus another. A metabolomic approach provides a comprehensive and holistic picture of plant defences, both constitutive and induced, and thus has the potential to give us a better understanding of the defensive strategy of plants.

Some of the benefits of a metabolomic approach derive at least in part from the intimate connections between primary and secondary metabolism initially revealed by genomic analysis – a substantial proportion of the genes responding to herbivore or pathogen attack are involved in primary metabolic processes such as photosynthesis (Schwachtje & Baldwin, 2008). Profiling all the metabolites produced by a plant simultaneously is the most effective way to assess the impacts of changes in both primary and secondary metabolism in response to attack by herbivores (Berenbaum, 1995). Using a metabolomic approach will greatly increase our understanding of the integrative role of PSMs in plant ecology and build on the insights that genomic methods have already provided.

Given the potential of this approach, it is unsurprising that the use of metabolomic technologies to identify marker biochemicals in plants has been rapidly increasing in the past decade. They have been used to identify agricultural traits beneficial for crop breeding, in the evaluation of genetically modified crops, and in food processing for identifying traits associated with taste preferences or degradation of products (Saito & Matsuda, 2010). Metabolomic analyses have also been used to investigate tolerance strategies of plants to environmental stress such as drought or salinity (Shao *et al.*, 2009) and to study metabolite pathways involved in plant defence responses to pathogenic bacteria and fungi (e.g. Sana *et al.*, 2010). Despite a wide range of applications across the plant sciences, the use of non-targeted metabolomic analyses in advancing the chemical ecology of plant herbivore or other invertebrate interactions has been limited as yet (Macel *et al.*, 2010). In addition, many of the more ecological studies have been targeted on only parts of the metabolome, either focusing on particular groups of specific metabolites or on the most abundant ones. It is timely to assess how state-of-the-art metabolomic techniques can advance our understanding of the interactions between plants and other organisms, and how chemical ecologists can exploit these techniques as effectively as other areas of plant science.

In this chapter we illustrate the benefits of using a metabolomic approach, using preliminary results from two ongoing studies where such an approach has provided new insights. We have selected two different study systems which investigate two contrasting aspects of plant defence against invading organisms: allocation of constitutive defences to tissue types with potentially different 'value' to the plant, and the induction of potential defensive compounds in plants in response to fungal infection. In the first study, metabolomic analysis is being used to assess how plants allocate PSMs between their vegetative and reproductive tissues, using the biennial plant ragwort (*Senecio jacobaea*) as a model system. Here, we aim to assess how the levels and types of constitutive PSMs differ between leaves and flowers and hence gain an insight into the factors governing defence allocation. In the second study, we investigate the chemical basis of interactions between plants and endophytic fungi, which live asymptomatically within plant tissues. Our model system for this study is the creeping thistle, *Cirsium arvense*, and we aim to characterise plant metabolic responses to these fungi, which are usually regarded as mutualists but may nevertheless cause defence induction.

11.2   Study objectives
11.2.1   Allocation of PSM to plant parts in *Senecio jacobaea*
How plants allocate limiting resources to optimise both their defence against attack and their reproductive fitness and competitive ability against their neighbours has long been a fundamental question in plant ecology (e.g. Herms & Mattson, 1992; Zangerl & Rutledge, 1996; Jones & Hartley, 1999). A key theory aiming to explain the distribution of a plant's defences within its

tissues is the optimal defence theory (ODT) first proposed by McKey (1974). To be 'optimally defended' a plant must distribute defences according to the perceived importance of the tissue, with important tissues being allocated the greatest defensive protection. Costs of defence prohibit the plant from allocating maximum defences throughout all tissues. It has been suggested that new leaves are more valuable than old leaves owing to a larger photo-synthetic potential, and flowers are more valuable than leaves owing to their importance for sexual reproduction (McKey, 1974). A recent meta-analysis of the ODT literature concluded that it has proved a useful framework for the prediction of the distribution of chemical defences (McCall & Fordyce, 2010).

Studies of the defensive strategy of *S. jacobaea* have traditionally focused almost exclusively on a single group of PSMs, collectively termed pyrrolizi-dine alkaloids (PAs). PAs have drawn most attention as they have been shown to be highly toxic to grazing livestock (Johnson, 1978) as well as to organisms of many different taxa, including: vertebrates (Cheeke & Pierson-Goeger, 1983), nematodes (Thoden *et al.*, 2009), fungi (Hol & van Veen, 2002) and insects (Macel *et al.*, 2005). In addition to acting as a feeding deterrent to a wide range of organisms, PAs are of interest to ecologists and evolutionary biologists because they provide specialist herbivores, such as the cinnabar moth (*Tyria jacobaeae*), with a mechanism for host recognition (Macel & Vrieling, 2003) and a source of chemicals that are metabolised for use in their own defence (Edgar *et al.*, 1980).

How PSMs are allocated between different plant parts in ragwort has been addressed in a few studies, all limited to studies of PAs alone. One glasshouse study found evidence supporting the optimal defence theory for the distribu-tion of PAs in ragwort rosettes. De Boer (1999) showed that in young rosette plants PA content decreased with leaf age, an effect not related to leaf size. To our knowledge no other study has tested ODT using ragwort as a model system, although some make brief mention of an increased level of defence in flowers when compared to leaves (Johnson *et al.*, 1985) and others note a difference in PA composition between these tissues (Witte *et al.*, 1992). Evidence for the costs of defence in ragwort is also limited to studies examin-ing PAs, and results are inconclusive. Vrieling and van Wijk did not detect any evidence of the costs of PA production in the greenhouse (1994a) or field (1994b), whereas in a recent field study Macel and Klinkhamer (2010) did.

Although PAs are the focus of the majority of studies of defence in ragwort, this plant contains a number of other significant PSMs. Kirk *et al.* (2005) report the presence of chlorogenic acid, kaempferol and jacaranone in *S. jacobaea*, all of which have been shown to have an effect on insect performance (Lajide *et al.*, 1996; Onyilagha *et al.*, 2004; Leiss *et al.*, 2009b). Leiss *et al.* (2009a) also identify kaempferol as an important chemical associated with thrip resist-ance in F_2 hybrids of *S. jacobaea* and *S. aquaticus*. Both of these studies use a holistic metabolomic approach, examining PAs in parallel to other defensive compounds. A metabolomic approach can assess the complete biochemical profiles of reproductive and non-reproductive plant parts simultaneously and thus gain an understanding of how plants allocate both primary and secondary metabolites to these tissues and whether their 'value' to the plant is reflected in how well defended they are (de Boer, 1999). Hence by using a metabolomic approach we can test the ODT in ragwort more effectively by examining whether it applies to all groups of PSMs in this plant or only to specific groups such as PAs. We can also look for trade-offs in the way different types of constitutive defences are allocated across tissue types by assessing the extent to which reproductive plant parts are better defended than vegetative plant parts and, if so, quantify the changes in relative concentrations of key metabolites between tissues.

11.2.2   Plant–endophyte interactions in *Cirsium arvense*

Endophytic fungi are present in most if not all plant species (Stone *et al.*, 2001). These fungi live asymptomatically within living plant tissues for some or all of their life cycle, and appear to occur ubiquitously in herbaceous plants and grasses, although far more is known about their behaviour in grass systems than herbaceous ones (Saikkonen *et al.*, 2006; Hartley & Gange, 2009; Rodriguez *et al.*, 2009). In grass systems, endophytes are associated with the production of toxic metabolites (Siegel & Bush, 1996), thought to defend the host against attack by herbivores (e.g. Wilkinson *et al.*, 2000); hence, these fungi are considered mutualistic. There is some evidence that they also protect their host against drought and other stresses (Bacon, 1993; Rodriguez *et al.*, 2009; but see Cheplick, 2004). However, the extent to which this relationship is truly mutualistic is debatable (Faeth, 2002; Sieber, 2007; Hartley & Gange, 2009), and indeed some authors consider endophytes parasitic (Faeth & Sullivan, 2003). There is less evidence for the role, mutualistic or otherwise, of endophytes in forbs, although some studies have demonstrated antagonistic effects of forb endophytes on herbivorous insects (Vidal, 1996; also see Gange *et al.*, Chapter 9).

We are using metabolomics to characterise the entire suite of biochemical changes within the plant in response to endophyte infection and so to inform this debate by identifying whether the plant's metabolic response to infection is similar to that following pathogen attack. Hence we may gain an insight into whether the plant 'views' these organisms as mutualists or pathogens at the biochemical level. *Cirsium arvense* (creeping thistle) was used as a model system in order to understand the biochemical basis of the interactions between a foliar endophyte and a herbaceous species, far less well-studied than those between endophytes and grasses. *Chaetomium cochlioides* was chosen as a model endophyte as it is a predominant species detected in *Cirsium arvense* in natural populations (Gange *et al.*, 2007).

11.3   Methods
11.3.1   Metabolomic methods

Plant tissues were snap frozen in liquid nitrogen to inhibit endogenous enzyme activity, and stored at –80 °C. An aliquot of ground frozen material was extracted in a pre-cooled ball mill using a multi-solvent composition of isopropanol : acetonitrile : water (3:3:2) developed by Sana *et al.* (2010). This solvent mixture extracts metabolites with a wide range of polarities more efficiently than the individual solvents alone or water–methanol mixtures. Deuterated analogues of internal standard compounds were added to the tissue extracts prior to solvent extraction to monitor sample recovery and enable chromatogram alignment. Tissues were extracted with a solvent to plant fresh weight ratio of 40:1, and an aliquot of the extract was dried down under vacuum and resuspended in 70% methanol in water, and filtered (0.2 μM) prior to mass spectrometry (MS) analyses. A summary of our approach is provided in Figure 11.1.

Plant extracts were profiled using ultra-high-performance liquid chromatography (UPLC) linked to time-of-flight (TOF) MS. Samples were analysed by electrospray ionisation (ESI) using a baffled lockspray source which allowed online calibration of mass ions. Mass spectral peaks were processed as described in Figure 11.1, and exported to SIMCA-P software for multivariate analyses (Umetrics UK Ltd, Windsor, UK). The data were analysed by principal component analysis (PCA) to detect and remove any outliers and then analysed by supervised partial least squares-discriminant analysis (PLS-DA) to find class separating differences in the datasets. Cross validation (CV) was used to determine the significant components of the models and thus minimise over-fitting. During each CV round, one sample from each class was excluded (in random order), and the classification of the excluded samples predicted from modelling the remaining dataset. The performance of the models was then described by the explained variation (R^2X for PCA and R^2Y for PLS-DA) and the predictive ability (Q^2X or Q^2Y) parameters of the models.

In order to identify the biochemicals which differed between the classes (i.e. between the different plant parts or between reference and endophyte-treated plants), the data were further analysed using orthogonal PLS (OPLS) to select data due only to class separation. The most discriminative variables (in terms of the RT $\times$ m/z signals) between the plant treatments were then visualised using an 'S'-plot, which is a scatter plot of covariance and correlation loading profiles resulting from the OPLS model (see Wiklund *et al.*, 2008 for further explanation). Significant metabolite markers were identified according to their *p*-value. However, the number of false positives (type 1 errors) in the statistical analysis of the data were likely to be high because of the abundance of variables being tested (Broadhurst & Kell, 2006), so an

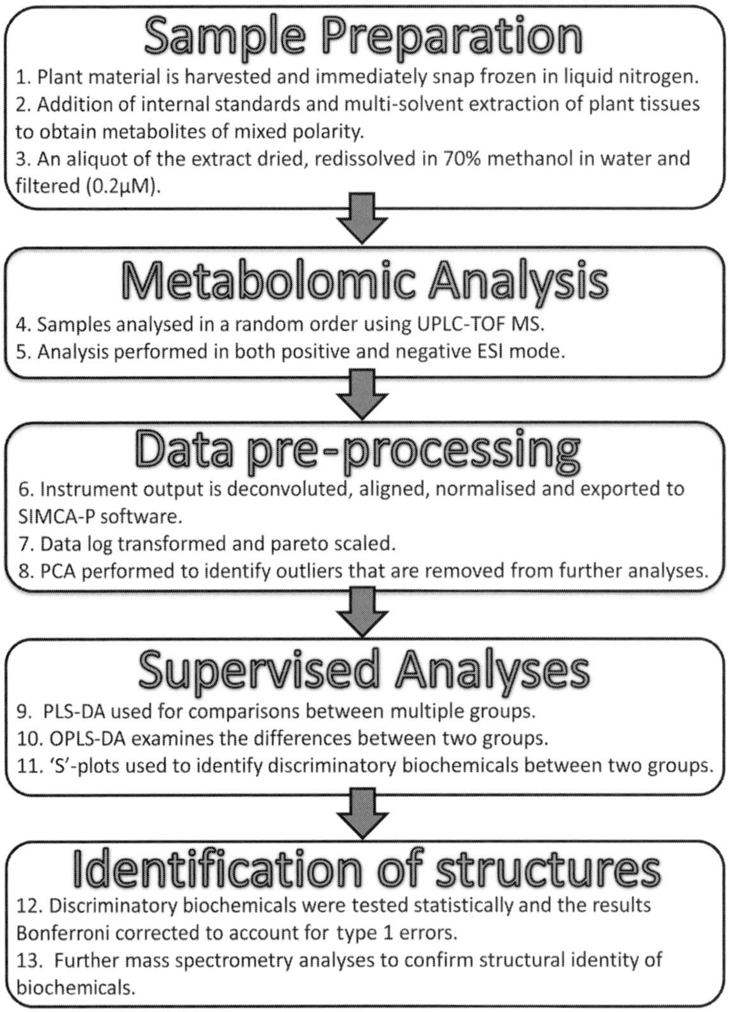

Sample Preparation

1. Plant material is harvested and immediately snap frozen in liquid nitrogen.
2. Addition of internal standards and multi-solvent extraction of plant tissues to obtain metabolites of mixed polarity.
3. An aliquot of the extract dried, redissolved in 70% methanol in water and filtered (0.2μM).

Metabolomic Analysis

4. Samples analysed in a random order using UPLC-TOF MS.
5. Analysis performed in both positive and negative ESI mode.

Data pre-processing

6. Instrument output is deconvoluted, aligned, normalised and exported to SIMCA-P software.
7. Data log transformed and pareto scaled.
8. PCA performed to identify outliers that are removed from further analyses.

Supervised Analyses

9. PLS-DA used for comparisons between multiple groups.
10. OPLS-DA examines the differences between two groups.
11. 'S'-plots used to identify discriminatory biochemicals between two groups.

Identification of structures

12. Discriminatory biochemicals were tested statistically and the results Bonferroni corrected to account for type 1 errors.
13. Further mass spectrometry analyses to confirm structural identity of biochemicals.

Figure 11.1 Schematic diagram outlining the metabolomic approach adopted throughout the presented studies. Abbreviations: UPLC-TOF MS, ultra-high-performance liquid chromatography/time-of-flight mass spectrometry; ESI, electrospray ionisation; PCA, principal component analysis; PLS-DA, partial least squares-discriminant analysis; OPLS, orthogonal PLS.

estimate of the false discovery rate (FDR) was made using the Bonferroni correction. Those biochemicals that were deemed to be significantly different between the classes after this correction have much lower p-values than the usual criterion of $p < 0.05$. The structural identities of the discriminatory biochemical markers were determined from the accurate mass data, and from further UPLC quadrupole-TOFMS (Q-TOFMS) analyses to obtain mass fragment information on each structure.

11.3.2   Plant sampling methods

Senecio jacobaea

To investigate the distribution of secondary metabolites within wild ragwort, plants were sampled at a site on the University of Sussex campus (Grid ref: TQ 348 098). Samples were collected in October 2008 from 15 flowering plants. From each plant, five flowers and the three most basal leaves (= old leaves) and the three uppermost leaves (= new leaves) from the flowering stem were collected. PAs (in both parent and N-oxide forms) and other metabolites were profiled; preliminary identifications were obtained by accurate mass and fragmentation data, and comparing their retention time with standards of senecionine and seneciphylline (Carl Roth, Germany). Differences in metabolite signal size between plant parts were examined using Kruskal–Wallis tests.

Cirsium arvense

Sixty *C. arvense* plants were grown from surface-sterilised seeds until 8 weeks old, when half the plants were sprayed with *C. cochlioides* spores harvested from a stock culture kept on agar. The spores were applied at a concentration of 35 000 per ml in a water/oil (1:1 v/v) suspension using a domestic vaporiser. During application, the soil was covered to prevent contamination. The untreated control plants were sprayed with an equal amount of water–oil mixture without spores. The plants were kept in a controlled-environment room (12 h day/night, 21 °C, 70% r.h.) and watered as needed. After a period of 2 weeks following infection, three old leaves originally sprayed with *C. cochlioides* spores and three newly emerged leaves from a part of the plant which had grown since spraying with spores were harvested from each infected plant. From control plants, which had been sprayed with the control oil–water mixture 2 weeks earlier, we also sampled three newly emerged unsprayed leaves in addition to three older sprayed leaves. All sampled leaves were extracted for metabolomic analyses. In addition, a disc from each of a further three old and three new leaves was removed from all the plants and surface sterilised using a method modified from Schulz *et al.* (1993). The discs were placed on agar to culture and identify emerging *C. cochlioides*, as well as any other endophytic fungi, and to determine rates of infection.

11.4   Results

11.4.1   Metabolite profiling in *Senecio jacobaea*

There were very clear separations between the different plant tissues within flowering *S. jacobaea* plants (Figure 11.2). The first axis of the figure shows the large separation between flowers and leaves, whilst the second axis illustrates the clear contrast between old and new leaves. Thus, flowers and leaves have a

Table 11.1 *Summary of marker biochemicals, not including PAs, that differ significantly between leaves and flowers of* Senecio jacobaea *when analysed in both positive and negative ESI mode. Results displayed remain significant after applying a Bonferroni correction. The total number of signals detected in the metabolome was between 10 000 and 12 000, representing 3000–4000 metabolites.*

		High in		
		Flowers	New leaves	Old leaves
Low in	Flowers	.	94	89
	New leaves	52	.	0
	Old leaves	53	7	.

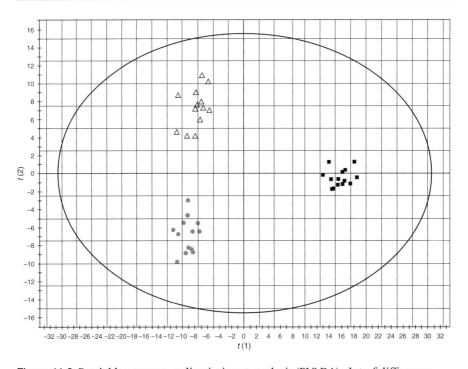

Figure 11.2 Partial least squares-discriminant analysis (PLS-DA) plot of difference between biochemical profiles of leaves and flowers of *Senecio jacobaea*. Squares = flowers; circles = new leaves; diamonds = old leaves. Profiles were obtained using UPLC-TOF MS in +ESI mode. The axes plot the first two latent variables (t_1 and t_2) that account for 49.8% and 45.4% of the modelled Y variation, respectively. The third latent variable, t_3, explained only 0.4% of the variation and is not shown.

different metabolic profile, as do leaves of different ages – together the factors of tissue type and leaf age explain 95.2% of the measured variation in tissue biochemicals. To examine which particular biochemical markers are driving these clear separations, further analyses were performed (see Figure 11.1 and

Figure 11.3). A total of 146 biochemical markers differed between flowers and new leaves, 94 of which were higher in new leaves (Table 11.1). Between flowers and old leaves, 142 markers differed, 89 of which were higher in old leaves. Between the old and new leaves, seven markers differed significantly, and all of these are found in higher amounts in new leaves.

One reason for the contrasting metabolomic profiles between flowers and leaves will of course be the inherent biochemical differences in these plant parts, for example, the presence of chlorophyll in the leaves but not flowers. The contribution of individual markers which reflect differences in leaf versus flower function, rather than differences in their defence allocation, to the patterns in Figure 11.2 is hard to estimate, particularly when many remain unidentified. One way forward is to determine the proportion of the markers that differ between flowers and either new or old leaves that are the same in both leaf types, suggesting they are closely associated with leaf (as opposed to flower) function. We found 183 markers elevated in either new (94) or old (89) leaves compared with flowers, and of these, 39 were markers present in both leaf types, perhaps indicating they are chemicals associated with leaf as opposed to flower function. Some of the other markers may reflect differences in secondary metabolite allocation, including PAs, between plant tissues.

A total of 18 PAs were found in the plant tissues of flowering ragwort plants. Using an analytical method targeted for quantification of PAs, Joosten *et al.* (2010) identified 24 structures in *S. jacobaea*, but some of these were present at very low concentrations and so would not have been detected in our analyses where less concentrated plant extracts were being profiled. A number of PAs differed significantly in concentration between plant parts (Kruskal–Wallis: $df = 2$; $p = 2.7 \times 10^{-4}$ to 1.6×10^{-6}): six of the PAs were highest in flowers, one was highest in new leaves, four were higher in both leaf types compared with flowers, and seven showed no difference in distribution between plant parts. The PA senecionine was found to be at 20- and 5-fold higher levels in flowers than in new or old leaves, respectively. The predominant PA present in leaves was identified as jaconine and its concentrations were 62- and 16-fold higher in new and old leaves, respectively than in the flowers. These results indicate significant differences between the distribution of specific PA structures in the plant parts, and reveal that plants allocate significant concentrations of defensive PAs to leaves.

Differences in classes of other PSMs and lipid metabolites were observed between plant parts of *S. jacobaea*. These included flavonoid mono- and diglucosides, flavonoid acetyl and malonyl monoglucosides, and free flavonoid structures, which were all at concentrations 100- to 800-fold higher in the flowers than in the new or old leaves. Mono- and trihydroxy metabolites of octadecadienoic acid and octadecenoic acids were detected at 5- to 30-fold

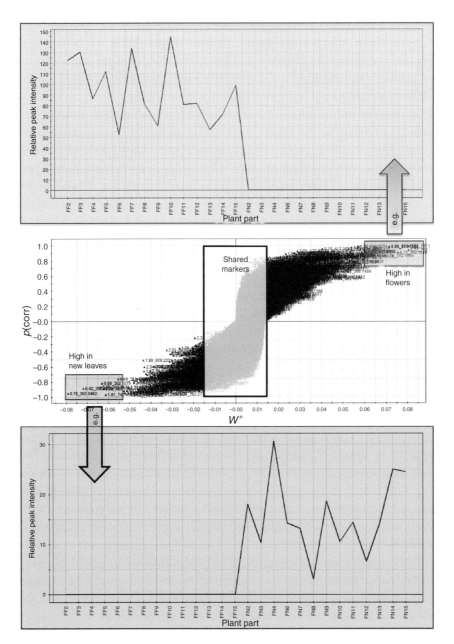

Figure 11.3 Example of 'S' plot of flowers and new leaves metabolomic datasets from *Senecio jacobaea*. Biochemical markers which differ significantly in leaves of different ages and between leaves and flowers are found at the extremities of the 'S', whilst those plotted around the centre do not differ between the plant parts examined. By selecting points on the plot, patterns for the individual biomarkers can be examined.

higher concentrations in flowers compared with leaves. In addition, concentrations of C18 esterified lysophosphatidic acids were 6- to 10-fold higher in flowers than leaves. Conversely chlorogenic acid and oxododecanoic acids were twice as high in leaves compared with the flowering parts.

11.4.2   Biochemical markers of endophyte infection
in *Cirsium arvense*

The infection frequency of *C. cochlioides* in the old (sprayed) and new (unsprayed) leaves of endophyte-treated plants 2 weeks after infection was 62.7 ± 5.6 and $91.1 \pm 3.6\%$, respectively. The presence of *C. cochlioides* in new leaves indicated movement of the endophyte from infected parts of the plant to sites of new growth, i.e. endophyte growth in this case was systemic, something which has not been demonstrated before in this system. In contrast, in the uninfected control plants no *C. cochlioides* was detected in old leaves and there was only a $2.4 \pm 1.7\%$ frequency of infection present in the new leaves. Although some of the leaf discs contained other endophytic fungi, the infection frequency was low and neither the number of fungal species nor their relative occurrence was related to the *C. cochlioides* treatment.

There were clear differences between the biochemical profiles of uninfected plants compared with those infected with *C. cochlioides* (Figure 11.4). The greatest difference in leaf biochemistry in response to endophyte infection appeared to reside in the younger unsprayed leaves harvested 2 weeks after infection; there was less separation between older infected and control leaves. The first latent variable was related to leaf age, namely biochemical differences between the old and new control leaves. The second and third latent variables were related to endophyte treatment, in the old and new leaves, respectively (Figure 11.4).

The main metabolite markers that were responsible for the discrimination between the control and endophyte-treated plants were a series of oxidised lipids. Key lipids that associated with endophyte infection included two trihydroxy fatty acids, as well as a class of 10 oxidised galactolipid structures. These types of galactolipids are termed 'arabidopsides' as they were previously only identified in *Arabidopsis thaliana*. The concentrations of some arabidopsides increased significantly in response to endophyte infection, whilst others decreased. For example, following infection, esters of dihydrojasmonic acid and 12-oxophytodienoic acid increased and decreased respectively (see Figure 11.5a and b). Both these compounds are common metabolites of the jasmonate pathway which controls plant responses to wounding. Interestingly, changes in response to infection were most pronounced in the new leaves of endophyte-treated plants: the ester of 12-oxophytodienoic acid, a precursor in the JA signalling pathway, was lower in new than in old leaves of endophyte-treated plants and lower than in the leaves of control plants,

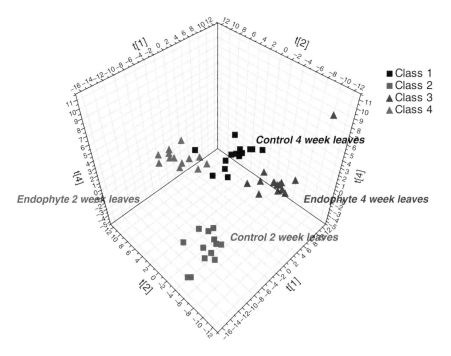

Figure 11.4 Partial least squares-discriminant analyses (PLS-DA) of changes in the metabolome of *Cirsium avenae* leaves infected with *Chaetomium cochlioides*. Biochemicals in the leaf extracts were profiled using UPLC-TOFMS (-ESI mode) and the datasets analysed by PLS-DA. Each symbol represents data from one plant. The axes represent the first three latent variables, t_1, t_2, t_3, of the PLS-DA model, and these account for 26.8%, 28.7% and 28.4 % of the Y variation. See plate section for colour version.

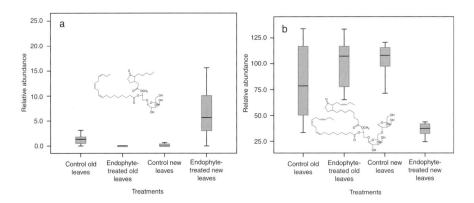

Figure 11.5 Changes in the relative abundance of arabidopsides in *Cirsium arvense* leaves in response to infection with *Chaetomium cochlioides*. Arabidopside structures were the most prominent discriminatory metabolites between the biochemical profiles of uninfected and endophyte-infected leaves. Their chemical structures were determined from Q-TOFMS fragmentation analyses of the parent ion.
(a) Dihydrojasmonic/linolenic acid ester of monogalactosyl diacylglycerol,
(b) 12-oxophytodienoic acid/linolenic acid ester of digalactosyl diacylglycerol.

while the ester of the signalling compound dihydrojasmonic acid showed the opposite response (Figure 11.5a and b). None of the arabidopside structures were detected in extracts of *C. cochlioides* that had been cultured on agar plates, confirming that the arabidopsides were of plant, not fungal, origin.

11.5 Discussion

11.5.1   Optimal defence in *Senecio jacobaea*

A metabolomic approach highlighted the very clear separation between the biochemical profile of leaves and flowers and of new and old leaves in ragwort. These contrasting metabolomes undoubtedly reflect differences in primary metabolism, at least in part: some metabolic pathways are obviously associated with leaf function (e.g. chlorophyll biosynthesis) and some with flower function (e.g. flower colour). However, even within a tissue type there are clear differences, with many biochemical markers significantly elevated or reduced in new leaves compared with old leaves.

Previous work (Leiss *et al.*, 2009a) used an NMR-based metabolomic analysis to compare *Senecio* F2 hybrids derived from tissue culture which were resistant or susceptible to thrips, but this is the first study to focus on wild plants of *S. jacobaea* and to compare the metabolomes of vegetative and reproductive tissues. Leiss *et al.* (2009a) found that developmental stage was the key factor in discrimination between leaves: metabolic differences between leaves of different ages were far greater than those between leaves from the resistant and susceptible hybrids. We also found significant differences in the profiles of young and old leaves, though differences between leaves and flowers were even more pronounced. Previous studies have found differences in the secondary compounds in plants of different ages (Barton & Koricheva, 2010; Koricheva & Barton, Chapter 3). In herbaceous plants like *S. jacobaea*, a meta-analysis (Barton & Koricheva, 2010) indicated that levels of defence increased during the seedling stage, and either increased or decreased within the juvenile stage according to type of compound, with terpenoids increasing, but alkaloids showing a tendency to decrease. Using metabolomic analysis to examine changes in many groups of secondary compounds simultaneously as plant age might be one way to gain a better understanding of these complex changes in allocation over plant development.

Within plants of a particular age, the ODT predicts that chemical defences will be allocated to plant tissues according to their value to the plant, generally assumed to be the cost of having that tissue removed (McCall & Fordyce, 2010). Hence younger leaves are predicted to have higher levels of defence than older leaves and flowers more than leaves. In a recent meta-analysis (McCall & Fordyce, 2010), younger leaves were indeed found to be better defended than older ones, but there was no evidence that flowers were better defended than leaves and concentrations of many types of secondary metabolite have been shown to be higher in young leaves than in mature ones (Koricheva & Barton, Chapter 3). We found that at least six PAs in *S. jacobaea* were significantly higher in flowers than in leaves, and one of these PAs was identified as senecionine. Higher levels of PAs in the flowers of *Senecio* spp. compared with leaves have been suggested by previous work (Johnson *et al.*, 1985), although the specific PA compounds responsible for these differences were not identified. Our results suggest that at least some PAs in *S. jacobaea* follow the predictions of the ODT by being higher in the more valuable reproductive tissue. However, further work is needed to confirm the identity of all the individual PAs in our study, particularly since our results suggest that different PAs appear to respond differently to the predictions of the ODT. We found that four PAs, one of which was identified as jaconine, were higher in both leaf types compared with flowers, and seven showed no difference in distribution between plant parts, patterns contradictory to the ODT. Some of the PA-like compounds we detected may be intermediate or breakdown products of PA pathways, so further work is needed to assess the biological significance of the different PA mixtures in leaves versus flowers, such as whether they result in different toxicities to herbivorous insects, or are sequestered to different extents within the herbivore.

PAs account for some of the chemical variation between the leaf and flower tissue types we found in our metabolomic analysis, but far from all of it – they were only 11 of the many other markers we found which differed between the plant parts (Table 11.1). Prominent markers that were associated with flowers included a variety of free or conjugated flavonoid structures. Flavonoids are known for their inhibitory effects on herbivores, and in hybrids of *Senecio* spp. have been associated with resistance to thrips (Leiss, 2009a). Conversely chlorogenic acid, which was detected at a two-fold higher concentration in leaves than flowers of *S. jacobaea* has also been shown to reduce thrip herbivory in plants (Leiss, 2009b). This data suggests that different plant tissues may adopt different chemical defence strategies to deter herbivory. This study highlights the value of adopting a metabolomic approach as opposed to targeted analyses of particular compounds because we can assess what other chemicals, including secondary metabolites, differ significantly between plant tissue types.

11.5.2 Plant responses to endophyte infection
in *Cirsium arvense*

Biochemical profiling using a non-targeted metabolomic approach has provided new information on the responses of plants to endophyte infection, which might have been missed if a more targeted analysis had been undertaken focusing only on one or two groups of metabolites (Jordon-Thaden & Louda, 2003). An array of oxidised galactolipids, known as arabidopsides, were identified, whose levels either decreased or increased as a result of endophyte infection, but the differences were mainly seen in the new leaves (Figure 11.5). Previously, arabidopsides had only been detected in *A. thaliana*, but here the use of a metabolomic approach has detected them in another, completely unrelated, species. Furthermore, our results suggest a role for these compounds in a defence response by the plant to endophyte infection. In *Arabidopsis*, levels of oxidised galactolipids have been shown to increase in the hypersensitivity response of the plant to avirulence proteins isolated from the bacterial pathogen *Pseudomonas syringae*, and *in vitro* they inhibited growth of bacterial pathogens (Andersson *et al.*, 2006). Some of these compounds, specifically dihydrojasmonic acid and 12-oxophytodienoic acid that were esterified as galactolipids, are common metabolites of the jasmonate pathway and so are closely associated with the defensive reactions of plants to wounding or insect attack.

The detection of these novel defensive compounds produced in *Cirsium* in response to endophyte infection suggests that plants may view endophytes as antagonists rather than mutualists. At least initially, *Cirsium* produces the same sort of chemicals in response to endophytes as it does in response to wounding, insect herbivory or pathogen invasion. However, their production appeared to be transitory, and further studies are needed to determine whether the endophyte overcomes this plant defence response, or whether the infection itself is time-limited because of these defensive changes in plant chemistry. Furthermore, although these compounds are not produced by the fungus when it is growing on an agar plate, it is difficult to rule out the possibility that it produces additional compounds when within a plant; it would be difficult to separate such compounds from plant-derived ones.

The distinct pattern of increases and decreases in the metabolites of the jasmonate pathway in new leaves of endophyte-treated plants reported here suggests that infection with *C. cochlioides* induced systemic defence signalling in *C. arvense*. Our observation that the invading organism reduces the levels of precursors of the jasmonate signalling pathway but increases the levels of signalling compounds is in line with the model of systemic plant signalling in response to wounding proposed by Wu and Baldwin (2010). Such a systemic response to wounding involves the production of jasmonic acid in undamaged leaves (Wu & Baldwin, 2010), allowing the plant to respond more effectively to a subsequent attack by natural enemies. Jasmonic acid plays a pivotal role in such induced systemic resistance in plants, protecting them against necrotrophic pathogens and controlling plant defence against herbivores (Reymond & Farmer, 1998; Wu & Baldwin, 2010). It appears likely that endophyte infection caused the plants to develop induced systemic resistance, something which might explain the larger difference between the metabolomes of new leaves of endophyte-treated and control plants than the difference between old leaves in these treatments (Figure 11.4).

That the responses to endophyte infection spread rapidly to new leaves in the plant suggests that the endophyte could increase resistance to other invading organisms. Resistance against pathogens inferred by foliar endophytes has previously been shown by Reuveni and Reuveni (2000), who found that application of the foliar non-pathogenic fungi *Alternaria cucumaria* or *Cladosporium fulvum* to the leaves of seedlings of cucumber plants caused induced systemic resistance and reduced infection by the powdery mildew *Sphaerotheca fuliginea* by 71% and 80%, respectively. Hence, although our results suggest that the plants perceived the endophytes as antagonists at the time of the infection, the impact of the infection may in the long term be more positive. Our results support recent ideas that to regard endophytes as either mutalistic or non-mutalistic may be too simplistic, and these definitions will change over time and under different environmental conditions (Bronstein, 1994; Davitt *et al.*, 2010), as noted in other symbiotic fungal interactions, such as mycorrhizae (Gange & Ayres, 1999).

11.6   Conclusions and future directions

The results from these two contrasting systems highlight the potential benefits of examining the role of PSMs in mediating interactions between organisms by using a metabolomic approach. New chemicals, such as the arabidopsides we found in *Cirsium* in response to infection by an endophyte, would not have been detected by conventional, more targeted chemical analysis. These compounds were previously only thought to occur in *Arabidopsis* and there was no reason to predict their presence in *Cirsium*, so they would certainly have been missed. Similarly the conventional focus on PAs when considering allocation to defence investment in *Senecio* neglects the other types of secondary metabolites that contribute to defence in this plant. The identity and function of the large numbers of metabolites which differ in abundance between the different ages of leaves and different tissue types await further more detailed investigation, but once complete, this will provide the basis for a much more detailed assessment of the factors controlling defence allocation in *Senecio* and hence a better test of the optimal defence theory. In fact, it could be argued that characterising the links between primary and secondary metabolism, something which can only be achieved using the metabolomic approach, is the best way to gain a more mechanistic understanding of defence allocation in plants (Berenbaum, 1995; Jones & Hartley, 1999; Gershenzon *et al.*, Chapter 4).

The metabolomic approach allows the simultaneous detection of a wide range of secondary metabolites, so it may provide a better understanding of the biochemical basis of resistance than other methods (Leiss *et al.*, 2009a). The methodological constraints of extracting and identifying specific groups of compounds separately may mean that some groups are missed altogether, a suggested explanation for the surprising absence of sesquiterpene lactones in studies of the herbivore-feeding deterrents in *Cirsium* (Jordon-Thaden & Louda 2003). A metabolomic technique can detect relatively subtle changes in plant responses over time, which can give valuable insights into the mechanistic basis of plant resistance. For example, in *Cirsium* we detected an endophyte-induced decrease in precursor compounds and an increase in the signalling compounds in the JA signal pathway. This highlights the signalling systems underpinning systemic resistances as well as showing the time scales over which such responses operate. We found responses to endophyte infection after 2 weeks, similar to the 3 weeks found by Reuveni and Reuveni (2000), but we still lack a good understanding of how long systemic responses to endophytes can last, or indeed how quickly they can begin. There is a need for a detailed study of the 'time-line' of plant metabolomic responses to invasion by attacking organisms in a way which characterises the changes in both primary and secondary metabolism as the interaction between the plant and the attacking organism develops. The complexity of ontogenetic variation in plant defence is just beginning to be realised (Barton & Koricheva, 2010; Tucker & Avila-Sakar, 2010; Koricheva & Barton, Chapter 3). Metabolomics will be a valuable tool in increasing our understanding of this complexity, as accurate 'snapshots' of plant defence allocation patterns across a range of PSMs simultaneously can be obtained.

The strength of metabolomics as a tool to unravel the complexities of plant responses to invading organisms lies partly in its ability to look across the whole metabolome and partly in the way it allows for larger replication and sample throughput compared with other '-omic' techniques, such as transcriptomics and proteomics. This maximises its potential for the discovery of new pathways and responses, such as the discovery of these arabidopside compounds in *Cirsium*, which might not have happened without the ability to assess the whole metabolome and highlight key compounds of interest relatively rapidly. Moving beyond characterising the responses of two or three 'lab rat' plant species to herbivores and pathogens will require methods such as metabolomics, which allow for rapid comparison of inter-specific differences in plant biochemistry and hence the characterising of unpredicted effects in other species, such as the discovery of arabidopsides in *Cirsium* highlighted here.

The identification of novel compounds produced in response to experimental treatments requires their structure to be determined, but this can be extremely time-consuming if there are many differences in response to treatments. Similarly, the metabolomic approach of characterising so many biochemical changes simultaneously generates very large datasets, which can be challenging to deal with and interpret (Prince & Pohnert, 2010). The development of improved multivariate techniques and better databases for marker identification are helping to mitigate this problem. The sensitivity of metabolomic techniques and the broad spectrum of plant biochemical responses they capture make it particularly important that protocols for sample collection, processing and analysis are followed precisely. This increases the likelihood of metabolomics detecting any small differences in key individual markers which may be driving the plants' responses to the experimental treatments imposed.

The benefits of the metabolomic approach are well established, and the potential applications of this technique and the insights it can generate are increasing rapidly. Collaborative interactions will extend these benefits to researchers who lack the expertise or resources to apply it themselves, and will allow the application of this technique to a broad range of ecological systems and questions. Assessing changes in the plant metabolome represents the analysis of the final end-product of gene expression; upstream analyses of transcripts or proteins are needed to understand fully how the plant responds to herbivores or pathogens, although metabolomic information can identify changes in phenotype that inform the search for transcriptome and genome differences associated with plant responses to environmental factors. Metabolomics is one of the most powerful of the new '-omics' technologies which are revolutionising our understanding of the role of PSMs in mediating plant responses to invading organisms.

Acknowledgements

We are grateful to Alan Gange and Adam Vanbergen for their input to the *Cirsium* and *Senecio* studies, respectively, and to Ali Abdul Sada for mass spectroscopy support. We also thank Brian Sutton for the identification of the endophytic fungi and the Natural Environment Research Council, the Centre for Ecology and Hydrology, and the University of Sussex for financial support.

References

Allwood, J. W., Ellis, D. I. and Goodacre, R. (2008) Biomarker metabolites capturing the metabolite variance present in a rice plant developmental period. *Physiologia Plantarum*, **132**, 117–135.

Andersson, M. X., Hamberg, M., Kourtchenko, O. et al. (2006) Oxylipin profiling of the hypersensitive response in *Arabidopsis thaliana* – formation of a novel oxo-phytodienoic acid-containing galactolipid, arabidopside E. *Journal of Biological Chemistry*, **281**, 31528–31537.

Bacon, C. W. (1993) Abiotic stress tolerances (moisture, nutrients) and photosynthesis in endophyte-infected tall fescue. *Agriculture Ecosystem Environment*, **44**, 123–141.

Barton, K. E. and Koricheva, J. (2010) The ontogeny of plant defense and herbivory: characterizing general patterns using meta-analysis. *American Naturalist*, **175**, 481–493.

Berenbaum, M. R. (1995) Turnabout is fair play: secondary roles for primary compounds. *Journal of Chemical Ecology*, **21**, 925–940.

Biere, A., Marak, H. B. and van Damme, J. M. M. (2004) Plant chemical defense against herbivores and pathogens: generalized defense or trade-offs? *Oecologia*, **140**, 430–441.

Broadhurst, D. I. and Kell, D. B. (2006) Statistical strategies for avoiding false discoveries in metabolomics and related experiments. *Metabolomics*, **2**, 171–196.

Bronstein, L. J. (1994) Conditional outcomes in mutualistic interactions. *Trends in Ecology and Evolution*, **9**, 214–217.

Bundy, J. G., Davey, M. P. and Viant, M. R. (2009) Environmental metabolomics: a critical review and future perspectives. *Metabolomics*, **5**, 3–21.

Cheeke, P. R. and Pierson-Goeger, M. L. (1983) Toxicity of *Senecio-jacobaea* and pyrrolizidine alkaloids in various laboratory-animals and avian species. *Toxicology Letters*, **18**, 343–349.

Cheplick, G. P. (2004) Recovery from drought stress in *Lolium perenne* (Poaceae): are fungal endophytes detrimental? *American Journal of Botany*, **91**, 1960–1968.

Dangl, J. L. and Jones, J. D. G. (2001) Plant pathogens and integrated defence responses to infection. *Nature*, **411**, 826–833.

Davitt, A. J., Stansberry, M. and Rudgers, J. A. (2010) Do the costs and benefits of fungal endophyte symbiosis vary with light availability? *New Phytologist*, **188**, 824–834.

de Boer, N. J. (1999) Pyrrolizidine alkaloid distribution in *Senecio jacobaea* rosettes minimises losses to generalist feeding. *Entomologia Experimentalis et Applicata*, **91**, 169–173.

Dicke, M. (2006) Chemical ecology from genes to communities – integrating 'omics' with community ecology. In M. Dicke and W. Takken (eds.) *Chemical Ecology: From Gene to Ecosystem*, Dordrecht: Springer, 175–189.

Edgar, J. A., Culvenor, C. C. J., Cockrum, P. A., Smith, L. W. and Rothschild, M. (1980) Callimorphine – identification and synthesis of the cinnabar moth metabolite. *Tetrahedron Letters*, **21**, 1383–1384.

Faeth, S. H. (2002) Are endophytic fungi defensive plant mutualists? *Oikos*, **98**, 25–36.

Faeth, S. H. and Sullivan, T. J. (2003) Mutualistic asexual endophytes in a native grass are usually parasitic. *American Naturalist*, **161**, 310–325.

Fiehn, O. (2002) Metabolomics – the link between genotypes and phenotypes. *Plant Molecular Biology*, **48**, 155–171.

Gange, A. C. and Ayres, R. L. (1999) On the relation between arbuscular mycorrhizal colonization and plant benefit? *Oikos*, **87**, 615–621.

Gange, A. C., Dey, S., Currie, A. F. and Sutton, B. C. (2007) Site- and species-specific differences in endophyte occurrence in two herbaceous plants. *Journal of Ecology*, **95**, 614–622.

Hartley, S. E. and Gange, A. C. (2009) Impacts of plant symbiotic fungi on insect herbivores: mutualism in a multitrophic context. *Annual Review of Entomology*, **54**, 323–342.

Herms, D. A. and Mattson, W. J. (1992) The dilemma of plants – to grow or defend. *Quarterly Review of Biology*, **67**, 283–335.

Hol, W. H. G. and van Veen, J. A. (2002) Pyrrolizidine alkaloids from *Senecio jacobaea* affect fungal growth. *Journal of Chemical Ecology*, **28**, 1763–1772.

Jansen, J. J., Alwood, J. W., Marsden-Edwards, E. *et al.* (2009) Metabolomic analysis of the interaction between plants and herbivores. *Metabolomics*, **5**, 150–161.

Johnson, A. E. (1978) Tolerance of cattle to tansy ragwort (*Senecio-jacobaea*). *American Journal of Veterinary Research*, **39**, 1542–1544.

Johnson, A. E., Molyneux, R. J. and Merrill, G. B. (1985) Chemistry of toxic range plants – variation in pyrrolizidine alkaloid content of *Senecio*, *Amsinckia*, and *Crotalaria* species. *Journal of Agricultural and Food Chemistry*, **33**, 50–55.

Jones, C. G. and Hartley, S. E. (1999) A protein competition model of phenolic allocation. *Oikos*, **86**, 27–44.

Joosten, L., Mulder, P. P. J., Vrieling, K., van Veen, J. A. and Klinkhamer, P. G. L. (2010) The analysis of pyrrolizidine alkaloids in *Jacobaea vulgaris*; a comparison of extraction and detection methods. *Phytochemical Analysis*, **21**, 197–204.

Jordon-Thaden, I. E. and Louda, S. M. (2003) Chemistry of *Cirsium* and *Carduus*: a role in ecological risk assessment for biological control of weeds? *Biochemical Systematics and Ecology*, **31**, 1353–1396.

Kant, M. R. and Baldwin, I. T. (2007) The ecogenetics and ecogenomics of plant–herbivore interactions: rapid progress on a slippery road. *Current Opinion in Genetics and Development*, **17**, 519–524.

Kessler, A. and Baldwin, I. T. (2002) Plant responses to insect herbivory: the emerging molecular analysis. *Annual Review of Plant Biology*, **53**, 299–328.

Kirk, H., Choi, Y. H., Kim, H. K., Verpoorte, R. and van der Meijden, E. (2005) Comparing metabolomes: the chemical consequences of hybridization in plants. *New Phytologist*, **167**, 613–622.

Lajide, L., Escoubas, P. and Mizutani, J. (1996) Cyclohexadienones-insect growth inhibitors from the foliar surface and tissue extracts of *Senecio cannabifolius*. *Experientia*, **52**, 259–263.

Leiss, K. A., Choi, Y. H., Abdel-Farid, I. B., Verpoorte, R. and Klinkhamer, P. G. L. (2009a) NMR metabolomics of thrips (*Frankliniella occidentalis*) resistance in *Senecio* hybrids. *Journal of Chemical Ecology*, **35**, 1–11.

Leiss, K. A., Maltese, F., Choi, Y. H., Verpoorte, R. and Klinkhamer, P. G. L. (2009b) Identification of chlorogenic acid as a resistance factor for thrips in chrysanthemum. *Plant Physiology*, **150**, 1567–1575.

Macel, M. and Klinkhamer, P. G. L. (2010) Chemotype of *Senecio jacobaea* affects damage by pathogens and insect herbivores in the field. *Evolutionary Ecology*, **24**, 237–250.

Macel, M. and Vrieling, K. (2003) Pyrrolizidine alkaloids as oviposition stimulants for the cinnabar moth, *Tyria jacobaeae*. *Journal of Chemical Ecology*, **29**, 1435–1446.

Macel, M., Bruinsma, M., Dijkstra, S. M. *et al.* (2005) Differences in effects of pyrrolizidine alkaloids on five generalist insect herbivore species. *Journal of Chemical Ecology*, **31**, 1493–1508.

Macel, M., van Dam, N. M. and Keurentjes, J. J. B. (2010) Metabolomics: the chemistry between ecology and genetics. *Molecular Ecology Resources*, **10**, 583–593.

Maleck, K. and Dietrich, R. A. (1999) Defense on multiple fronts: how do plants cope with diverse enemies? *Trends in Plant Science*, **4**, 215–219.

McCall, A. C. and Fordyce, J. A. (2010) Can optimal defence theory be used to predict the distribution of plant chemical defences? *Journal of Ecology*, **98**, 985–992.

McKey, D. (1974) Adaptive patterns in alkaloid physiology. *American Naturalist*, **108**, 305–320.

Onyilagha, J. C., Lazorko, J., Gruber, M. Y., Soroka, J. J. and Erlandson, M. A. (2004) Effect of flavonoids on feeding preference and development of the crucifer pest *Mamestra configurata* Walker. *Journal of Chemical Ecology*, **30**, 109–124.

Paul, N. D., Hatcher, P. E. and Taylor, J. E. (2000) Coping with multiple enemies: an integration of molecular and ecological perspectives. *Trends in Plant Science*, **5**, 220–225.

Prince, E. K. and Pohnert, G. (2010) Searching for signals in the noise: metabolomics in chemical ecology. *Analytical and Bioanalytical Chemistry*, **396**, 193–197.

Reymond, P. and Farmer, E. E. (1998) Jasmonate and salicylate as global signals for defence gene expression. *Current Opinion in Plant Biology*, **1**, 404–411.

Reuveni, M. and Reuveni, R. (2000) Prior inoculation with non-pathogenic fungi induces systemic resistance to powdery mildew on cucumber plants. *European Journal of Plant Pathology*, **106**, 633–638.

Rodriguez, R. J., White, J. F. Jr, Arnold, A. E. and Redman, R. S. (2009) Fungal endophytes: diversity and functional roles. *New Phytologist*, **182**, 314–330.

Saikkonen, K., Lehtonen, P., Helander, M., Koricheva, J. and Faeth, S. (2006) Model systems in ecology: dissecting the endophyte-grass literature. *Trends in Plant Science*, **11**, 428–443.

Saito, K. and Matsuda, F. (2010) Metabolomics for functional genomics, systems biology, and biotechnology. *Annual Review of Plant Biology*, **61**, 463–489.

Sana, T. R., Fischer, S., Wohlgemuth, G. *et al.* (2010) Metabolomic and transcriptomic analysis of the rice response to the bacterial blight pathogen *Xanthomonas oryzae pv. oryzae*. *Metabolomics*, **6**, 451–465.

Schulz, B., Wanke, U., Draeger, S. and, Aust, H.-J. (1993) Endophytes from herbaceous plants and shrubs: effectiveness of surface sterilization methods. *Mycological Research*, **97**, 1447–1450.

Schwachtje, J. and Baldwin, I. T. (2008) Why does herbivore attack reconfigure primary metabolism? *Plant Physiology*, **146**, 845–851.

Shao, H. B., Chu, L. Y., Jaleel, C. A. *et al.* (2009) Understanding water deficit stress-induced changes in the basic metabolism of higher plants – biotechnologically and sustainably improving agriculture and the ecoenvironment in arid regions of the globe. *Critical Reviews in Biotechnology*, **29**, 131–151.

Sieber, T. (2007) Endophytic fungi in forest trees: are they mutualists? *Fungal Biology Reviews*, **21**, 75–89.

Siegel, M. R. and Bush, L. P. (1996) Defensive chemicals in grass-fungal endophyte associations. In J. T. Romeo, J. A. Saunders and P. Barbosa (eds.) *Phytochemical Diversity and Redundancy in Ecological Interactions*. New York and London: Plenum Press, 81–119.

Stone, J., Bacon, C. and White, J. (2001) An overview of endophytic microbes: endophytism defined. In C. W. Bacon and J. White (eds.) *Microbial Endophytes*. New York: Marcel Dekker, 3–29.

Thaler, J. S., Karban, R., Ullman, D. E., Boege, K. and Bostock, R. M. (2002) Cross-talk between jasmonate and salicate plant defence pathways: effects on several plant parasites. *Oecologia*, **131**, 227–235.

Thoden, T. C., Boppre, M. and Hallmann, J. (2009) Effects of pyrrolizidine alkaloids on the performance of plant-parasitic and free-living nematodes. *Pest Management Science*, **65**, 823–830.

Tucker, C. and Avila-Sakar, G. (2010) Ontogenetic changes in tolerance to herbivory in Arabidopsis. *Oecologia*, **164**, 1005–1015.

van der Putten, W. H. (2003) Plant defense belowground and spatiotemporal processes in natural vegetation. *Ecology*, **84**, 2269–2280.

Verpoorte, R., Choi, Y. H. and Kim, H. K. (2010) Metabolomics: what's new? *Flavour and Fragrance Journal*, **25**, 128–131.

Vidal, S. (1996) Changes in suitability of tomato for whiteflies mediated by a non-pathogenic endophytic fungus. *Entomologia Experimentalis et Applicata*, **80**, 272–274.

Vrieling, K. and van Wijk, C. A. M. (1994a) Cost assessment of the production of pyrrolizidine alkaloids in ragwort (*Senecio-jacobaea* L). *Oecologia*, **97**, 541–546.

Vrieling, K. and van Wijk, C. A. M. (1994b) Estimating costs and benefits of the pyrrolizidine alkaloids of *Senecio-jacobaea* under natural conditions. *Oikos*, **70**, 449–454.

Wiklund, S., Johansson, E., Sjostrom, L. *et al.* (2008) Visualization of GC/TOF-MS-based metabolomics data for identification of biochemically interesting compounds using OPLS class models. *Analytical Chemistry*, **80**, 115–122.

Wilkinson, H. H., Siegel, M. R., Blankenship, J. D. *et al.* (2000) Contribution of fungal loline alkaloids to protection from aphids in a grass-endophyte mutualism. *Molecular Plant–Microbe Interactions*, **13**, 1027–1033.

Witte, L., Ernst, L., Adam, H. and Hartmann, T. (1992) Chemotypes of 2 pyrrolizidine alkaloid-containing *Senecio* species. *Phytochemistry*, **31**, 559–565.

Wu, J. and Baldwin I. T. (2010) New insights into plant responses to the attack from insect herbivores. *Annual Review of Genetics*, **44**, 1–24.

Zangerl, A. R. and Rutledge, C. E. (1996) The probability of attack and patterns of constitutive and induced defense: a test of optimal defense theory. *American Naturalist*, **147**, 599–608.

Zheng, S. J. and Dicke, M. (2008) Ecological genomics of plant–insect interactions: from gene to community. *Plant Physiology*, **146**, 812–817.

CHAPTER TWELVE

Integrating the effects of PSMs on vertebrate herbivores across spatial and temporal scales

BEN D. MOORE

Ecological Sciences Group, The James Hutton Institute

JANE L. DEGABRIEL

School of Biological Sciences, University of Aberdeen

12.1 Introduction

Since Fraenkel (1959) proposed a leading role for plant secondary metabolites (PSMs) in the interactions between plants and herbivores, science has achieved broad insight into the diversity of PSMs and herbivores' counter-adaptations to them (Freeland & Janzen, 1974; Foley *et al.*, 1999; Foley & Moore, 2005). However, the more we learn about the distributions and functions of PSMs in natural systems, the sharper the limitations of our understanding become. In countless plant–herbivore interactions, ecologists have identified PSMs that act as feeding deterrents, toxins, digestibility reducers, feeding or oviposition cues, and signals for communicating to neighbouring plants and natural enemies of herbivores. However, most studies focus on the interaction between single species of herbivore and plant, usually with observations of captive animals fed diets containing PSMs under highly simplified conditions. Although such approaches are a necessary first step in isolating and characterising the actions of PSMs, they greatly oversimplify the complex interactions that occur between wild herbivores and plants. The next challenge for ecologists is to 'scale up' the roles of PSMs in plant–herbivore interactions, as we understand them from controlled experiments at small temporal scales, to predict ecological interactions at greater temporal and spatial extents.

A captive herbivore may commonly eat less as PSM concentrations in its food increase, but can this predict the foraging decisions of a wild animal within its home range, or, ultimately the distributions and abundances of plant and herbivore species and genotypes? Scaling up has an obvious spatial component, because wild animals forage more extensively than do captive animals, but it also has a temporal component. Experiments usually describe plant–herbivore interactions over very short time intervals, but in

The Ecology of Plant Secondary Metabolites: From Genes to Global Processes, eds. Glenn R. Iason, Marcel Dicke and Susan E. Hartley. Published by Cambridge University Press. © British Ecological Society 2012.

nature they are continuous and the effects of PSMs can be long-lasting (Cheeke, 1998). Animal feeding preferences are dynamic and often change with season or reproductive state, or through the ongoing process of refinement of conditioned flavour aversions (Provenza, 1996). With increasing spatial extent and finer spatial grain size comes greater complexity in the interactions between plants and animals; PSMs are rarely distributed evenly throughout landscapes, and understanding how this influences plant–animal interactions requires approaches adopted from resource ecology, foraging theory and spatial ecology and often an extensive, high-resolution picture of the foodscapes within which animals forage (van Langevelde & Prins, 2008).

   We consider that the major impediments to scaling up studies of vertebrate herbivore–PSM interactions are: integrating the effects of multiple PSMs and nutrients to obtain a holistic measure of plant nutritional quality; identifying the scales (grain and extent) at which herbivores interact with PSMs; and observing/describing herbivore behaviours and plant communities at sufficient resolution at the relevant scales. We will discuss both theoretical and practical approaches to the challenge of 'scaling up' and conclude with a case study in which we used an understanding derived from captive experiments to produce a palatability map for a freely foraging dietary specialist, the koala (*Phascolarctos cinereus*).

12.2   What feeding experiments with captive animals do and don't tell us

In contrast to many feeding experiments with insects, experiments with captive vertebrates are usually conducted over a single, or at most several feeding bouts, e.g. overnight (Moore *et al.*, 2005; Marsh *et al.*, 2006; Wiggins *et al.*, 2006a). In no-choice experiments, herbivores only decide whether and how much to eat from the food presented. However, the foraging decisions of wild herbivores are more complex, starting with the decision whether to eat a food or to keep searching for a preferred one. Over longer time frames, wild herbivores' foraging goals must balance the intake of multiple nutrients and toxins while assembling a diet from many alternative feeding options.

12.2.1   PSM concentrations and intake

Two conceptual models are often used to describe the amount eaten by herbivores as a function of PSM concentration in their food: (1) consumption is constant across a range of low concentrations, but then declines curvilinearly beyond a threshold concentration (Figure 12.1a, b); (2) consumption declines linearly as PSM concentrations increase from zero (Figure 12.1c, d). Theoretical discussions of these often neglect to define the relevant timescale, but they are most useful when considering intake over the course of one or several feeding bouts. Model 1 is known as the 'regulation model': a herbivore's tolerance of a PSM is determined by its rate and extent of absorption of the PSM from the gut, distribution of the PSM throughout different physiological compartments, and the rates of and capacity for metabolism (detoxification) and excretion of the PSM and its metabolites (absorption, detoxification, metabolism, excretion (ADME), McLean & Duncan, 2006; Sorensen *et al.*, 2006), as well as the herbivore's tolerance of PSMs at active sites. At very low or negligible PSM concentrations, food intake can be maximised subject to the constraints of handling time and gut capacity. At and above a certain dietary PSM concentration, intake is limited by the capacity of ADME mechanisms to keep plasma PSM concentrations below a harmful level, and consumption declines along a convex curve towards an asymptote at zero, while PSM intake is held constant. In contrast, for a linear, non-threshold response such as model 2 (Figure 12.1c, d), the curve describing PSM intake across an increasing range of PSM concentrations is dome-shaped, which is consistent with the idea that herbivores trade off the amount of toxin

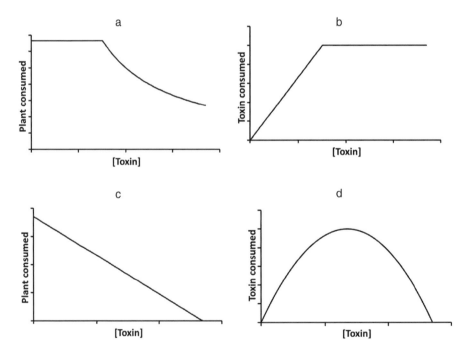

Figure 12.1 Consumption of food and of PSMs as a function of PSM concentration in food, as envisaged by two conceptual models. (a, b) The 'regulation model', in which maximum food intake is maintained until a threshold PSM intake is reached; (c, d) an alternative model, in which food intake declines linearly with PSM concentration, and which forms the basis of the functional response of Li *et al.* (2006).

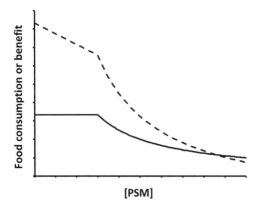

Figure 12.2 Food intake across a range of PSM concentrations under the regulation model (solid line) and the associated benefit to the herbivore after accounting for the nutritional or detoxification costs of PSM ingestion (dashed line). Under choice conditions, the relative preference of an 'ideal' herbivore for plants varying only in PSM concentrations should most closely reflect the benefit curve.

they are willing to ingest against the amount of food (and hence nutrients) consumed.

Although the regulation model offers a mechanistic explanation for why PSMs cause herbivores to regulate their consumption of plants during a single feeding bout, it may not explain the preferences of wild herbivores if these reflect the net benefit received from each food, because the model does not incorporate the costs associated with PSM ingestion. In reality, herbivores are unlikely to consume infinitesimally small amounts of very toxic plants (except in the context of sampling), because the benefit is negligible and the costs are high. The costs of ingesting PSMs are varied but can often be high: for Stephen's woodrats (*Neotoma stephensi*) feeding on juniper foliage, the energetic investment required for detoxification may be similar to that of reproduction (Sorensen *et al.*, 2005) and for ruffed grouse (*Bonasa umbellus*) feeding on aspen, energetic loss to excreted conjugates is 10% of intake (Jakubas *et al.*, 1993). Lagomorph and rodent species can incur a possibly unsustainable sodium cost after eating many PSMs (Iason & Palo, 1991), and tolerance of cyanogenic glycosides incurs a protein cost (Gleadow & Woodrow, 2002). Various curves have been proposed to describe the dose–response relationships for different xenobiotics (Raubenheimer & Simpson, 2009), but for simplicity, we can describe the costs of PSM ingestion with a linear, non-threshold curve and combine this with the regulation model to produce a concentration–benefit curve (Figure 12.2). In situations where a herbivore has had sufficient experience with a plant and its PSMs to assess the net benefit or cost returned from eating it, optimal food consumption should closely match this or a similar benefit function.

12.2.2   Temporal scaling up – from one meal to a lifetime

Herbivores' fitness and ultimately, their distribution and abundance, should reflect their interactions with PSMs over long periods of times, such as seasons and animal lifetimes. However, while captive feeding experiments provide valuable insight into the interactions of herbivores with PSMs, they are generally conducted over short time periods, such as a single meal or day/night, and a number of caveats should be considered before extrapolating these observations to predict herbivore responses over longer timescales.

First, some captive experiments may not include an acclimation period or be of sufficient duration for herbivores to accurately integrate the benefits and detrimental consequences of plant foods. This may lead to an overestimate of wild animals' feeding preferences if captive animals cannot assess the long-term costs of PSM ingestion (for examples, see Cheeke, 1998), or an underestimate if the animals learn to tolerate higher PSM concentrations after longer exposure, for example via the induction of microsomal enzymes involved in detoxification (Pass *et al.*, 2001; McLean *et al.*, 2008). Furthermore, common brushtail possums will increase their tolerance of 1,8-cineole (McLean *et al.*, 2008) and the bitter alcoholic β-glucoside, salicin (Pass & Foley, 2000) over and above that explained by enzyme induction after prolonged exposure.

Other caveats are associated with the no-choice protocol used in many feeding experiments to isolate the pairwise plant–herbivore interaction for study. Where no-choice experiments are conducted overnight, herbivores may ingest several small meals containing high PSM concentrations, obscuring differences in individual meal size from plants of different toxin concentrations (Figure 12.3). Wild herbivores, however, might completely reject the same plants or just eat a single meal, then search for an alternative food. Captive herbivores are also exempt from constraints on foraging behaviour such as minimising predation risk and may be prepared to tolerate concentrations of PSM which would cause toxicosis in wild herbivores and make them more vulnerable to predation (Schmidt, 2000).

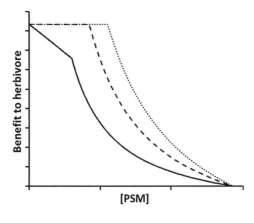

Figure 12.3 Total daily benefit accrued to a herbivore as a function of PSM concentration when allowed to eat a single meal (solid line), two meals (long dashes) or three meals (short dashes), assuming the herbivore can meet its daily requirements in a single meal from an undefended plant. The figure demonstrates how measuring food intake over multiple meals can obscure some herbivore feeding preferences.

Finally, the timescales over which PSMs are ingested and detoxified may affect various herbivore species differently according to their foraging strategies. For example, wapiti (*Cervus elaphus*; Wilmshurst *et al.*, 1995; Fryxell, 2008) exercise short-term energy maximisation, whereas cattle and wood bison (*Bison bison athabascae*) maximise energy intake at a daily scale (WallisDeVries & Daleboudt, 1994; Fryxell, 2008), so we can expect that PSMs might be more likely to affect foraging outcomes for wapiti than for cattle.

12.3   Patchy PSMs in the landscape

The interaction between a plant and its herbivores can be strongly affected by the plant's spatial context. The nutritional and PSM composition of plants is variable both within and between species, creating heterogeneous, often patchy, foraging environments (where 'patchiness' is shorthand for structured heterogeneity which can vary from unambiguous, discrete patches to continuously varying aggregation patterns). Patchiness can be beneficial to herbivores if it concentrates particular plant species or low-PSM genotypes in space. However, it can also decrease the ability of herbivores to ameliorate the effects of PSMs by eating mixed diets (Freeland & Janzen, 1974) when patches of plants defended by different PSMs are too widely separated (Wiggins *et al.*, 2006b). Crucially, however, the significance of spatial heterogeneity for a herbivore's ecology depends upon (1) the herbivore's anatomy and powers of perception, which limit the minimum spatial grain size at which heterogeneity is relevant and (2) the herbivore's mobility and size, which constrain the spatial extent at which it can forage.

Phenotypic patchiness often reflects an underlying genetic structure which can arise neutrally when gene flow is limited (Andrew *et al.*, 2007) and which can be strengthened by spatially uneven selection by other biotic agents. One fine-grained example is that slug browsing is greatest near adult sycamore trees (Pigot & Leather, 2008). At broader scales, fire drives the abundance of early successional birch trees across North America, and subsequent browsing of these by snowshoe hares selects for defensive PSMs (Bryant *et al.*, 2009). Abiotic causes of spatial structure in PSM concentrations can include underlying edaphic, altitudinal, climatic or fire-regime heterogeneity and even patchy light and shade in forests (Moore *et al.*, 2004; Brenes-Aguedas & Coley, 2005; Hakes & Cronin, 2011). The grain size of spatial autocorrelation in PSM concentrations can vary from centimetres to thousands of kilometres; however, patchiness occurs at multiple scales simultaneously, creating what has been described as a nested hierarchy of aggregated resources (Kotliar & Wiens, 1990; Searle *et al.*, 2005).

12.3.1   What does patchiness mean for herbivores?

Herbivores do not perceive individual plants in isolation, but make feeding decisions across a hierarchy of scales, from individual bites to entire home ranges. Therefore, the rate at which herbivores encounter food items is influenced by their foraging strategy, selectivity and morphology, as well as the spatial arrangement of plants (Searle *et al.*, 2007). From a herbivore perspective, a patch can be defined as a collection of feeding stations separated in space by distance that is sufficient to cause a change in intake rate (Hobbs *et al.*, 2003) and PSMs might determine patch quality in a number of ways. For example, foraging might be influenced by the average PSM concentration in a patch, the proportion of plants or feeding stations which fall below some threshold concentration at which plants are rejected or at which feeding bouts are restricted, or by the likelihood of encountering extremely toxic plants. A more plant-centric definition of a patch, however, is a spatial cluster of plants which are more similar to each other than they are to plants outside the patch.

12.3.2   Identifying the scales of selection in heterogeneous landscapes

Many studies have addressed the influence of multi-scale spatial heterogeneity in food resources for herbivores, particularly in the context of understanding associational resistance, where plant susceptibility to herbivores is influenced by the quality and proximity of neighbouring plants (Atsatt & O'Dowd, 1976; Barbosa *et al.*, 2009). PSMs are a common modifier of food quality in these studies (e.g. Iason, 2005; Bergvall *et al.*, 2006). If herbivores make foraging decisions at the scale of a patch of plants, plants can gain associational resistance to herbivores if neighbouring plants are unpalatable or suffer associational susceptibility if the neighbours are palatable. However, unpalatable neighbours can also produce associational susceptibility via a 'neighbour contrast effect' when herbivores forage at an individual-plant scale (Bergvall *et al.*, 2006). This mechanism has a temporal as well as spatial component, because it relies upon the herbivore making either a simultaneous contrast between (for example) high- and low-PSM foods or retaining a memory of the initial food at least until the next plant is encountered and the contrast can be assessed (Bergvall & Balogh, 2009). Associational effects are herbivore-specific because they depend upon the scale of the herbivore's foraging movements (Milchunas & Noy-Meir, 2002) and can operate from a scale smaller than the herbivore's smallest cropping size (McNaughton, 1978) in mixed swards up to large movement scales (e.g. van Beest *et al.*, 2010).

In experimental arenas with mule deer (*Odocoileus hemionus*) and with grizzly bears (*Ursus arctos*), Searle *et al.* (2006) investigated the influence of food quantity in patches comprising several feeding stations, as well as in surrounding patches and within- and inter-patch distances on patch residence time. Herbivore behaviour was always explained by more than one level of heterogeneity in their patch hierarchy, and they argued that their empirical statistical models provided a useful alternative to optimal foraging models. In those experiments, herbivores were able to assess surrounding and more distant patches visually, because it was food quantity that was varied. Where patch quality is mediated by PSMs or plant nutrition, however, the scale of animal perception is often likely to be smaller. This limitation can be overcome where animals remain resident in established home ranges and develop a spatial memory of resource quality (Bailey & Provenza, 2008; Fryxell *et al.*, 2008) and heterogeneity.

To consider the scales of herbivore foraging decisions in context, we must address heterogeneity in natural landscapes. One approach is to understand and model the processes which create heterogeneity and structure the spatial distribution of plant species and phenotypes. An empirical approach demands high-resolution measurement of habitats across wide areas. Increasingly, remote sensing offers a realistic prospect of achieving this as it becomes feasible to map the presence of vegetation types and even plant phenotypes at scales as fine as individual trees or shrubs (Skidmore & Ferwerda, 2008; Kokaly *et al.*, 2009). Across smaller study areas, habitat quality can be described from the ground, and here technologies such as near-infrared spectroscopy can be invaluable (see case study below for an example).

12.4   Theoretical approaches to scaling up

The central process in the assemblage of herbivore diets is foraging (van Langevelde & Prins, 2008), and a rich literature of predominantly mathematical foraging theory has accumulated. Until very recently, however, this theory has mostly ignored the role of PSMs (Belovsky, 1990; Swihart *et al.*, 2009).

12.4.1   Patch departure models

Patch departure models, such as the marginal value theorem (Charnov, 1976) are probably the least relevant branch of optimal foraging theory to our discussion. Optimal foraging approaches based around patch use or patch departure rules have been criticised from a number of perspectives (Kotliar & Wiens, 1990; Searle *et al.*, 2005), in particular because they typically only study processes at a single scale, despite ample evidence that herbivores make foraging decisions across numerous scales, often simultaneously.

12.4.2   Linear and dynamic linear programming approaches

Linear programming is often used to solve for optimal diet composition subject to linear constraints such as daily energy or other nutritional

requirements, available feeding time and digestive capacity. Where constraints are non-linear, such as where toxin–toxin or toxin–nutrient combinations may act synergistically, non-linear programming solutions are
required. Most linear programming models are considered in the currency
of a single nutrient (often energy maximisation or time minimisation) to be
optimised, but there is no reason that the currency to be optimised cannot be
the maximum ingestion of multiple nutrients that must be consumed in ratio
and/or the minimal ingestion of multiple PSMs (Belovsky, 1990).

Maximisation of energy intake may be a common goal for many animals
and linear programming models produce good agreement with observed diet
selection for a number of species (Belovsky, 1994). However, studies conducted within the geometric framework of nutrition (Felton *et al.*, 2009)
have sometimes found that neither nutrient maximisation nor PSM minimisation, but instead 'protein leveraging' (Simpson & Raubenheimer, 2005)
explains foraging behaviour of wild animals, which may limit the usefulness
of linear programming.

12.4.3 Functional response models

Traditional functional response models describe the amount of food eaten by
foraging animal(s) as a function of the abundance of food items. Difference
equations can be used to consider herbivore and plant functional responses
together to model both herbivore and plant (predator–prey) population
dynamics. Li *et al.* (2006) recently formulated a toxin-determined functional
response model (TDFRM) which permits herbivores to regulate their consumption of toxic PSMs, accounting for the negative impacts of these metabolites on herbivore growth. They first took a conventional Holling type II
functional response describing food intake, where N is the abundance of the
plant food resource, e is the resource encounter rate, h is the handing time for
each food item, σ is the proportion of food items (or mouthfuls) encountered
that are ingested by the herbivore:

$$f(N) = \frac{e\sigma N}{1 + he\sigma N}$$

They then modified this functional response to produce the following TDFRM
describing consumption (C) as a function of N:

$$C(N) = f(N)\left(1 - \frac{f(N)}{4G}\right)$$

in which G is the ratio M/T, and M is the maximum amount of toxic PSM per
unit time that the herbivore can tolerate – exceeding this would be fatal – and
T is the toxin concentration in the plant (G can also be thought of as the
herbivore's tolerance of the plant given its toxin concentration, T). The

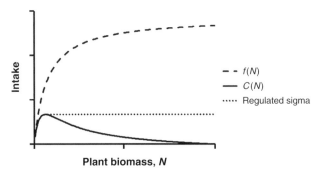

Intake

Plant biomass, *N*

- - *f*(N)
— *C*(N)
····· Regulated sigma

Figure 12.4 The toxin determined functional response model of Li *et al.* (2006). The dashed line, *f*(N), describes food consumption by herbivores when constrained by searching and handling time, but not by PSM. The solid line, *C*(N), describes herbivore consumption of food when limited by the effects of high PSM concentrations. Beyond the maximum consumption rate, intake declines because the toxic effects of the PSM force the herbivore to stop feeding. However, the model allows the herbivore to eat a smaller proportion (denoted by σ) of encountered food items beyond that point, thus maintaining the maximum allowable consumption rate (dotted line).

negative effects of PSMs on consumption by the herbivore are described by the second factor in this equation (Swihart *et al.*, 2009). As the conventional functional response grows with increasing plant abundance, so too does the herbivores' potential to ingest PSMs and the difference between *f*(N) and *C*(N) (Figure 12.4).

Feng *et al.* (2008) show that under circumstances where a herbivore can tolerate high consumption of the plant (because *T* is low and/or *M* is high), *C*(N) is monotonically increasing, but where *G* is lower, *C*(N) is unimodal, reaching a maximum less than *G* at plant abundance N_m. If, in the latter case, the herbivore does not modify σ, the proportion of encountered plants which it ingests, then it experiences negative physiological consequences which forcibly reduce its consumption of the plant resource beyond N_m. However, most vertebrate herbivores do regulate their intake of PSM below physiologically safe levels, and the TDFRM also allows for herbivores to do this by reducing σ at levels of *N* greater than N_m to maintain consumption of the resource (Figure 12.4).

This TDFRM does not explicitly model the effects of toxic PSM on herbivores, but simply postulates a negative influence on the functional response that increases with toxin intake (Feng *et al.*, 2008, p. 450). Part of its appeal lies in accounting for both the direct costs of PSMs (avoidance) and the negative effects of toxins on herbivore growth. The PSM concentration–intake curve follows the linear, non-threshold model discussed earlier when σ is not regulated and becomes convex when regulation of intake commences (Figure 12.5). A limitation of this model is that it assumes that the anti-feedant

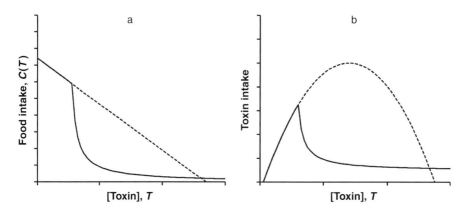

Figure 12.5 (a) Food intake and (b) PSM intake as a function of PSM concentration in plants when the proportion of encountered food items is not regulated (dashed line) and when it is (solid line), under the TDFRM of Li *et al.* (2006).

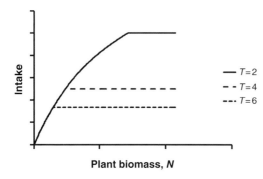

Figure 12.6 Holling type II functional response model constrained by a fixed herbivore detoxification capacity at three different toxin (T) concentrations.

effects of PSMs scale directly to the cost of detoxification, which need not be the case. However, in the context of predicting herbivore impacts on plant populations, the failure of this TDFRM to permit plants with low PSM concentrations to be treated equally below a threshold level could be problematic because linear threshold dose–responses are common. Nonetheless, this model has been extensively analysed, giving insight into the role of PSMs in controlling cyclic plant and herbivore populations (Feng *et al.*, 2008) and herbivore-mediated invasion processes in plant associations (Feng *et al.*, 2009), and no doubt further advances will emerge.

An alternative approach to Liu's TDFRM would be to retain the traditional Holling type II functional response to describe off-take of plant biomass, but with the constraint

$$f(N) \leq \frac{E}{T}$$

where E is the maximum toxin elimination rate (Figure 12.6). The net benefit of consumption experienced by herbivores can then be calculated by subtracting the digestibility-reducing and detoxification costs associated with PSM consumption from $f(N)$. This functional response corresponds to the regulation model of PSM intake described earlier, and intake across a range of PSM concentrations follows the form of Figure 12.1a, b.

12.4.4 Limitations to theoretical approaches

Linear programming and functional response models usually only deal with average plant attributes, ignoring common within-population variation in plant quality; they also treat resource abundance and encounter rates as constant across space. The intake rate calculated for a landscape which is assumed to have an evenly distributed food resource will be greater than the average intake rate experienced by herbivores in a landscape possessing the same overall resource abundance but in which the resource is distributed in patches of low and high abundance, resulting in a variable encounter rate. This occurs because the functional response, $f(N)$, is concave increasing with respect to plant abundance (dashed line, Figure 12.4) and can be attributed to Jensen's inequality (DeGroot, 1970). Thus, if animals forage randomly, theoretical models will overestimate their intake in heterogeneous landscapes. However, many vertebrate herbivores do not forage randomly, because they are familiar with the spatial distribution of resources and PSMs in their home range and so actively forage in more rewarding patches. When this occurs, theoretical models are instead likely to underestimate the functional response if we assume the average landscape values of N or T. Both of these scenarios lend support to the hypothesis that functional responses are related to resource heterogeneity as well as resource quality (Laca, 2008).

12.4.5 Foodscapes

Recently, Searle *et al.* (2007) have promoted the 'foodscape' concept. The foodscape is a herbivore-specific construct which recognises that the value of food resources in a landscape at any given time is a product of both the landscape and the foraging animal (Searle *et al.*, 2007), and thus can be characterised by relevant behavioural metrics across multiple scales. Animal behaviour is plastic, and changes in animal behaviour (and hence foodscapes) can indicate changing forage quality. For example, when forage quality declines, large ungulate herbivores can choose one of two, potentially coexisting, foodscapes: they either adjust their behaviour to maintain high diet quality or they accept a lower quality diet but adjust their behaviour to improve the efficiency of digestion.

Can the foodscape framework also offer a useful perspective from which to understand how PSMs affect the value of food resources for herbivores in a landscape? Perhaps, but the selection of appropriate behavioural indices is critical. Some candidate indices discussed by Searle *et al.* (2007) are bite size, bite rate, bites and time per plant or feeding station. In the absence of PSMs, reducing bite size can allow herbivores to select smaller and rarer plant parts, often young growing tips, achieving increased dietary quality at the expense of rate of intake. However, as PSMs are often concentrated in young growing leaves (Lambdon & Hassall, 2005), browsers may instead be forced to accept larger, older plant parts when PSMs are present, permitting larger bite sizes and greater instantaneous rates of intake, but sacrificing nutritional quality. In ruminants, the lower quality may slow down passage rates and reduce intake on a daily timescale.

In the context of PSMs, we would suggest additional indices commonly used in captive animal studies, such as meal size (where a meal might include one or more than one feeding station) and inter-meal interval (IMI, Torregrossa & Dearing, 2009), that are more targeted towards understanding the constraints imposed by PSMs. IMI can reflect a detoxification time constraint, but care must be taken to distinguish it from between-meal intervals brought about by satiation or by searching. Note that increases in IMI can be considered analogous to (and might potentially be confused with) increases in time devoted to rumination where ungulates are forced to digest less nutritious forage more thoroughly.

12.5   Integrating PSM effects on diet and habitat quality to predict herbivore distributions

PSMs can influence habitat quality for herbivores in two ways: they deter herbivores from eating some plants, effectively reducing the overall quantity of food available, and they reduce plants' nutritional quality, either by reducing their digestibility or by imposing significant detoxification costs. Thus, to predict animal performance in a given habitat, and ultimately distribution and abundance, it is necessary to know which plants herbivores eat (i.e. diet selection) and to integrate the effects of ingested PSMs into a measure of nutritional benefit gained from that diet. However, knowing exactly what wild herbivores consume is an enormous challenge, particularly when PSMs can cause the quality of individual plants, or even individual leaves on a single plant, to differ significantly. Herbivore-feeding preferences can vary greatly between individuals, populations and seasons (DeGabriel *et al.*, 2009b). Specialist herbivores can specialise on different plants in different habitats, and generalists can have broad diets or, perhaps commonly, specialise on one or a small number of locally available plants (Shipley *et al.*, 2009). This complexity is intrinsically difficult to capture.

The geometric framework of nutrition (Simpson & Raubenheimer, 1999) suggests that organisms' foraging is guided by clear intake targets, comprising an optimal mixture and amount of nutrients, rather than the maximisation of one or all nutrients. Consequently, a herbivore might eat a plant which offers little energetic reward and a substantial toxin cost if it still offers another nutrient reward that could help it to reach its nutritional target, provided other plants are available to make up the required blend of nutrients. Many tropical plants that are defended by PSMs such as alkaloids and cyanogens are nonetheless rich protein or energy sources and can form a valuable part of herbivore diets (Chapman & Chapman, 2002). Behmer et al. (2002) showed that the response of herbivores to PSMs can depend on the associated nutrient profile of a food, with PSMs more effective at deterring herbivores from carbohydrate-rich foods than from protein-rich foods.

Two recent studies have successfully linked concentrations of digestibility-reducing PSMs to the reproductive performance of wild common brushtail possums and moose (DeGabriel et al., 2009a; McArt et al., 2009). In both cases, fecundity was best explained by the concentration of digestible protein in the animals' diets, and this was strongly influenced by concentrations of condensed tannins. The nutritional cost of consumed PSMs, but not their anti-feedant effects, probably play a key role in structuring population sources and sinks at a scale greater than the size of these herbivores' home ranges.

12.6   Case study: linking a toxic PSM to patch use by koalas

Because many aspects of plant nutrition and chemistry vary between species, the ecological role of PSMs can be best addressed by studying how herbivores, particularly specialists, respond to quantitative variation in PSM concentrations within a plant population. Several studies have successfully used this approach to demonstrate that PSMs influence the choice by herbivores of individual plants from a population (Farentinos et al., 1981; Bryant et al., 1983), but fewer have tried to understand whether larger-scale PSM patchiness affects herbivores similarly.

Koalas (*Phascolarctos cinereus*) are highly specialised arboreal folivores of *Eucalyptus*. Formylated phloroglucinol compound (FPC) concentrations are particularly variable amongst trees within eucalypt populations and cause koalas to markedly reduce their intake of foliage in no-choice captive feeding experiments (Moore et al., 2005; Marsh et al., 2007). Moore and Foley (2005) and Moore et al. (2010) describe studies of a koala population at Phillip Island, Australia in which they scaled up observations from captive experiments to predict patch use at a scale of several hectares. Methodological details of the studies which are summarised here can be found in those papers.

12.6.1   PSMs in the landscape

Near-infrared spectroscopy (NIRS, Foley et al., 1998) was used to determine FPC and nitrogen concentrations in all available trees and predict palatability of eucalypt foliage to koalas, using no-choice feeding trials with captive koalas to calibrate the models. This approach is not exempt from the caveats outlined earlier, and the variability in lengths of visits of wild koalas to trees at Phillip Island (from 1 to 5625 minutes; K. J. Marsh & B. D. Moore, unpublished data) means that captive feeding trials probably artificially restricted the true range of variation in palatability. Spatial patchiness in FPC concentrations occurred at two scales: (1) strong autocorrelation to a distance of 150 m, attributable to the woodland community structure (i.e. patches of tree species) and (2) weaker autocorrelation within each eucalypt species to 30 m, probably explainable by genetic patchiness resulting from isolation by distance (for a similar example and detailed explanation see Andrew *et al.*, 2007).

12.6.2   Koalas in the landscape

Koala use of individual trees increased with tree size and with NIRS-predicted palatability, but decreased at high FPC concentrations (Moore & Foley, 2005; Moore *et al.*, 2010). Neighbourhood quality (the mean predicted number of koala visits to surrounding trees within a 25 m radius of each tree) also had a strong influence on the likelihood of koala visits, with trees in both 'good' and 'bad' neighbourhoods experiencing more visits than trees in 'average neighbourhoods'. The former was attributed to associational susceptibility and the latter to a neighbour contrast effect where trees become relatively more appealing because the immediately available alternatives are so poor. A smoothed 'palatability map', based on the model predictions, showed heterogeneity in predicted habitat quality throughout the woodland (Moore *et al.*, 2010).

12.6.3   A caution about herbivore and site-specific attributes of habitat quality

The modelling approach of Moore *et al.* could easily be extended into woodland beyond the confines of the reserve, but its predictions are very specific to this study population. The predicted koala densities on the map are scaled to the local koala density (here monitored and controlled by wildlife managers). In other areas, where large-scale koala movements are less constrained, areas similar to ones occupied by koalas in the study woodland may be unoccupied. Tree size was the strongest factor influencing tree use at Phillip Island, but koalas can thrive in woodlands with much smaller trees, so it can be deduced that this factor is less likely to influence dispersal and home range selection. In environments where koala-feeding trees occur at much lower densities, such as in a matrix of other tree species, and koalas' between-tree movements increase accordingly, then how will the scale of associational effects and within-home range patch selection be affected?

In *E. globulus* and *E. microcorys* (another species favoured by the koala), mean FPC concentrations can vary more than two-fold among sites (Moore *et al.*, 2004; O'Reilly-Wapstra *et al.*, 2004), alongside other leaf properties. Far greater variation in the types and amounts of PSMs occurs among the hundreds of species of eucalypts, even if only those used by koalas are considered (DeGabriel *et al.*, 2008; Brophy *et al.*, 2009; Wallis *et al.*, 2010). As a result, foraging is plastic even for this dietary specialist; home range sizes can vary from 3 ha to more than 100 ha. Little is known about inter-population differences in tolerance of PSMs by koalas, but the requirements of lactation and thermoregulation and efficiency of mastication related to tooth wear create large differences in the intake of PSMs by koalas different in sex, reproductive status and age, and across seasons (Logan & Sanson, 2002; Krockenberger & Hume, 2007). These sorts of between-site and between-season differences in the spatial scale of plant heterogeneity, and the plasticity of herbivore responses are enormous challenges in understanding the part played by PSMs in plant–herbivore interactions at large scales.

12.7   Conclusions and future directions

Ultimately, if ecologists are to understand the influence of PSMs on herbivore distribution and abundance, they need to understand and to integrate the deterrent effects of PSMs with their effects on herbivore performance. More research is needed to extrapolate relationships between PSM concentration and intake derived from captive feeding experiments to foraging behaviour of wild herbivores. This is because relatively little evidence exists to demonstrate the true form of these relationships, and because captive and wild herbivores often make feeding decisions at different spatial and temporal resolutions. A framework for understanding how deterrence by PSMs shapes wild herbivore diets needs to incorporate the role of associational effects and to consider the relative value of different food options in a way that integrates both nutritional and detrimental aspects of plant chemistry. PSM concentration–benefit curves demonstrating the consequences of PSMs for lifetime fitness are easily obtainable for monophagous insects, but understanding the long-term consequences of toxic PSM consumption for generalist herbivores is likely to remain elusive. In contrast, integrating the effects of digestibility-reducing PSMs is achievable and has provided evidence that tannins can reduce reproductive success in herbivores.

If theoretical models are to play a useful role in 'scaling up' the role of PSMs, they must be reconciled with detailed empirical data addressing questions about animal foraging in realistic environments such as: how random or directed is animal foraging? How important is spatial memory? At what strength and spatial extent do herbivores respond to spatial structure in PSM distributions? All of these are substantial challenges, because ecological interactions in real world ecosystems are complex, but the application of accrued knowledge of PSMs' effects to questions of biodiversity and animal conservation in the face of global environmental change will no doubt prove rewarding.

Acknowledgements

Thanks to Stuart McLean, Jack Lennon, Jennifer Forbey, Bill Foley and Glenn Iason for helpful discussions and comments on the manuscript.

References

Andrew, R. L., Peakall, R., Wallis, I. R. and Foley, W. J. (2007) Spatial distribution of defense chemicals and markers and the maintenance of chemical variation. *Ecology*, **88**, 716–728.

Atsatt, P. R. and O'Dowd, D. J. (1976) Plant defense guilds. *Science*, **193**, 24–29.

Bailey, D. W. and Provenza, F. D. (2008) Mechanisms determining large-herbivore distribution. In H. H. T. Prins and F. van Langevelde (eds.) *Resource Ecology: Spatial and Temporal Dynamics of Foraging*. Dordrecht: Springer, 7–28.

Barbosa, P., Hines, J., Kaplan, I. *et al.* (2009) Associational resistance and associational susceptibility: having right or wrong neighbors. *Annual Review of Ecology, Evolution and Systematics*, **40**, 1–20.

Behmer, S. T., Simpson, S. J. and Raubenheimer, D. (2002) Herbivore foraging in chemically heterogeneous environments: nutrients and secondary metabolites. *Ecology*, **83**, 2489–2501.

Belovsky, G. E. (1990) How important are nutrient constraints in optimal foraging models or are spatial/temporal factors more important? In *Behavioural Mechanisms of Food Selection*. Berlin: Springer-Verlag, 255–278.

Belovsky, G. E. (1994) How good must models and data be in ecology? *Oecologia*, **100**, 475–480.

Bergvall, U. A. and Balogh, A. C. V. (2009) Consummatory simultaneous positive and negative contrast in fallow deer: implications for selectivity. *Mammalian Biology*, **74**, 238–241.

Bergvall, U. A., Rautio, P., Kesti, K., Tuomi, J. and Leimar, O. (2006) Associational effects of plant defences in relation to within- and between-patch food choice by a mammalian herbivore: neighbour contrast susceptibility and defence. *Oecologia*, **147**, 253–260.

Brenes-Aguedas, T. and Coley, P. D. (2005) Phenotypic variation and spatial structure of secondary chemistry in a natural population of a tropical tree species. *Oikos*, **108**, 410–420.

Brophy, J. J., Forster, P. I., Goldsack, R. J., Hibbert, D. B. and Punruckvong, A. (2009) Essential oil variation in *Eucalyptus crebra, E. melanophloia* (Myrtaceae) and their hybrids. *Australian Journal of Botany*, **57**, 425–431.

Bryant, J. P., Wieland, G. D., Reichardt, P. B., Lewis, V. E. and McCarthy, M. C. (1983) Pinosylvin methyl ether deters snowshoe hare feeding on green alder. *Science*, **222**, 1023–1025.

Bryant, J. P., Clausen, T. P., Swihart, R. K. *et al.* (2009) Fire drives transcontinental variation in tree birch defense against browsing by snowshoe hares. *American Naturalist*, **174**, 13–23.

Chapman, C. A. and Chapman, L. J. (2002) Foraging challenges of red colobus monkeys: influence of nutrients and secondary compounds. *Comparative Biochemistry and Physiology A – Molecular and Integrative Physiology*, **133**, 861–875.

Charnov, E. L. (1976) Optimal foraging, the marginal value theorem. *Theoretical Population Biology*, **9**, 129–136.

Cheeke, P. R. (1998) *Natural Toxicants in Feeds, Forages and Poisonous Plants*. Danville, IL: Interstate Publishers.

DeGabriel, J. L., Wallis, I. R., Moore, B. D. and Foley, W. J. (2008) A simple, integrative assay to quantify nutritional quality of browses for herbivores. *Oecologia*, **156**, 107–116.

DeGabriel, J. L., Moore, B. D., Foley, W. J. and Johnson, C. N. (2009a) The effects of plant defensive chemistry on nutrient availability predict reproductive success in a mammal. *Ecology*, **90**, 711–719.

DeGabriel, J. L., Moore, B. D., Shipley, L. A. *et al.* (2009b) Inter-population differences in the tolerance of a marsupial folivore to plant secondary metabolites. *Oecologia*, **161**, 539–548.

DeGroot, M. H. (1970) *Optimal Statistical Decisions*. New York: McGraw-Hill.

Farentinos, R. C., Capretta, P. J., Kepner, R. E. and Littlefield, V. M. (1981) Selective herbivory in tassel-eared squirrels: role of monoterpenes in ponderosa pines chosen as feeding trees. *Science*, **213**, 1273–1275.

Felton, A. M., Felton, A., Wood, J. T. *et al.* (2009) Nutritional ecology of *Ateles chamek* in lowland Bolivia: how macronutrient balancing influences food choices. *International Journal of Primatology*, **30**, 675–696.

Feng, Z. L., Liu, R. S. and DeAngelis, D. L. (2008) Plant–herbivore interactions mediated by plant toxicity. *Theoretical Population Biology*, **73**, 449–459.

Feng, Z. L., Liu, R. S., DeAngelis, D. L. *et al.* (2009) Plant toxicity, adaptive herbivory, and plant community dynamics. *Ecosystems*, **12**, 534–547.

Foley, W. J., McIlwee, A., Lawler, I. *et al.* (1998) Ecological applications of near infrared reflectance spectroscopy: a tool for rapid, cost-effective prediction of the composition of plant and animal tissues and aspects of animal performance (review). *Oecologia*, **116**, 293–305.

Foley, W. J., Iason, G. R. and McArthur, C. (1999) Role of plant secondary metabolites in the nutritional ecology of mammalian herbivores: how far have we come in 25 years? In *Nutritional Ecology of Herbivores: Proceedings of the Vth International Symposium on the Nutrition of Herbivores*. Savoy, IL: American Society of Animal Science, 130–209.

Foley, W. J. and Moore, B. D. (2005) Plant secondary metabolites and vertebrate herbivores – from physiological regulation to ecosystem function. *Current Opinion in Plant Biology*, **8**, 430–435.

Fraenkel, G. S. (1959) The *raison d'être* of secondary plant substances. *Science*, **129**, 1466–1470.

Freeland, W. J. and Janzen, D. H. (1974) Strategies in herbivory by mammals: the role of plant secondary compounds. *American Naturalist*, **108**, 269–287.

Fryxell, J. M. (2008) Predictive modelling of patch use by terrestrial herbivores. In H. H. T. Prins and F. van Langevelde (eds.) *Resource Ecology. Spatial and Temporal Dynamics of Foraging*. Dordrecht: Springer, 105–123.

Fryxell, J. M., Hazell, M., Borger, L. *et al.* (2008) Multiple movement modes by large herbivores at multiple spatiotemporal scales. *Proceedings of the National Academy of Sciences USA*, **105**, 19114–19119.

Gleadow, R. M. and Woodrow, I. E. (2002) Constraints on effectiveness of cyanogenic glycosides in herbivore defense {review}. *Journal of Chemical Ecology*, **28**, 1301–1313.

Hakes, A. S. and Cronin, J. T. (2011) Environmental heterogeneity and spatiotemporal variability in plant defense traits. *Oikos*, **120**, 452–462.

Hobbs, N. T., Gross, J. E., Shipley, L. A., Spalinger, D. E. and Wunder, B. A. (2003) Herbivore functional response in heterogeneous environments: a contest among models. *Ecology*, **84**, 666–681.

Iason, G. (2005) The role of plant secondary metabolites in mammalian herbivory: ecological perspectives. *Proceedings of the Nutrition Society*, **64**, 123–131.

Iason, G. R. and Palo, R. T. (1991) Effects of birch phenolics on a grazing and a browsing mammal – a comparison of hares. *Journal of Chemical Ecology*, **17**, 1733–1743.

Jakubas, W. J., Karasov, W. H. and Guglielmo, C. G. (1993) Ruffed grouse tolerance and biotransformation of the plant secondary metabolite coniferyl benzoate. *Condor*, **95**, 625–640.

Kokaly, R. F., Asner, G. P., Ollinger, S. V., Martin, M. E. and Wessman, C. A. (2009) Characterizing canopy biochemistry from imaging spectroscopy and its application to ecosystem studies. *Remote Sensing of Environment*, **113**, S78–S91.

Kotliar, N. B. and Wiens, J. A. (1990) Multiple scales of patchiness and patch structure – a hierarchical framework for the study of heterogeneity. *Oikos*, **59**, 253–260.

Krockenberger, A. K. and Hume, I. D. (2007) A flexible digestive strategy accommodates the nutritional demands of reproduction in a free-living folivore, the koala (*Phascolarctos cinereus*). *Functional Ecology*, **21**, 748–756.

Laca, E. A. (2008) Foraging in a heterogeneous environment: intake and diet choice. In H. H. Prins and F. van Langevelde (eds.) *Resource Ecology: Spatial and Temporal Dynamics of Foraging*. Dordrecht: Springer, 81–100.

Lambdon, P. W. and Hassall, M. (2005) How should toxic secondary metabolites be distributed between the leaves of a fast-growing plant to minimize the impact of herbivory? *Functional Ecology*, **19**, 299–305.

Li, Y., Feng, Z. L., Swihart, R., Bryant, J. and Huntly, N. (2006) Modeling the impact of plant toxicity on plant–herbivore dynamics. *Journal of Dynamics and Differential Equations*, **18**, 1021–1042.

Logan, M. and Sanson, G. D. (2002) The effect of tooth wear on the feeding behaviour of free-ranging koalas (*Phascolarctos cinereus* Goldfuss). *Journal of Zoology*, **256**, 63–69.

Marsh, K. J., Wallis, I. R., McLean, S., Sorensen, J. S. and Foley, W. J. (2006) Conflicting demands on detoxification pathways influence how common brushtail possums choose their diets. *Ecology*, **87**, 2103–2112.

Marsh, K. J., Wallis, I. R. and Foley, W. J. (2007) Behavioural contributions to the regulated intake of plant secondary metabolites in koalas. *Oecologia*, **154**, 283–290.

McArt, S. H., Spalinger, D. E., Collins, W. B. *et al.* (2009) Summer dietary nitrogen availability as a potential bottom-up constraint on moose in south-central Alaska. *Ecology*, **90**, 1400–1411.

McLean, S. and Duncan, A. J. (2006) Pharmacological perspectives on the detoxification of plant secondary metabolites: implications for ingestive behavior of herbivores. *Journal of Chemical Ecology*, **32**, 1213–1228.

McLean, S., Brandon, S., Boyle, R. R. and Wiggins, N. L. (2008) Development of tolerance to the dietary plant secondary metabolite 1,8-cineole by the brushtail possum (*Trichosurus vulpecula*). *Journal of Chemical Ecology*, **34**, 672–680.

McNaughton, S. J. (1978) Serengeti ungulates – feeding selectivity influences effectiveness of plant defense guilds. *Science*, **199**, 806–807.

Milchunas, D. G. and Noy-Meir, I. (2002) Grazing refuges, external avoidance of herbivory and plant diversity. *Oikos*, **99**, 113–130.

Moore, B. D. and Foley, W. J. (2005) Tree use by koalas in a chemically complex landscape. *Nature*, **435**, 488–490.

Moore, B. D., Wallis, I. R., Wood, J. and Foley, W. J. (2004) Foliar nutrition, site quality and temperature affect foliar chemistry of tallowwood (*Eucalyptus microcorys*). *Ecological Monographs*, **74**, 553–568.

Moore, B. D., Foley, W. J., Wallis, I. R., Cowling, A. and Handasyde, K. A. (2005) A simple understanding of complex chemistry explains feeding preferences of koalas. *Biology Letters*, **1**, 64–67.

Moore, B. D., Lawler, I. R., Wallis, I. R., Beale, C. M. and Foley, W. J. (2010) Palatability mapping: a koala's eye view of spatial variation in habitat quality. *Ecology*, **91**, 3165–3176.

O'Reilly-Wapstra, J. M., McArthur, C. and Potts, B. M. (2004) Linking plant genotype, plant defensive chemistry and mammal browsing in a *Eucalyptus* species. *Functional Ecology*, **18**, 677–684.

Pass, G. J. and Foley, W. J. (2000) Plant secondary metabolites as mammalian feeding deterrents: separating the effects of the taste of salicin from its post-ingestive consequences in the common brushtail possum (*Trichosurus vulpecula*). *Journal of Comparative Physiology B: Biochemical, Systemic and Environmental Physiology*, **170**, 185–192.

Pass, G. J., McLean, S., Stupans, I. and Davies, N. (2001) Microsomal metabolism of the terpene 1,8-cineole in the common brushtail possum (*Trichosurus vulpecula*), koala (*Phascolarctos cinereus*), rat and human. *Xenobiotica*, **31**, 205–221.

Pigot, A. L. and Leather, S. R. (2008) Invertebrate predators drive distance-dependent patterns of seedling mortality in a temperate tree *Acer pseudoplatanus*. *Oikos*, **117**, 521–530.

Provenza, F. D. (1996) Acquired aversions as the basis for varied diets of ruminants foraging on rangelands. *Journal of Animal Science*, **74**, 2010–2020.

Raubenheimer, D. and Simpson, S. J. (2009) Nutritional PharmEcology: doses, nutrients, toxins, and medicines. *Integrative and Comparative Biology*, **49**, 329–337.

Schmidt, K. A. (2000) Interactions between food chemistry and predation risk in fox squirrels. *Ecology*, **81**, 2077–2085.

Searle, K. R., Hobbs, N. T. and Shipley, L. A. (2005) Should I stay or should I go? Patch departure decisions by herbivores at multiple scales. *Oikos*, **111**, 417–424.

Searle, K. R., Vandervelde, T., Hobbs, N. T., Shipley, L. A. and Wunder, B. A. (2006) Spatial context influences patch residence time in foraging hierarchies. *Oecologia*, **148**, 710–719.

Searle, K. R., Hobbs, N. T. and Gordon, I. J. (2007) It's the 'foodscape', not the landscape: using foraging behavior to make functional assessments of landscape condition. *Israel Journal of Ecology and Evolution*, **53**, 297–316.

Shipley, L. A., Forbey, J. S. and Moore, B. D. (2009) Revisiting the dietary niche: when is a mammalian herbivore a specialist? *Integrative and Comparative Biology*, **49**, 274–290.

Simpson, S. J. and Raubenheimer, D. (1999) Assuaging nutritional complexity: a geometrical approach. *Proceedings of the Nutrition Society*, **58**, 779–789.

Simpson, S. J. and Raubenheimer, D. (2005) Obesity: the protein leverage hypothesis. *Obesity Reviews*, **6**, 133–142.

Skidmore, A. K. and Ferwerda, J. G. (2008) Resource distribution and dynamics. In H. H. T. Prins and F. van Langevelde (eds.) *Resource Ecology: Spatial and Temporal Dynamics of Foraging*. Dordrecht: Springer, 57–77.

Sorensen, J. S., McLister, J. D. and Dearing, M. D. (2005) Plant secondary metabolites compromise the energy budgets of specialist and generalist mammalian herbivores. *Ecology*, **86**, 125–139.

Sorensen, J. S., Skopec, M. M. and Dearing, M. D. (2006) Application of pharmacological approaches to plant–mammal interactions. *Journal of Chemical Ecology*, **32**, 1229–1246.

Swihart, R. K., DeAngelis, D. L., Feng, Z. and Bryant, J. P. (2009) Troublesome toxins: time to re-think plant–herbivore interactions in vertebrate ecology. *BMC Ecology*, **9**, 5.

Torregrossa, A. M. and Dearing, M. D. (2009) Nutritional toxicology of mammals: regulated intake of plant secondary compounds. *Functional Ecology*, **23**, 48–56.

van Beest, F.M., Mysterud, A., Loe, L.E. and Milner, J.M. (2010) Forage quantity, quality and depletion as scale-dependent mechanisms driving habitat selection of a large browsing herbivore. *Journal of Animal Ecology*, **79**, 910–922.

van Langevelde, F. and Prins, H.H.T. (2008) Introduction to resource ecology. In H.H. Prins and F. van Langevelde (eds.) *Resource Ecology: Spatial and Temporal Dynamics of Foraging*. Dordrecht: Springer, 1–6.

Wallis, I.R., Nicolle, D. and Foley, W.J. (2010) Available and not total nitrogen in leaves explains key chemical differences between the eucalypt subgenera. *Forest Ecology and Management*, **260**, 814–821.

WallisDeVries, M.F. and Daleboudt, C. (1994) Foraging strategy of cattle in patchy grassland. *Oecologia*, **100**, 98–106.

Wiggins, N.L., Marsh, K.J., Wallis, I.R., Foley, W.J. and McArthur, C. (2006a) Sideroxylonal in *Eucalyptus* foliage influences foraging behaviour of an arboreal folivore. *Oecologia*, **147**, 272–279.

Wiggins, N.L., McArthur, C., Davies, N.W. and McLean, S. (2006b) Spatial scale of the patchiness of plant poisons: a critical influence on foraging efficiency. *Ecology*, **87**, 2236–2243.

Wilmshurst, J.F., Fryxell, J.M. and Hudson, R.J. (1995) Forage quality and patch choice by wapiti (*Cervus elaphus*). *Behavioral Ecology*, **6**, 209–217.

CHAPTER THIRTEEN

Plant secondary metabolite polymorphisms and the extended chemical phenotype

GLENN R. IASON, BEN D. MOORE, JACK J. LENNON, JENNI A. STOCKAN, GRAHAM H. R. OSLER, COLIN D. CAMPBELL, DAVID A. SIM and JOAN R. BEATON
Ecological Sciences Group, The James Hutton Institute
JOANNE R. RUSSELL
The James Hutton Institute

13.1 Introduction

As it was originally proposed, the extended phenotype comprised 'all effects of a gene upon the world' (Dawkins, 1989) and portrayed how the effects of a gene borne by an organism influenced its biotic and abiotic environments. The consideration of indirect genetic effects, in which an organism's phenotype becomes part of the selective environment of conspecifics (Wolf *et al.*, 1998), was developed rigorously in the population genetics context and the concept subsequently extended to include effects on heterospecifics (Whitham *et al.*, 2003). The extended phenotype concept has been adopted as a framework by some evolutionary biologists and ecologists to study the roles of plant secondary metabolites (PSMs) since Whitham *et al.* (2003) used heritable variation in tissue tannin concentrations among *Populus* species and hybrids to develop the concept of community and ecosystem genetics (Antonovics, 1992).

Many studies of how genetically determined variation in plant traits, including PSMs, drive associated community phenotypes and processes, have been based on differences between hybrids (Dungey *et al.*, 2000; Hochwender & Fritz, 2004; Bailey *et al.*, Chapter 14). Fewer studies have investigated the effects on extended phenotypes of continuously varying PSMs or between known genotypes within a species (Whitham *et al.*, 2006; Schweitzer *et al.*, 2008; Barbour *et al.*, 2009; O'Reilly-Wapstra *et al.*, Chapter 2). A convenient approach to identification and utilisation of genotypic variation for the study of multiple effects of PSMs is provided by the use of genetic polymorphisms. A polymorphism can be defined as occurring when a trait such as a morphological or biochemical character exists in two or more distinct forms in a randomly mating population within a species (Ford, 1975).

The Ecology of Plant Secondary Metabolites: From Genes to Global Processes, eds. Glenn R. Iason, Marcel Dicke and Susan E. Hartley. Published by Cambridge University Press. © British Ecological Society 2012.

The approach is particularly useful in species that cannot be readily cloned. Here, we review examples of how intra-specific variation in a particular group of PSMs, the monoterpenes, has informed our understanding of how PSMs can play multiple ecological roles and mediate the extended phenotype of plants. The monoterpenes are a group of low-molecular-weight, volatile terpenoids which form a very diverse group in terms of number of compounds, structure and function (Gershenzon & Dudareva, 2007). We use variation within species which are polymorphic for concentrations or presence of monoterpenes to provide an insight into their ecological ramifications and larger-scale consequences, against the background of intra-specific variation in other traits. This approach can also be used to inform strategies of selection or transgenic manipulation for commercial purposes or to provide additional tools to reveal ecological function (Bohlmann *et al.*, 2004; Aharoni *et al.*, 2005; Schuman & Baldwin, Chapter 15; Dicke *et al.*, Chapter 16).

The monoterpenes have multiple functions and in many species, including the main examples used in this chapter, are also under strong genetic control. Their heritabilities are known to be very high, and the genomic and molecular backgrounds to their syntheses are well understood (Hiltunen, 1976; Thompson *et al.*, 2003; Keeling & Bohlmann, 2006; Degenhardt *et al.*, 2009). They can therefore potentially form a bridge between genes and ecosystems. We use examples of polymorphisms in monoterpenes found in common thyme (*Thymus vulgaris*) and Scots pine (*Pinus sylvestris*), to examine, respectively, (1) the ecological significance of different monoterpenes produced by the same species, and (2) the multiple effects of a single compound produced by a species and how it influences different components of the extended phenotype. We briefly review the known ecological functions of monoterpenes more broadly, and discuss the biotic and abiotic interactions that might be mediated by them in these two species. These reviewed examples extend from pairwise interactions with other species to the determination of community composition via competition and influence on ecosystem processes and function.

13.2 Ecological functions of monoterpenes

13.2.1 Monoterpene properties and their effects at the cellular level

The biological activity of monoterpenes is largely due to their lipophilicity, which means they can impair membrane integrity and function (Cox & Markham, 2007; Uribe-Carvajal *et al.*, 2008), including in chloroplasts (Klingler *et al.*,1991). For plants, monoterpene autotoxicity is therefore a possibility, but can be averted by the storage of the compounds in vacuoles, trichomes and resin canals (McKey, 1979). The study of allelopathic effects of either individual or mixtures of monoterpenes against other plants or microbes has provided useful information, but this research is often con-ducted in the absence of a rigorous framework, and rarely allows modes of action to be identified or even, in some cases, identification of the compounds involved (Vokou, 2007). However some notable attempts to identify cellular-level mechanisms of action of monoterpenes have occurred in the context of allelopathic effects. Specifically, the common monoterpene α-pinene, abun-dant in many woody plants including *Artemisia scoparia*, leads to cytotoxicity and oxidative damage due to the generation of reactive oxygen species that results in inhibition of seed germination and shoot and root growth of poten-tially competitive weed species (Singh *et al.*, 2006, 2009; Jin *et al.*, 2010). The oxygenated monoterpenes appear to produce the most potent allelopathic effects (Kordali *et al.*, 2007), but other aspects of molecular structure also play a role. For example, the antimicrobial activity of carvacrol is attributable both to its high lipophilicity and also to the presence of a system of delocalised electrons, which allow the molecule to act as a proton exchanger, reducing the gradient across the microbial cytoplasmic membrane (Ben Arfa *et al.*, 2006). Other effects including antiviral properties are of possible medical significance (Astani *et al.*, 2010), but are as yet unexplored outwith the cellular environment and in natural ecosystems. These examples illustrate that mono-terpenes have multiple structure-based modes of action at the cellular level.

13.2.2   Other ecological effects attributed to monoterpenes

Many monoterpenes are nematicidal (Kong *et al.*, 2007), acaricidal (Macchioni *et al.*, 2002; Cetin *et al.*, 2010), molluscicidal and repellent against molluscs (Frank *et al.*, 2002; Abdelgaleil, 2010), toxic and repellent to insects (Schlyter *et al.*, 2004; Kordali *et al.*, 2007; Salle & Raffa, 2007) and mammalian herbivores (Duncan *et al.*,1994; Iason, 2005), and mycotoxic (Nakahara *et al.*, 2003; Ludley *et al.*, 2008). They can also act as attractants for insect pollinators (Beker *et al.*, 1989). However, owing to co-evolutionary relationships between specialist insect herbivores and their host plants, volatiles such as monoterpenes can be interpreted as signals acting as attractants for feeding or oviposition by herbivores, phagostimulants or aggregation promoters (Feeney, 1991; Carisey & Bauce, 1997), all of which are of potential disadvantage to the plant.

13.2.3   Monoterpene effects on community composition and multi-trophic interactions

The induction and release of volatiles including monoterpenes, by plants attacked by herbivores, can attract enemies of the herbivores, mediating multi-trophic interactions in both the above- and belowground environment (Kessler & Baldwin, 2004; Rasmann *et al.*, 2005; Gershenzon *et al.*, Chapter 4; Schuman & Baldwin, Chapter 15; Dicke *et al.*, Chapter 16). Integration of these above- and belowground processes occurs via plants' internal vertical chemical communication (van Dam, Chapter 10). The mechanisms of interactions involving monoterpenes that produce broader community effects can be complex, e.g. with aggregation pheromones of bark beetles mediating their efficacy (Erbilgin & Raffa, 2000) and combinations rather than single monoterpenes influence numerous other plant–insect interactions (Cates, 1996; Gershenzon *et al.*, Chapter 4). The chemical diversity of needle monoterpenes is positively correlated with the species richness and composition of ground flora beneath individual Scots pine trees (Iason *et al.*, 2005; Pakeman *et al.*, 2006). The monoterpenes as a group influence the composition and structure of belowground fungal communities because of the differential susceptibility of the fungal species to growth inhibition by different monoterpenes (Ludley *et al.*, 2008).

13.2.4   The effects of monoterpenes on soil function

Terpenoids form a large component of the volatile organic compounds emitted by the shoots and roots of plants, which have been most extensively studied in conifer-dominated systems. They enter the soil system via fallen litter or from plant roots (Asensio *et al.*, 2008; Maurer *et al.*, 2008; Ludley *et al.*, 2009a) and their volatility favours diffusion through the gaseous phase (Hiltpold & Turlings, 2008). Litter degradation releasing monoterpenes is a slow process taking several months (Kainulainen & Holopainen, 2002; Ludley *et al.*, 2009b), and monoterpenes can reach high concentrations in the soil atmosphere (Smolander *et al.*, 2006) where they influence belowground processes. Monoterpenes can be rapidly metabolised by soil microorganisms (Owen *et al.*, 2007) including bacteria (Misra *et al.*, 1996), some of which use them as a carbon source (Smolander *et al.*, 2006; Maurer *et al.*, 2008). Conversely, the rate of respiration of some saprotrophic fungi, and biomass growth of both saprophytic and ectomycorrhizal (ECM) fungi was depressed by monoterpenes, although the rate of colonisation of root tips by ECM fungi was greater in the presence of monoterpenes (Ludley *et al.*, 2009b). The most consistent effect of monoterpenes in soil systems is their inhibition of nitrification and generally negative effects on soil nitrogen cycling (White, 1991; Paavolainen *et al.*, 1998; Smolander *et al.*, 2006; Uusitalo *et al.*, 2008), possibly through direct inhibition of nitrifying bacteria (Ward *et al.*, 1997) and/or immobilisation of labile N utilised by organisms degrading the carbon-rich monoterpenes (White, 1994).

We conclude that monoterpenes can have numerous ecological functions, and it should be of no surprise that monoterpenes as a diverse group of compounds, and as individual compounds, influence a wide range of functions within a single system that contains multiple receptor organisms. However, despite monoterpenes being very well studied, coordinated evidence for their multi-functionality within single systems is scant. We review evidence of their effects within two relatively well-studied systems: (1) common thyme, in which a monoterpene polymorphism defines a number of distinct chemotypes, and (2) Scots pine, the two main chemotypes of which either possess or lack a specific monoterpene, δ-3-carene.

13.3   Polymorphic variation and the ecological roles of monoterpenes in common thyme

Although there are many naturally occurring phytochemical polymorphisms reflecting intra-specific genetic variability, few if any have been as deeply investigated as that in common thyme. Natural populations of common thyme are polymorphic and comprise different frequencies of at least six chemotypes, each of which has a dominant monoterpene resulting from a few epistatically interacting loci (Linhart & Thompson, 1999). The genetic and phytochemical polymorphism has been intensively and quantitatively studied, including the occurrence and correlations among the monoterpene and other terpenoid constituents (Thompson *et al.*, 2003), and aspects of gene flow and genetic differentiation among and within populations (Tarayre *et al.*, 1997). These exemplary studies of thyme demonstrate the diverse and far-reaching ecological causes and consequences of the phytochemical polymorphism.

The chemotypes segregate among habitats (Thompson *et al.*, 1998) and have differential survival in relation to climatic conditions, as they vary in frequency at both the local landscape scale of a few kilometres, and over broader geographical scales. Particular chemotypes survive either summer drought conditions or early winter frosts (Thompson *et al.*, 2007), the latter being the result of a specific trichome-related adaptation which releases volatiles during freezing (Amiot *et al.*, 2005). In addition to the maintenance of the polymorphism by these adaptations to the physical environment, it is also maintained by biotic interactions that vary between chemotypes, as a result of their monoterpene composition. It was concluded that a combination of both differential selective herbivory by insect, mollusc and mammalian herbivores, and differential allelopathic activity of the chemotypes that inhibit seed germination of a commonly competing grass species (*Brachypodium phoenicoides*), also contribute to maintenance of the polymorphism (Tarayre *et al.*, 1995; Linhart & Thompson, 1999). Additionally, an elegant experimental approach demonstrated that the flowering success of the thyme chemotypes was differentially negatively influenced by aphids in the absence of their predators. The strength of this interaction was influenced by competition from another grass species, *Bromus madritensis* (Linhart *et al.*, 2005). The local soil chemical environment, influenced by the monoterpene composition of the thyme chemotypes, can also shape the associated plant communities via modification of interactions with co-occurring species. Germination, growth and reproductive effort of *Bromus erectus*, a co-occurring grass species, responded according to the chemotype of thyme that had been previously growing on the soil medium. These responses, however, also varied with whether or not the *Bromus* seed itself was collected from plants associated with the same or different thyme chemotype as the growing medium (Ehlers & Thompson, 2004). This approach has been extended to other chemo-typically varying thyme species and their associated species (Grondahl & Ehlers, 2008). The response of some thyme-associated species to soil containing the thyme phenolic-monoterpene carvacrol varied from negative through to positive, according to whether the species' site of origin contained thyme (Jensen & Ehlers, 2010). This work confirms the importance of co-adaptation of associated species to the thyme chemotypes and firmly implicates the monoterpene polymorphisms in thyme in the co-evolution and dynamics of the associated plant community.

13.4   Recent studies of the variation and ecological roles of δ-3-carene in Scots pine woodland

Scots pine is a coniferous tree with a widespread distribution ranging from the Mediterranean to the Arctic in Europe, and from western Scotland across the high temperate regions of Asia to the Okhotsk Sea in eastern Siberia. It is characterised by substantial concentrations of monoterpenes in its tissues, the monoterpene composition of which is very variable among individuals within a population (Kinloch *et al.*, 1986). In Scots pine the broad sense heritabilities of the monoterpenes are 70–98% (Hiltunen, 1976) and they are relatively unresponsive to environmental manipulation and difficult to induce by simulated herbivore damage (Honkanen *et al.*, 1999; Iason *et al.*, 2011). The longevity of the trees, and the year-to-year stability of the genetically determined constitutive monoterpenes (Thoss *et al.*, 2007), mean that a Scots pine tree will influence its local chemical environment consistently over many years. We therefore hypothesise that the chemical composition of the tree may be a driver of the tree's extended phenotype, including interactions with other organisms, community composition of associated organisms and local parameters of ecosystem function, particularly those associated with the soil, which receives chemical inputs via roots and needle litter (Ludley *et al.*, 2009a). Here, we examine the consequences of a specific chemotypic difference among individual pine trees, that of the presence or absence of δ-3-carene in the needles, on plant–vertebrate herbivore interactions, on the plant–invertebrate interaction and how this is modified by wood ants, and on impacts on soil mite community composition and decomposition.

Previous studies have reported bi-modal or tri-modal distributions of the frequency of the proportion of total monoterpenes represented by δ-3-carene in the cortico-resin of Scots pine (Kinloch *et al.*, 1986; Muona *et al.*, 1986).

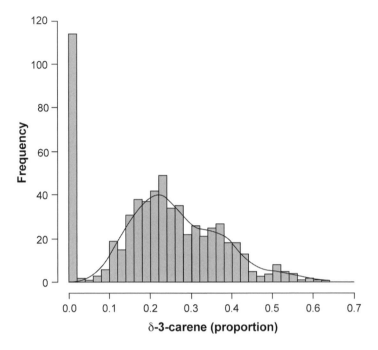

Figure 13.1 Histogram and associated density function illustrating the percentage of δ-3-carene in the needle monoterpenes of 630 Scots pine trees drawn from 21 populations across Scotland.

Our own data from 21 populations across Scotland show that trees without δ-3-carene are readily distinguishable and strongly differentiated from those with the compound (Figure 13.1). This is the starkest contrast among monoterpenes in Scots pine and suggests a simple inheritance at a single locus (Iason *et al.*, 2011), although the distribution of the concentration and the proportion of δ-3-carene in the total monoterpenes is likely to arise from a more complex inheritance. When it is present, δ-3-carene is the second-most abundant monoterpene in Scots pine, comprising about 25% of the total; this proportion is of course partially dependent on the concentration of the other monoterpenes, which themselves may be epistatically controlled (Kinloch *et al.*, 1986). These factors, and its lack of enantiomeric variation, makes δ-3-carene a good model system in which to study the ecological effects of a monoterpene (Thoss *et al.*, 2007).

13.4.1   Monoterpenes and high-impact herbivores

Work on foraging by high-impact herbivores on Scots pine was recently summarised to examine the evolutionary mechanisms underlying the diversity of monoterpenes (Iason *et al.*, 2011). In experiments on selection of Scots pine trees by red deer (*Cervus elaphus*), under extensive grazing conditions in

0.25 ha plots containing trees superimposed in pots onto a vegetation back-ground of moorland dominated by *Calluna vulgaris*, we showed a weak but significant effect of δ-3-carene in reducing the damage inflicted on the trees, whilst no other monoterpenes explained herbivory. In contrast, an experi-mental study of bank voles (*Myodes glareolus*) on their selection of 1200 Scots pine seedlings found no effects of the monoterpene composition on the probability or extent of herbivory, and an inductive field study of capercaillie (*Tetrao urogallus*), a forest galliform bird which browses Scots pine needles, found that monoterpenes other than δ-3-carene were associated negatively with the use of the trees by this bird. A study of the mortality inflicted by slugs (*Arion ater*) on very young Scots pine seedlings less than 3 weeks old showed that a greater proportion of seedlings that were eaten were of the zero δ-3-carene chemotype (Figure 13.2). But positive correlations among the con-centrations of monoterpenes present meant that α-pinene and β-pinene were also positively associated with survival of the seedlings. Both δ-3-carene and α-pinene also significantly reduced the intake of an artificial diet by slugs (O'Reilly-Wapstra *et al.*, 2007). From this compiled work, it was concluded that some but by no means all of the diversity of the monoterpenes in Scots pine could be explained by defensive functions against these high-impact herbi-vores, and that δ-3-carene seems to be associated with seedling survival by reducing the risk of attack by slugs, but has only a weak effect as a defence against red deer. In all studies, other plant parameters also influenced herbi-vore foraging either independently or in conjunction with the monoterpene,

Figure 13.2 Scots pine seedlings eaten by the large black slug (*Arion ater*) in a feeding experiment contained a higher percentage of plants of the zero δ-3-carene chemotype than did trees that were not eaten (O'Reilly-Wapstra *et al.*, 2007). The numbers of seedlings in each class are shown. Needle δ-3-carene concentration explained a significant proportion of the deviance in a logistic regression model ($\chi^2 = 6.88$, $P = 0.0087$) of whether or not the seeding was eaten by slugs. Stem diameter, needle nitrogen concentration and needle α-pinene concentration were also significant in alternative models.

namely seedling location, stem diameter and nitrogen concentration (slugs), associated plant community and whether it had been mown (voles), needle length and stem diameter (red deer), and needle magnesium concentration (capercaillie).

13.4.2   Interactions between δ-3-carene, crown invertebrates and ants

Wood ants are known to have a high impact on the ecology of forest systems including trophic relationships and nutrient transfers, which they influence on a strongly spatial basis (Mooney, 2007; Domisch *et al.*, 2009). The ants mediate several types of trophic interaction (Mooney & Tillberg, 2005), e.g. facultatively tending or predating aphids, and predation of herbivores and competing predatory invertebrates. These interactions and their impact on invertebrate communities are likely to vary with the characteristics of the host-plant species, including plant defensive chemistry against herbivores and chemical defence by herbivores against their predators including ants (Sipura, 2002; Lindstedt *et al.*, 2006). We investigated the basis for selection of Scots pine trees by ants and showed that after the main environmental effect reflecting habitat characteristics was removed, δ-3-carene was the main determinant of whether the tree was used by ants, with ants using the zero δ-3-carene chemotype significantly more than the δ-3-carene chemotype (Figure 13.3). Of the randomly pre-selected sample of pine trees used for this work undertaken at Ballochbuie Wood, northeast Scotland, eight were of the zero δ-3-carene chemotype. Of those eight, 87.5% were used by ants as compared with only 48.6% of the δ-3-carene trees ($\chi^2 = 4.283$, $df = 1$, $P = 0.038$), indicating a strong preference for the zero δ-3-carene chemotype. The use of the tree by ants was negatively associated with the number of non-aphid phytophagous arthropods in the crown ($F = 11.91$, $df = 1,44$, $P < 0.001$, $r^2 = 19.5\%$) and positively associated with the number of aphids ($F = 21.58$, $df = 1,44$, $P < 0.001$, $r^2 = 31.4$) trapped in five 30×20 cm funnel traps mounted at 0.75 m height beneath the crown of each of 45 Scots pine trees, from April 2002 to October 2003. A fine mesh filter in the base of each funnel allowed retention of arthropod material but not water. Traps were emptied 32 times; 4-weekly during the winter months and otherwise at 2-weekly intervals. The glossy finish of the vertical upper wall of the funnels prevented escapes. These results indicate the strong importance of ants as causal agents of crown invertebrate populations (Mooney & Tillberg, 2005). An experiment using 30 of the initial 46 trees to experimentally exclude ants from branches found that the only tree that was of the zero δ-3-carene chemotype and that was not used by ants had a large population of crown non-aphid invertebrates (197 trapped) that was greater than the upper 95% confidence interval calculated for the remaining ant-used trees that were zero δ-3-carene (mean: 25.1

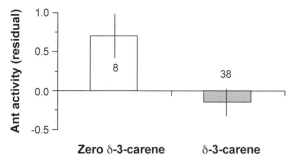

Figure 13.3 The residual effect (mean and SD) of whether or not the trees were zero δ-3-carene chemotype on the activity of ants ($F = 4.77$, df 1,43, $P < 0.05$), after adjusting for the altitude of the tree. The numbers of trees in each sample are shown. Ant activity was determined during a 30 minute observation of each of the 45 study trees in each month from June 2002 until October 2003 in which the total number of descending ants was recorded. If no ants were recorded after 5 minutes (e.g. during winter months) the count was abandoned. The data were transformed by $\log_{10}$ [1 + number of descents] and analysed according to a general linear model with a normal distribution which entered altitude and chemotype as explanatory variables. The main effect of altitude explained significant variation ($F = 12.60$, $df = 1$, 43, $P < 0.001$, regression coefficient = 0.0106, SE = 0.0032) and this variable represents the environment of the tree, as it correlates negatively with local tree density and the northing of the tree's location.

individuals, upper 95% CI: 108.9). However, the small sample of zero δ-3-carene trees in that study means that further experimental work is being undertaken to ascertain the underlying reason for selection of trees by ants, and to test whether this is related to the aphid and non-aphid invertebrate populations, independently of the direct negative feedback effects of ants on these assemblages.

13.4.3   The role of δ-3-carene in shaping soil mite communities and decomposition

Increasing attention is being paid to the effects of phenotypic variation in plants on their associated belowground communities. Phenotypic variation in leaf litter chemistry is known to affect decomposition rates (e.g. Driebe & Whitham, 2000; Madritch & Hunter, 2002), and plant-induced defences are known to affect belowground communities (van Dam *et al.*, 2003), although Madritch and Hunter (2005) found no effect of different oak phenotypes on the abundance of different microarthropod orders. However, they speculated that this level of taxonomic resolution might have been insufficient to detect differences (Madritch & Hunter, 2005). Although the monoterpene concentration of needle litter decreases during decomposition, α- and β-pinene, camphene and δ-3-carene can remain at relatively high concentrations for at least 6 months (Kainulainen & Holopainen, 2002). Given the wide-ranging influences of pine needle monoterpenes on soil processes (reviewed above), we therefore tested the hypothesis that Scots pine trees of different δ-3-carene chemotype in their needles would affect the oribatid mite assemblages at their bases.

Without inclusion of soil moisture, on a per gram of soil basis, there were 65% more species per gram (1.78 ± 0.18 species g^{-1}) associated with the low δ-3-carene trees compared with the high δ-3-carene trees (1.08 ± 0.08 species g^{-1}, $P = 0.025$). Analysis of co-variance indicated that the number of species per soil core removed from beneath the two chemotypes did not respond equally to differences in soil moisture content ($P < 0.005$). When soil moisture was included as a co-variate, a trend for greater species richness in the low δ-3-carene trees on a per core basis persisted ($P = 0.057$), with low δ-3-carene trees in five of the six pairs of trees having greater species richness (Figure 13.4). Our results indicate that less litter material is supporting a greater number of species under the low δ-3-carene trees, but that it does not increase abundance of mites.

An associated experiment using the same study trees showed that decomposition rates of a standard pool of Scots pine litter in nylon litter bags with a 2 mm mesh (16 replicates per tree), located at the soil surface, decomposed at a faster rate under the zero δ-3-carene trees. The mass loss after 12 weeks was 31.9% under the δ-3-carene trees compared with 35.2% under the zero δ-3-carene trees ($P = 0.074$, *least significant difference* 3.6). After 36 weeks the mass loss was 47.4% under the δ-3-carene trees compared with 50.9% under the zero δ-3-carene trees ($P = 0.034$, *least significant difference* 3.0). The difference in species richness of mites per core may not be directly determined by the needle chemistry but may instead be driven by factors that co-vary with needle chemistry and have indirect effects, such as rates of input of needles, decomposition rates or microbial population size or composition. Manipulative experiments are required to elucidate the factors that may drive this difference in the oribatid mite communities, but the experiments reported here indicate that monoterpene chemotypes influence both associated soil invertebrate communities and the decomposition of litter, an important measure of ecosystem function.

13.4.4 Is large- and small-scale spatial variation in δ-3-carene chemotypes driven by natural selection?

A large-scale analysis of variation of the proportions of monoterpenes in the cortico-resin among 40 of the highly fragmented Scots pine populations in Scotland showed low G_{st} values (0.045) indicating large intra-population versus smaller inter-population variability (Kinloch *et al.*, 1986). We collected needle samples from 30 individuals in 21 Scots pine populations to analyse needle monoterepnes, since we considered these to be more relevant to a

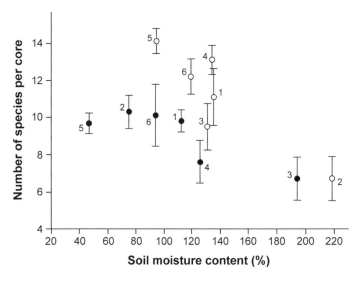

Figure 13.4 The effects of soil moisture (%) and δ-3-carene chemotype on the species richness of soil oribatid mites (mean ± SE, n = 10 for each point). Six zero δ-3-carene trees (open circles) were paired with six δ-3-carene trees (closed circles) in similar locations, aspect and soil type for comparison. The paired trees are indicated by the same numbers. Ten soil cores were taken randomly within 50 cm of the base of the 12 trees, where virtually no ground vegetation was present, in October 2004. The soil cores were 3.5 cm in diameter and 5 cm deep, and were cooled on collection, before being slowly increased to 40 °C over 7 days in modified Tullgren funnels for soil animal collection into 70% ethanol. All adult oribatids were sorted using microscopy to species level except the Brachychthoniidae and Suctobelbidae, which were grouped by family. Analyses with and without these groups yielded similar results. Soil moisture was determined by collecting three cores of the same size as the microarthropod cores randomly from the tree base. The whole core was separated by hand, and moisture determined gravimetrically after oven drying at 105 °C for 48 hr; hence the dry soil mass represented the bulk density of the cores. Oribatid community structure was described as the number of individuals, number of species and Shannon species diversity on a whole tree basis and on a per core basis, presented here (Kaneko *et al.*, 2005). Analysis of variance (ANOVA) and analysis of co-variance (ANCOVA) were conducted, with the treatment being δ-3-carene level (zero 3-carene, 3-carene) and tree pairs as blocks (six pairs), with soil moisture as a co-variate.

broader spectrum of ecological interactions. We confirmed the relatively high frequency of zero δ-3-carene chemotypes in the Shieldaig population in the Wester Ross area, in the northwest of Scotland where 76.7% of trees were zero δ-3-carene, as compared with only 15.7% in the remaining 20 populations (χ^2 = 84.49, df = 1, P < 0.0001) (Figure 13.1). This was interpreted to represent this population's origin from a different glacial refugium from the other Scottish populations, which are considered genetically more similar to those on the

European mainland (Kinloch *et al.*, 1986). There was also an increase from the south to the north of Finland in the frequency of the zero δ-3-carene chemotype (Muona *et al.*, 1986). However, this discontinuity of δ-3-carene chemotypes between the Scottish populations, and the north–south cline in northern Europe, may represent differences in selection for δ-3-carene. At a smaller spatial scale, most variability in monoterpenes including δ-3-carene lies within rather than between populations. We therefore sought, for the first time, to quantify the spatial variability in δ-3-carene chemotypes, within a single population of Scots pine, at Ballochbuie in northeast Scotland, in which only 17.5% of mature trees were the zero δ-3-carene chemotype. We tested the spatial variation in δ-3-carene within and between seven 1 ha plots within the overall population of Scots pine. We hypothesised that such spatial variation would, if discovered, be consistent with spatially varying natural selection for the monoterpene, if a set of neutral genetic markers (six simple sequence repeat nuclear microsatellite DNA markers) in the same trees showed no concomitant spatial variation. There was significant variation among plots in needle concentrations of δ-3-carene, independently of any genotypic variation in (presumably) selectively neutral microsatellites (Figure 13.5). This variation is also independent of any spatial variation within the hectare plots. This indicates distinct spatial structure in δ-3-carene chemotypes within the forest which is not duplicated among neutrally varying DNA markers. This strongly suggests that the spatial variation in δ-3-carene chemotypes of Scots pine is not due to any gene flow, founder effects or genetic drift, but rather is the result of spatially varying selection for or against δ-3-carene.

13.5 Conclusion and future directions

The monoterpenes influence multiple ecological processes from above-ground interactions of plants to soil processes, including those driven by plant litter quality which relates directly to shoot chemical quality. They therefore have the potential to drive or influence the outcomes of many ecological processes including herbivory, allelopathy and ecosystem functions such as decomposition. As a diverse group of chemical compounds, the monoterpenes can be expected to influence multiple species within any trophic group, such as herbivores, decomposers or primary producers. However, monoterpenes, such as carvacrol in thyme and δ-3-carene in Scots pine, can also individually influence other species including different taxa within a trophic group that feed on different stages of the plant's life cycle. With the notable exception of studies on aspen hybrids and species (Whitham *et al.*, 2006), and studies of the effects of volatiles of the Brassicaceae and *Nicotiana* on associated communities (Kessler & Baldwin, 2004; Schuman & Baldwin, Chapter 15; Dicke *et al.*, Chapter 16), there exist relatively few

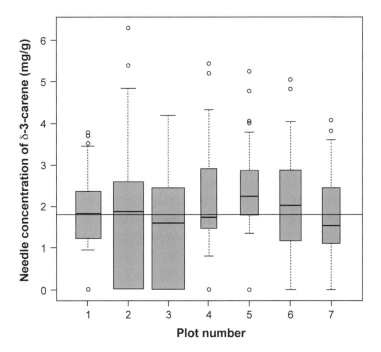

Figure 13.5 The concentrations of needle δ-3-carene of all trees in each of seven 100 m × 100 m plots, within Ballochbuie Wood, Scotland, shown as boxplots: median, interquartile range, 2 and 98 percentile (whiskers) and outlying individual datapoints. The mean spatial distance between the plot centroids was 986 m (SD: 438 m). We collected needles from all mature trees ($n = 442$) and analysed them for monoterpene concentration (Thoss *et al.*, 2007). The needle monoterpene concentration was entered as the response variable using 'multiple regression on distance matrices' (Lichstein, 2007). Explanatory variables were: genetic distance (calculated from six nuclear satellite DNA markers as a matrix of Loiselle genetic distances among trees which incorporates and removes genetic isolation/drift), spatial distance among trees (entered as 5 m distance classes between 5 m and 100 m, which removes the effect of small-scale within-plot spatial variation), and plot identification. The overall model was statistically significant ($F = 29.18$, $P = 0.003$), within which only the effect of plot identification (the effect shown), after adjusting for other model parameters, was statistically significant.

examples where several different functions of the same compound, or groups of compounds, have been investigated within a single system. The long-term studies described here move us towards this goal of understanding the multiple roles of monoterpenes. Given the roles of volatile organic compounds including isoprene and monoterpenes as precursors to ozone production in the troposphere (Bagnoli *et al.*, Chapter 6), these example systems also have the potential to connect the genes for their production with global-scale processes (Lerdau & Gray, 2003). This is especially likely in the case of the Taiga or boreal forests, dominated by Scots pine or similar needle-leaved coniferous trees, which cover approximately 11% of the land surface of the Earth (http://www.borealforest.org/: last accessed 27 April 2009).

There are numerous examples of how plant traits other than PSMs influence the biotic communities associated with them (Wimp *et al.*, 2005; Johnson *et al.*, 2009). In most of the interactions mediated by or involving monoterpenes, these compounds do not drive the interactions in isolation, but act in addition to or in association with some other component of the biotic or abiotic environment. Examples of their effects acting in association with abiotic factors are soil moisture in the case of soil mite communities, and climate and microclimate influencing survival of thyme seedlings in relation to their chemotype. Examples of monoterpenes acting via interactions among biotic components of the environment include their likely effects on Scots pine crown invertebrates in conjunction with a profound influence of wood ants, and in thyme, the differential effect of aphids on different chemotypes being modified by competition from a grass species. We advocate that the ecological roles of PSMs should be considered in an integrated way with non-PSM traits, and note that experimental approaches, rather than reliance on inductive field studies, have shed more light on these complex inter-relationships (Linhart *et al.*, 2005; Lankau & Strauss, 2008).

A genetic basis for between-plant differences in associated soil microbial biomass, microbial community composition and nitrogen cycling has been demonstrated in *Populus* and is mediated by genetic differences in their PSMs (Schweitzer *et al.*, 2004). Differences in nutrient fluxes during decomposition have also been reported among *Quercus* genotypes (Madritch & Hunter, 2002). There are several reasons to suspect that genetically determined monoterpenes would also influence decomposition and soil nutrient dynamics, particularly in coniferous trees and eucalypts in which they are prevalent (O'Reilly-Wapstra *et al.*, Chapter 2). Not only do monoterpenes persist in the litter (Kainulainen & Holopainen, 2002), but they also affect fungal and bacterial decomposers (Langenheim, 1994) and soil mite communities (Figure 13.4). Although chemotypic differentiation on the basis of δ-3-carene in Scots pines appears to materially affect rates of litter decomposition (this study), broader evidence for genotypically varying soil processes in coniferous trees is lacking.

A suitable means to determine the relative effectiveness of a particular PSM, or a group of PSMs that act on many different processes within the same system, is to consider its impact on the fitness of the plant that bears the trait. The general point concerning the correlation between ecological functions of plant traits, including plant secondary chemistry, was previously made by Grime *et al.* (1996) who drew attention to the correlation across species of palatability to herbivores and decomposition. The positive effects on fitness of a defence against herbivores may be offset by negative consequences for the plant, if the trait slows the rate of decomposition, which may render nutrients less available to the plant. However, the ways in which plant litter quality, to which PSMs are a contributory factor, may feed back via soil processes to affect fitness of the bearer of the PSMs are diverse (Northup *et al.*, 1998); they may be positive or negative, and their feedbacks to fitness definitely require further investigation.

Arising from the contrasting multiple environmental factors that provide selective pressures on the monoterpenes (Langenheim, 1994) is the possibility that these operate at different spatial scales, and generate spatial pattern in the distribution of the chemotypes. This is illustrated here for the spatial distribution of thyme chemotypes in relation to climatic variables, and at a smaller scale, for δ-3-carene chemotypes of Scots pine (Figure 13.5). For example, the impact of slug herbivory on Scots pine seedlings is likely to generate selective forces that vary at a scale that contrasts with that of more widely ranging mammalian herbivores. As slugs select damp habitats, the contribution of PSMs to the plant resistance against them is also likely to co-vary with soil moisture. But as yet there has been relatively little study of the causes and consequences of spatial heterogeneity in PSM concentrations and their ecological effects (see Moore & DeGabriel, Chapter 12).

When relating genes to ecosystems, the effects of the PSMs as gene products are potentially numerous and may feed back to affect the fitness of the bearer of the gene positively, negatively or indeed not at all (Biernaskie & Tyerman, 2005; Whitham *et al.*, 2005). We contend that in order to understand the roles of PSMs in ecosystems, we must understand their roles in the full range of simple paired and multiple interactions, in a multi-trophic context, in relation to the feedbacks from ecosystem modification and processes, and from the physical environment. The requirement now is for in-depth studies of multiple compounds, to reveal how they mediate interactions at different trophic levels and their effects on ecosystem processes, and the fitness of the plants producing them. Such a task is not trivial and is likely to require collaborative pooling of multi-disciplinary resources, probably in the context of the common garden approach to isolate genetic from environmental variation. This approach will not only tell us the full ramifications of individual and suites of PSMs, but will also elucidate the individual and combined selective forces generating and maintaining them.

Acknowledgements
Thanks to Robin Pakeman and two anonymous reviewers for helpful comments. Duncan White, Anastasia Koryczynska, Pete Hollingsworth, Richard Ennos and Allan Booth all assisted with discussion, comments or technical skills. We thank Peter Dennis for initiation of the ant studies, the Scottish Government Rural and Environment Research and Analysis Directorate for funding, and forest owners, particularly Balmoral Estate, for their cooperation.

References

Abdelgaleil, S. A. M. (2010) Molluscicidal and insecticidal potential of monoterpenes on the white garden snail, *Theba pisana* (Muller) and the cotton leafworm, *Spodoptera littoralis* (Boisduval). *Applied Entomology and Zoology*, **45**, 425–433.

Aharoni, A., Jongsma, M. A. and Bouwmeester, H. J. (2005) Volatile science? Metabolic engineering of terpenoids in plants. *Trends in Plant Science*, **10**, 594–602.

Amiot, J., Salmon, Y., Collin, C. and Thompson, J. D. (2005) Differential resistance to freezing and spatial distribution in a chemically polymorphic plant *Thymus vulgaris*. *Ecology Letters*, **8**, 370–377.

Antonovics, J. (1992) Toward community genetics. In R. S. Fritz and E. L. Simms (eds.) *Plant Resistance to Herbivores and Pathogens; Ecology Evolution and Genetics*. Chicago, IL: University of Chicago Press, 426–449.

Asensio, D., Owen, S. M., Llusia, J. and Peñuelas, J. (2008) The distribution of volatile isoprenoids in the soil horizons around *Pinus halepensis* trees. *Soil Biology and Biochemistry*, **40**, 2937–2947.

Astani, A., Reichling, J. and Schnitzler, P. (2010) Comparative study on the antiviral activity of selected monoterpenes derived from essential oils. *Phytotherapy Research*, **24**, 673–679.

Barbour, R. C., O'Reilly-Wapstra J. M., De Little, D. W. *et al.* (2009) A geographic mosaic of genetic variation within a foundation tree species and its community-level consequences. *Ecology*, **90** 1762–1772.

Beker, R., Dafni, A., Eisikowitch, D. and Ravid, U. (1989) Volatiles of two chemotypes of *Majorana syriaca* (Labiatae) as olfactory cues for the honeybee. *Oecologia*, **79**, 446–451.

Ben Arfa, A., Combes, S., Preziosi-Belloy, L., Gontard, N. and Chalier, P. (2006) Antimicrobial activity of carvacrol related to its chemical structure. *Letters in Applied Microbiology*, **43**, 149–154.

Biernaskie, J. M. and Tyerman, J. G. (2005) The overextended phenotype. *Ecoscience*, **12**, 3–4.

Bohlmann, J., Martin, D., Miller, B. and Huber, D. (2004) Terpenoid synthases in conifers and poplars. In *Plantation Forest Biotechnology for the 21st Century*, 181–201.

Carisey, N. and Bauce, E. (1997) Balsam fir foliar chemistry in middle and lower crowns and Spruce budworm growth, development, food and nitrogen utilization. *Journal of Chemical Ecology*, **23**, 1963–1978.

Cates, R. G. (1996) The role of mixtures and variation in the production of terpenoids in conifer–insect–pathogen interactions. In J. T. Romeo, J. A. Saunders and P. Barbosa (eds.) *Phytochemical Diversity and Redundancy in Ecological Interactions*. New York: Plenum, 179–216.

Cetin, H., Cilek, J. E., Oz, E. *et al.* (2010) Acaricidal activity of *Satureja thymbra* L. essential oil and its major components, carvacrol and gamma-terpinene against adult *Hyalomma marginatum* (Acari: Ixodidae). *Veterinary Parasitology*, **170**, 287–290.

Cox, S. D. and Markham, J. L. (2007) Susceptibility and intrinsic tolerance of *Pseudomonas aeruginosa* to selected plant volatile compounds. *Journal of Applied Microbiology*, **103**, 930–936.

Dawkins, R. (1989) *The Extended Phenotype: The Long Reach of the Gene*, 2nd edn. Oxford: Oxford University Press.

Degenhardt, J., Kollner, T. G. and Gershenzon, J. (2009) Monoterpene and sesquiterpene synthases and the origin of terpene skeletal diversity in plants. *Phytochemistry*, **70**, 1621–1637.

Domisch, T., Finer, L., Neuvonen, S. *et al.* (2009) Foraging activity and dietary spectrum of wood ants (*Formica rufa* group) and their role in nutrient fluxes in boreal forests. *Ecological Entomology*, **34**, 369–377.

Driebe, E. M. and Whitham, T. G. (2000) Cottonwood hybridization affects tannin and nitrogen content of leaf litter and alters decomposition. *Oecologia*, **123**, 99–107.

Duncan, A. J., Hartley, S. E. and Iason, G. R. (1994) The effect of monoterpene concentrations in Sitka spruce (*Picea sitchensis*) on the browsing behavior of red deer (*Cervus elaphus*). *Canadian Journal of Zoology-Revue Canadienne de Zoologie*, **72**, 1715–1720.

Dungey, H. S., Potts, B. M., Whitham, T. G. and Li, H. F. (2000) Plant genetics affects arthropod community richness and composition: evidence from a synthetic eucalypt hybrid population. *Evolution*, **54**, 1938–1946.

Ehlers, B. K., and Thompson, J. (2004) Do co-occurring plant species adapt to one another? The response of *Bromus erectus* to the presence of different *Thymus vulgaris* chemotypes. *Oecologia*, **141**, 511–518.

Erbilgin, N. and Raffa, K. F. (2000) Opposing effects of host monoterpenes on responses by two sympatric species of bark beetles to their aggregation pheromones. *Journal of Chemical Ecology*, **26**, 2527–2548.

Feeney, P. (1991) The evolution of chemical ecology: contributions from the study of herbivorous insects. In G. A. Rosenthal and M. R. Berenbaum (eds.) *Herbivores: Their Interaction with Plant Secondary Metabolites*. San Diego, CA: Academic Press, 1–44.

Ford, E. B. (1975) *Ecological Genetics*, 4th edn. London: Chapman and Hall.

Frank, T., Bieri, K. and Speiser, B. (2002) Feeding deterrent effect of carvone, a compound from caraway seeds, on the slug *Arion lusitanicus*. *Annals of Applied Biology*, **141**, 93–100.

Gershenzon, J. and Dudareva, N. (2007) The function of terpene natural products in the natural world. *Nature Chemical Biology*, **3**, 408–414.

Grime, J. P., Cornelissen, J. H. C., Thompson, K. and Hodgson, J. G. (1996) Evidence of a causal connection between anti-herbivore defence and the decomposition rate of leaves. *Oikos*, **77**, 489–494.

Grondahl, E. and Ehlers, B. K. (2008) Local adaptation to biotic factors: reciprocal transplants of four species associated with aromatic *Thymus pulegioides* and *T. serpyllum*. *Journal of Ecology*, **96**, 981–992.

Hiltpold, I. and Turlings, T. C. J. (2008) Belowground chemical signaling in maize: when simplicity rhymes with efficiency. *Journal of Chemical Ecology*, **34**, 628–635.

Hiltunen, R. (1976) On variation, inheritance and chemical interrelationships of monoterpenes in Scots pine (*Pinus sylvestris* L.). *Annales Academiae Scientiarium Fennicae*, **208**, 1–54.

Hochwender, C. G., and Fritz, R. S. (2004) Plant genetic differences influence herbivore community structure: evidence from a hybrid willow system. *Oecologia*, **138**, 547–557.

Honkanen, T., Haukioja, E. and Kitunen, V. (1999) Responses of *Pinus sylvestris* branches to simulated herbivory are modified by tree sink/source dynamics and by external resources. *Functional Ecology*, **13**, 126–140.

Iason, G. (2005). The role of plant secondary metabolites in mammalian herbivory. *Proceedings of the Nutrition Society*, **64**, 123–131.

Iason, G. R., Lennon, J. J., Pakeman, R. J. *et al.* (2005). Does chemical composition of individual Scots pine trees determine the biodiversity of their associated ground vegetation? *Ecology Letters*, **8**, 364–369.

Iason, G. R., O'Reilly-Wapstra, J., Brewer, M. J., Summers, R. W. and Moore, B. D. (2011) Do multiple herbivores maintain the chemical diversity of Scots pine monoterpenes? *Philosophical Transactions of the Royal Society B: Biological Sciences*, **366**, 1337–1345.

Jensen, C. G. and Ehlers, B. K. (2010) Genetic variation for sensitivity to a thyme monoterpene in associated plant species. *Oecologia (Berlin)*, **162**, 1017–1025.

Jin, K. S., Bak, M. J., Jun, M. *et al.* (2010) Alpha-pinene triggers oxidative stress and related signalling pathways in A549 and HepG2 cells. *Food Science and Biotechnology*, **19**, 1325–1332.

Johnson, M. T. J., Vellend, M. and Stinchcombe, J. R. (2009) Evolution in plant populations as a driver of ecological changes in arthropod communities. *Philosophical Transactions of the Royal Society B: Biological Sciences*, **364**, 1593–1605.

Kainulainen, P. and Holopainen, J. K. (2002) Concentrations of secondary compounds in Scots pine needles at different stages of decomposition. *Soil Biology and Biochemistry*, **34**, 37–42.

Kaneko, N., Sugawara, Y., Miyamoto, T., Hasegawa, M. and Hiura, T. (2005) Oribatid mite community structure and tree species diversity: a link? *Pedobiologia*, **49**, 521–528.

Keeling, C. I. and Bohlmann, J. (2006) Genes, enzymes and chemicals of terpenoid diversity in the constitutive and induced defence of conifers against insects and pathogens. *New Phytologist*, **170**, 657–675.

Kessler, A. and Baldwin, I. T. (2004) Herbivore-induced plant vaccination. Part I. The orchestration of plant defenses in nature and their fitness consequences in the wild tobacco *Nicotiana attenuata*. *Plant Journal*, **38**, 639–649.

Kinloch, B. B., Westfall, R. D. and Forrest, G. I. (1986) Caledonian Scots pine – origins and genetic-structure. *New Phytologist*, **4**, 703–729.

Klingler, H., Frosch, S. and Wagner, E. (1991) *In vitro* effects of monoterpenes on chloroplast membranes. *Photosynthesis Research*, **28**, 109–118.

Kong, J. O., Park, I. R., Choi, K. S., Shin, S. C. and Afini, Y. (2007) Nematicidal and propagation activities of thyme red and white oil compounds toward *Bursaphelenchus xylophilus* (Nematoda: Parasitaphelenchidae). *Journal of Nematology*, **39**, 237–242.

Kordali, S., Cakir, A. and Sutay, S. (2007) Inhibitory effects of monoterpenes on seed germination and seedling growth. *Zeitschrift für Naturforschung C: A Journal of Biosciences*, **62**, 207–214.

Langenheim, J. H. (1994) Higher-plant terpenoids – a phytocentric overview of their ecological roles. *Journal of Chemical Ecology*, **20**, 1223–1280.

Lankau, R. A. and Strauss, S. Y. (2008) Community complexity drives patterns of natural selection on a chemical defense of *Brassica nigra*. *American Naturalist*, **171**, 150–161.

Lerdau, M. and Gray, D. (2003) Ecology and evolution of light-dependent and light-independent phytogenic volatile organic carbon. *New Phytologist*, **157**, 199–211.

Lichstein, J. W. (2007) Multiple regression on distance matrices: a multivariate spatial analysis tool. *Plant Ecology*, **188**, 117–131.

Lindstedt, C., Mappes, J., Paivinen, J. and Varama, M. (2006) Effects of group size and pine defence chemicals on Diprionid sawfly survival against ant predation. *Oecologia*, **150**, 519–526.

Linhart, Y. B. and Thompson, J. D. (1999) Thyme is of the essence: biochemical polymorphism and multi-species deterrence. *Evolutionary Ecology Research*, **1**, 151–171.

Linhart, Y. B., Keefover-Ring, K., Mooney, K. A., Breland, B. and Thompson, J. D. (2005) A chemical polymorphism in a multitrophic setting: thyme monoterpene composition and food web structure. *American Naturalist*, **166**, 517–529.

Ludley, K., Jickells, S., Chamberlain, P., Whitaker, J. and Robinson, C. (2009a) Distribution of monoterpenes between organic resources in upper soil horizons under monocultures of *Picea abies, Picea sitchensis* and *Pinus sylvestris*. *Soil Biology and Biochemistry*, **41**, 1050–1059.

Ludley, K., Robinson, C., Jickells, S., Chamberlain, P. and Whitaker, J. (2009b) Potential for monoterpenes to affect ectomycorrhizal and saprotrophic fungal activity in coniferous forests is revealed by novel experimental system. *Soil Biology and Biochemistry*, **41**, 117–124.

Ludley, K. E., Robinson, C. H., Jickells, S., Chamberlain, P. M. and Whitaker, J. (2008) Differential response of ectomycorrhizal and saprotrophic fungal mycelium from coniferous forest soils to selected monoterpenes. *Soil Biology and Biochemistry*, **40**, 669–678.

Macchioni, F., Cioni, P. L., Flamini, G. *et al.* (2002) Acaricidal activity of pine essential oils and their main components against *Tyrophagus putrescentiae*, a stored food mite. *Journal of Agricultural and Food Chemistry*, **50**, 4586–4588.

Madritch, M.D. and Hunter, M.D. (2002) Phenotypic diversity influences ecosystem functioning in an oak sandhills community. *Ecology*, **83**, 2084–2090.

Madritch, M.D. and Hunter, M.D. (2005) Phenotypic variation in oak litter influences short- and long-term nutrient cycling through litter chemistry. *Soil Biology and Biochemistry*, **37**, 319–327.

Maurer, D., Kolb, S., Haumaier, L. and Borken, W. (2008) Inhibition of atmospheric methane oxidation by monoterpenes in Norway spruce and European beech soils. *Soil Biology and Biochemistry*, **40**, 3014–3020.

McKey, D. (1979) The distribution of secondary compounds within plants. In G.A. Rosenthal and D.H. Janzen (eds.) *Herbivores: Their Interaction with Plant Secondary Metabolites*. New York: Academic Press, 135–160.

Misra, G., Pavlostathis, S.G., Perdue, E.M. and Araujo, R. (1996) Aerobic biodegradation of selected monoterpenes. *Applied Microbiology and Biotechnology*, **45**, 831–838.

Mooney, K.A. (2007) Tritrophic effects of birds and ants on a canopy food web, tree growth, and phytochemistry. *Ecology*, **88**, 2005–2014.

Mooney, K.A. and Tillberg, C.V. (2005) Temporal and spatial variation to ant omnivory in pine forests. *Ecology*, **86**, 1225–1235.

Muona, O., Hiltunen, R., Shaw, D.V. and Morén, E. (1986) Analysis of monoterpene variation in natural stands and plustrees of *Pinus sylvestris* in Finland. *Silva Fennica*, **20**, 1–8.

Nakahara, K., Alzoreky, N.S., Yoshihashi, T., Nguyen, H.T.T. and Trakoontivakorn, G. (2003) Chemical composition and antifungal activity of essential oil from *Cymbopogon nardus* (citronella grass). *Jarq – Japan Agricultural Research Quarterly*, **37**, 249–252.

Northup, R.R., Dahlgren, R.A. and McColl, J.G. (1998) Polyphenols as regulators of plant–litter–soil interactions in northern California's pygmy forest: a positive feedback? *Biogeochemistry*, **42**, 189–220.

O'Reilly-Wapstra, J.M., Iason, G.R. and Thoss, V. (2007) The role of genetic and chemical variation of *Pinus sylvestris* seedlings in influencing slug herbivory. *Oecologia*, **152**, 82–91.

Owen, S.M., Clark, S., Pompe, M. and Semple, K.T. (2007) Biogenic volatile organic compounds as potential carbon sources for microbial communities in soil from the rhizosphere of *Populus tremula*. *Fems Microbiology Letters*, **268**, 34–39.

Paavolainen, L., Kitunen, V. and Smolander, A. (1998) Inhibition of nitrification in forest soil by monoterpenes. *Plant and Soil*, **205**, 147–154.

Pakeman, R.J., Beaton, J.K., Thoss, V. *et al.* (2006) The extended phenotype of Scots pine *Pinus sylvestris* structures the understorey assemblage. *Ecography*, **29**, 451–457.

Rasmann, S., Kollner, T.G., Degenhardt, J. *et al.* (2005) Recruitment of entomopathogenic nematodes by insect-damaged maize roots. *Nature*, **434**, 732–737.

Salle, A. and Raffa, K.F. (2007) Interactions among intraspecific competition, emergence patterns, and host selection behaviour in Ips pini (Coleoptera: Scolytinae). *Ecological Entomology*, **32**, 162–171.

Schlyter, F., Smitt, O., Sjodin, K., Hogberg, H.E. and Lofqvist, J. (2004) Carvone and less volatile analogues as repellent and deterrent antifeedants against the pine weevil, *Hylobius abietis*. *Journal of Applied Entomology*, **128**, 610–619.

Schweitzer, J.A., Bailey, J.K., Rehill, B.J. *et al.* (2004) Genetically based trait in a dominant tree affects ecosystem processes. *Ecology Letters*, **7**, 127–134.

Schweitzer, J.A., Bailey, J.K., Fisher, D.J. *et al.* (2008) Plant–soil–microorganism interactions: heritable relationship between plant genotype and associated soil microorganisms. *Ecology*, **89**, 773–781.

Singh, H.P., Batish, D.R., Kaur, S., Arora, K. and Kohli, R.K. (2006) Alpha-pinene inhibits growth and induces oxidative stress in roots. *Annals of Botany*, **98**, 1261–1269.

Singh, H. P., Kaur, S., Mittal, S., Batish, D. R. and Kohli, R. K. (2009) Essential oil of *Artemisia scoparia* inhibits plant growth by generating reactive oxygen species and causing oxidative damage. *Journal of Chemical Ecology*, **35**, 154–162.

Sipura, M. (2002) Contrasting effects of ants on the herbivory and growth of two willow species. *Ecology*, **83**, 2680–2690.

Smolander, A., Ketola, R. A., Kotiaho, T. *et al.* (2006) Volatile monoterpenes in soil atmosphere under birch and conifers: effects on soil N transformations. *Soil Biology and Biochemistry*, **38**, 3436–3442.

Tarayre, M., Thompson, J. D., Escarre, J. and Linhart, Y. B. (1995) Intra-specific variation in the inhibitory effects of *Thymus vulgaris* (Labiatae) monoterpenes on seed-germination. *Oecologia*, **101**, 110–118.

Tarayre, M., Saumitou-Laprade, P., Cuguen, J., Couvet, D. and Thompson, J. D. (1997) The spatial genetic structure of cytoplasmic (cpDNA) and nuclear (allozyme) markers within and among populations of the gynodioecious *Thymus vulgaris* (Labiatae) in southern France. *American Journal of Botany*, **84**, 1675–1684.

Thompson, J. D., Manicacci, D. and Tarayre, M. (1998) Thirty-five years of thyme: a tale of two polymorphisms. *BioScience*, **48**, 805–815.

Thompson, J. D., Chalchat, J. C., Michet, A., Linhart, Y. B. and Ehlers, B. (2003) Qualitative and quantitative variation in monoterpene co-occurrence and composition in the essential oil of *Thymus vulgaris* chemotypes. *Journal of Chemical Ecology*, **29**, 859–880.

Thompson, J. D., Gauthier, P., Amiot, J. *et al.* (2007) Ongoing adaptation to Mediterranean climate extremes in a chemically polymorphic plant. *Ecological Monographs*, **77**, 421–439.

Thoss, V., O'Reilly-Wapstra, J. and Iason, G. R. (2007) Assessment and implications of intraspecific and phenological variability in monoterpenes of Scots pine (*Pinus sylvestris*) foliage. *Journal of Chemical Ecology*, **33**, 477–491.

Uribe-Carvajal, S., Guerrero-Castillo, S., King-Diaz, B. and Hennsen, B. L. (2008) Allelochemicals targeting the phospholipid bilayer and the proteins of biological membranes. *Allelopathy Journal*, **21**, 1–23.

Uusitalo, M., Kitunen, V. and Smolander, A. (2008) Response of C and N transformations in birch soil to coniferous resin volatiles. *Soil Biology and Biochemistry*, **40**, 2643–2649.

van Dam, N. M., Harvey, J. A., Wackers, F. L. *et al.* (2003) Interactions between aboveground and belowground induced responses against phytophages. *Basic and Applied Ecology*, **4**, 63–77.

Vokou, D. (2007) Allelochemicals, allelopathy and essential oils: a field in search of definitions and structure. *Allelopathy Journal*, **19**, 119–133.

Ward, B. B., Courtney, K. J. and Langenheim, J. H. (1997) Inhibition of *Nitrosomonas europaea* by monoterpenes from coastal redwood (*Sequoia sempervirens*) in whole-cell studies. *Journal of Chemical Ecology*, **23**, 2583–2598.

White, C. S. (1991) The role of monoterpenes in soil-nitrogen cycling processes in Ponderosa pine – results from laboratory bioassays and field studies. *Biogeochemistry*, **12**, 43–68.

White, C. S. (1994) Monoterpenes – their effects on ecosystem nutrient cycling. *Journal of Chemical Ecology*, **20**, 1381–1406.

Whitham, T. G., Young, W. P., Martinsen, G. D. *et al.* (2003) Community and ecosystem genetics: a consequence of the extended phenotype. *Ecology*, **84**, 559–573.

Whitham, T. G., Lonsdorf, E., Schweitzer, J. A. *et al.* (2005) 'All effects of a gene on the world': extended phenotypes, feedbacks, and multi-level selection. *Ecoscience*, **12**, 5–7.

Whitham, T. G., Bailey, J. K., Schweitzer, J. A. *et al.* (2006) A framework for community and ecosystem genetics: from genes to ecosystems. *Nature Reviews Genetics*, **7**, 510–523.

Wimp, G. M., Martinsen, G. D., Floate, K. D.,
    Bangert, R. K. and Whitham, T. G. (2005)
    Plant genetic determinants of arthropod
    community structure and diversity.
    Evolution, **59**, 61–69.

Wolf, J. B., Brodie, E. D., Cheverud, J. M.,
    Moore, A. J. and Wade, M. J. (1998)
    Evolutionary consequences of indirect
    genetic effects. *Trends in Ecology and Evolution*,
    13, 64–69.

CHAPTER FOURTEEN

From genes to ecosystems: emerging concepts bridging ecological and evolutionary dynamics

JOSEPH K. BAILEY, JENNIFER A. SCHWEITZER,
FRANCISCO ÚBEDA, BENJAMIN M. FITZPATRICK, MARK
A. GENUNG and CLARA C. PREGITZER
Department of Ecology and Evolutionary Biology, University of Tennessee
MATTHEW ZINKGRAF, THOMAS G. WHITHAM and
ARTHUR KEITH
Department of Biological Sciences, Northern Arizona University
JULIANNE M. O'REILLY-WAPSTRA and BRADLEY M. POTTS
School of Plant Science, University of Tasmania
BRIAN J. REHILL
Department of Chemistry, US Naval Academy
CARRI J. LEROY and DYLAN G. FISCHER
Environmental Studies Program, The Evergreen State College

14.1 Introduction

Relatively little is understood about the extent to which evolution in one species can result in changes to associated communities and ecosystems, the potential mechanisms responsible for those changes (genetic drift, gene flow or natural selection), the phenotypes or candidate genes that may link ecological and evolutionary dynamics, or the role of rapid evolution and feedbacks. However, linking genes and ecosystems in this manner is fundamental to placing community structure and ecosystem function in an evolutionary framework. This is not an easy endeavour as the field of community genetics is multi-disciplinary (Whitham *et al.*, 2006), and ecological and evolutionary dynamics occur at different spatial and temporal scales. Recent reviews show that plant genetic variation can have extended consequences at the community and ecosystem level (extended phenotype; Whitham *et al.*, 2003) affecting arthropod diversity, soil microbial communities, trophic interactions, carbon dynamics and soil nitrogen availability (Whitham *et al.*, 2006; Johnson & Stinchcombe, 2007; Hughes *et al.*, 2008; Bailey *et al.*, 2009a). Its effects are not limited to single systems or even foundation species, but are common across broadly distributed plant and animal systems, and can have effects at the

The Ecology of Plant Secondary Metabolites: From Genes to Global Processes, eds. Glenn R. Iason, Marcel Dicke and Susan E. Hartley. Published by Cambridge University Press. © British Ecological Society 2012.

community and ecosystem level of similar magnitude to traditional ecological factors, such as differences among species (Bailey *et al.*, 2009a, b).

Theory in the fields of community genetics (Shuster *et al.*, 2006; Whitham *et al.*, 2006) and co-evolution (Thompson, 2005) also supports the connection between evolutionary and ecological dynamics (Johnson *et al.*, 2009). Multiple investigators argue that community and ecosystem phenotypes represent complex traits related to variation in the fitness consequences of inter-specific indirect genetic effects (IIGEs) (Thompson, 2005; Shuster *et al.*, 2006; Whitham *et al.*, 2006; Tetard-Jones *et al.*, 2007). In their most basic form, IIGEs occur when the genotype of one individual affects the phenotype and fitness of an associated individual of a different species (Moore *et al.*,1997; Agrawal *et al.*, 2001; Shuster *et al.*, 2006; Wade, 2007). Such interactions are important in the geographic mosaic theory of co-evolution (Thompson, 2005), the development of community heritability (Shuster *et al.*, 2006) and non-additive responses of community structure, biodiversity and ecosystem function (Bailey *et al.*, 2009a). Empirical evidence for the effects of plant genetic variation on communities and ecosystems, paired with growing theoretical models explaining evolutionary mechanisms for these results, provides a solid foundation for understanding how evolutionary processes, such as drift and selection, may affect community structure and ecosystem function.

Placing community structure and ecosystem function in an evolutionary framework represents an important emerging frontier in science, with implications for how we understand the broad consequences of environmental problems (Parker *et al.*, 2006; Bradley & Pregitzer, 2007; Evans *et al.*, 2008; Cavender-Bares *et al.*, 2009; Ryan *et al.*, 2009; Genung *et al.*, 2011). For example, genetically based thresholds for photosynthetic responses to elevated CO_2 can define rates of biomass accumulation and hence carbon-flux through an ecosystem, and determine a plant's response to climate change (Bradley & Pregitzer, 2007). In another example, Genung *et al.* (2010) showed that floral visitation by pollinators demonstrated a non-additive response associated with flowering phenology. This response was positively related to increased genotypic diversity in *Solidago altissima*, which has implications for understanding the relationship between reductions in genetic diversity and the global decline in pollinators as well as agricultural productivity and sustainability. Given the relatively widespread findings of a genetic basis to community and ecosystem phenotypes in multiple systems (Bailey *et al.*, 2009a), such a perspective may reveal greater consequences of environmental perturbations than previously considered.

Taking a comparative approach, we explore emerging concepts bridging the genetic basis to community and ecosystem ecology, and the feedbacks between ecology and evolution. Here we expand on previous work and explore the generality of the extended phenotype and the role of plant secondary metabolites (PSMs), shown to be correlated with a diverse group of interacting species and ecosystem properties among *Populus* hybrids (Section 14.2). The logical extension of this work suggests that interacting species and the ecosystem processes they mediate may respond to underlying evolutionary dynamics. Using results from three systems we investigate how genetic differentiation in plants can affect communities and ecosystem function (Section 14.3). Because evolution can affect community structure and ecosystem function, there must be genes or gene regions in the host plant driving these extended phenotypes. Using a model *Populus* system, we link plant genomes (candidate genes) to phenotypes and ultimately community structure (Section 14.4). We then use an example of ecological and evolutionary feedbacks (eco–evo dynamics) to show how genetically based linkages may reinforce ecological and evolutionary dynamics (Section 14.5). Secondary metabolites tend to be correlated with plant growth, which does appear to be related to variation at the community and ecosystem level (Section 14.5). The synthesis of this work indicates that if the evolutionary dynamic in a system changes, then the ecological dynamic is also likely to change (and vice versa) through feedbacks, with consequences that can extend from the phenotype of an individual to affect the ecosystem. As the field of community and ecosystem genetics is currently focused on documenting the extended effects of genetic variation in a focal species, this review provides a broad conceptual framework for understanding how evolution may affect communities and ecosystems.

14.2   The generality of the extended phenotype

Much recent empirical work has shown that genetic variation in one species can have extended consequences for associated species and that these interactions can result in significant variation in biodiversity as well as patterns of energy flow and nutrient cycling (see reviews by Whitham *et al.*, 2006; Johnson & Stinchcombe, 2007; Hughes *et al.*, 2008; Schweitzer *et al.*, 2008b; Bailey *et al.*, 2009b). Such extended phenotypes are a consequence of the effects of genes at levels higher than the population (Whitham *et al.*, 2003). For example, in a well-studied riparian forest system of *Populus angustifolia* × *P. fremontii*, genetic variation between the two species and their hybrids for foliar condensed tannin concentration was related to differences in terrestrial and aquatic macroinvertebrate communities, fungal endophyte infection, rates of decomposition and nitrogen mineralisation (Whitham *et al.*, 2006). Although the mechanisms are still being worked out, genetic variation within *P. angustifolia* also explains approximately 45% of the variation in associated communities and ecosystem function (Figure 14.1). These include traits potentially important for many functions/processes, including: (1) phytochemistry, clearly important among the hybrids; (2) whole-tree productivity and belowground carbon storage, which is vital for the mitigation of rising atmospheric CO_2 (Lojewski *et al.*, 2009); (3) soil microbial composition and soil nitrogen availability, which mediates plant growth; (4) top-down regulation of

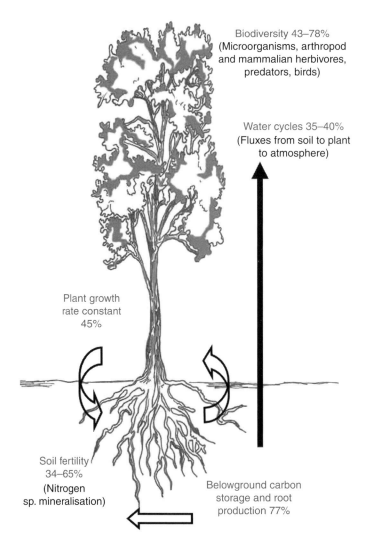

Figure 14.1 Across diverse plant systems, a large proportion of the variation in biodiversity and ecosystem function in common environment trials can be explained by plant genetic factors. For example, in *Populus angustifolia*, plant genetic factors can affect soil nutrient availability and belowground carbon storage as well as species interactions with arthropods, avian predators, mammalian herbivores like beaver and even soil microbes. See plate section for colour version.

herbivores by avian predators; and (5) overall biodiversity, as different tree genotypes support different communities of canopy arthropods, fungal endophytes and soil microbes.

Although plant genetic variation clearly affects the services provided by this representative *Populus* ecosystem, this example is not unique. Similar results have been found across broadly distributed plant and animal systems (see reviews by Hughes *et al.*, 2008; Bailey *et al.*, 2009a, 2009b; Post & Palkovaks, 2009). For example, genetic variation in a dominant old-field plant, *Solidago altissima*, can explain up to 60% of the among-plant variation in associated community structure and ecosystem function including: (1) variation in the abundance, richness and composition of floral visitors; and (2) variation in plant biomass (Genung *et al.*, 2010).

Note that the percentages of these community and ecosystem phenotypes explained by plant genetics in these experimental gardens are similar even though these species differ strongly in their life history and ecological influence (e.g. a tree vs. perennial herb). Moreover, increasing numbers of plant and animal systems demonstrate such effects (Whitham *et al.*, 2006; Johnson & Stinchcombe, 2007; Post & Palkovacs, 2008, 2009; Bailey *et al.*, 2009b; Harmon *et al.*, 2009; Palkovacs *et al.*, 2009).

Based on theory outlined by Shuster *et al.* (2006), genetically based variation in community phenotypes provides evidence that: (1) fitness effects arise from IIGEs among the interacting species in a community where the genotype of one individual affects the phenotype and fitness of another individual (or many individuals); and (2) through differential fitness consequences of indirect genetic effects, selection can occur in a community context. Moreover, recent studies show that arthropod population dynamics may vary based upon the plant genotype with which the arthropods are associated (McIntyre & Whitham, 2003; Tetard-Jones *et al.*, 2007; Johnson, 2008) which can lead to genetic change in the population or even speciation. For example, Evans *et al.* (2008) found cryptic speciation in a galling mite associated with hybridisation in a *P. angustifolia* × *P. fremontii* complex. Because natural selection in a community context is implicit in the measurement of community and ecosystem phenotypes, it is reasonable to expect that genetic differentiation and evolution among populations in one species may alter their extended phenotypes to affect associated communities and ecosystem function (i.e. evolution leads to ecological change), particularly for foundation species.

14.3 Cascading ecological consequences of evolution

Evolution can occur through a variety of mechanisms including mutation, natural selection and genetic drift, all of which represent important pathways of population genetic change. Feedbacks between ecological and evolutionary dynamics have long been discussed, but studies are only now beginning to emerge demonstrating that evolution in one species can have community and ecosystem consequences (Stireman *et al.*, 2005, 2006; Post & Palkovacs, 2008, 2009; Barbour *et al.*, 2009a, b, c; Harmon *et al.*, 2009; Palkovacs *et al.*, 2009). For example, Palkovacs *et al.* (2009) experimentally compared the effects of two fish species (guppies, *Poecilia reticulata*; and killifish, *Rivulus hartii*) that had evolved in either the presence or absence of predators (i.e. evolutionary diversity) and under different contexts of sympatry (i.e. co-evolutionary diversity) on aquatic macroinvertebrates and algal biomass. Their results demonstrated that populations which evolved under these different conditions differentially influenced both invertebrate and algal biomass in mesocosms. Similarly, the adaptive radiation of three-spine sticklebacks (*Gasterosteus aculeatus*) has repeatedly resulted in specialised benthic and limnetic 'trophic morphs' (Schluter & Nagel, 1995; Schluter, 1996). In an experimental mesocosm experiment, Harmon *et al.* (2009) showed that genetic divergence in trophic niche has strong effects on biodiversity and ecosystem function, with consequences for prey community structure, total primary production and dissolved organic materials that can affect light attenuation in this aquatic system.

Identifying and understanding genetic differentiation along specific environmental gradients (Endler, 1977; Storfer, 1999; Foster *et al.*, 2007) provides an important tool for addressing the hypothesis that evolution in one species can lead to change at the community and ecosystem level. Elevational and latitudinal gradients help show how environmental factors can affect ecological properties of biodiversity and ecosystem function (Hodkinson, 2005). For example, temperature, barometric pressure, precipitation, wind and a suite of edaphic factors all vary along elevational gradients (Hodkinson, 2005). These physical selective forces may affect the underlying genetic variation for plant traits which convey fitness advantages; however, few data exist on the extended consequences of this on higher levels of organisation. In the few studies of the extended effects of evolution in plants, Barbour *et al.* (2009a, b, c) found that genetic differentiation amongst *Eucalyptus globulus* trees collected from different provenances in Southeastern Australia, and grown in a common environment trial, supported different canopy, trunk and litter communities of arthropods and fungi. Using ion-exchange resins to estimate differences in soil inorganic nitrogen (due either to differential nutrient turnover or differential nutrient/microbial uptake; Binkley & Hart, 1989), provenance-level variation in soil nitrate (NO_3^-) was also found (Figure 14.2a). These studies provide another example of extended community and ecosystem phenotypes and support the hypothesis that broad-scale genetic differentiation and evolution in a foundation plant species, such as trees, may contribute to the changes we see in community composition, species interactions and ecosystem properties across the landscape.

Results similar to those found in the *Eucalyptus globulus* system have been observed in studies of *Populus angustifolia* where elevational gradients have been used to identify genetically differentiated populations of plants. In comparative common garden studies planted with *Populus angustifolia* genotypes

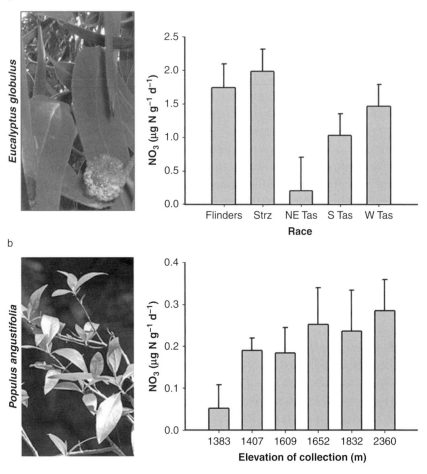

Figure 14.2 Evolution in host plants can alter biodiversity and ecosystem function. Common garden experiments indicate that genetic differentiation along elevational and latitudinal gradients can alter soil nutrients. (a) Race-level genetic differentiation in *Eucalyptus globulus* alters soil nitrate (Family (race): $F = 0.12$, $P = 0.4$, race: $F = 3.8$, $P = 0.011$, diameter at breast height: $F = 4.42$, $P = 0.039$). For specific methods on the number of families and provenances in the Eucalypt trial see Barbour *et al.* (2009a). (b) There was also evidence for genetic differentiation in P. *angustifolia* altering soil nitrate, which varied in common garden experiments according to the elevation of the site of origin of the trees (generalised linear model (GLM) – genotype: $F = 0.09$, $P = 0.4$, site: $F = 2.88$, $P = 0.028$). *Populus angustifolia* source trees for this common garden were randomly selected from six sites at different elevations along the Weber River that were <1 km to >100 km apart, and which differed in mean annual precipitation by 200 mm and in mean annual temperature by 3 °C, and are representative of the genetic variation one may find naturally occurring in this river system (Martinsen *et al.*, 2001). Within each site, the numbers of genotypes collected varied, but there were at least five genotypes. The trees are now 15 years old and planted in a randomised design with 1–7 replicates per genotype. Bars represent population level means (+1 SE). See plate section for colour version.

collected from six sites of different elevation in Utah, USA, there was greater soil NO_3^- under trees originating from sites at high elevation than low elevation (Figure 14.2b). We examined the potential mechanistic link between phytochemistry and soil nutrient availability and found no correlation, suggesting that differences in soil nutrient availability were due in part to differences in growth rate (*sensu* Lojewski *et al.*, 2009) and subsequent nutrient uptake. While we do not have the statistical power to examine whether the change at the community and ecosystem-level phenotypes across elevational gradients in *P. angustifolia* was due to natural selection or genetic drift, the experimental results using common gardens indicate that evolution through population-level genetic differentiation in plants can alter community and ecosystem phenotypes, including arthropod community composition and soil nitrogen content.

Across both animal and plant systems, evolution can drive change in multiple community and ecosystem phenotypes (Tetard-Jones *et al.*, 2007; Post & Palkovacs, 2008, 2009; Bailey *et al.*, 2009a, 2009b; Harmon *et al.*, 2009; Palkovacs *et al.*, 2009). In plant systems, these results suggest that patterns of arthropod and microbial diversity in the landscape may not be driven solely by the geographic distribution of host species or abiotic factors, but the geographic distribution of a species may be a *response to evolutionary change in their host plants* (Pennings & Silliman, 2005). Because plant genetic variation can have predictable extended effects at the community and ecosystem levels, one would expect that functional genes may underlie these effects.

14.4   Community and ecosystem genomics

As the above sections have demonstrated, extended phenotypes occur across multiple plant and animal systems, and evolution can act to alter communities and ecosystems. However, linking genes to ecosystems is fundamental to understanding the genetic mechanisms for these patterns and the genes upon which selection may act (Whitham *et al.*, 2008). With a few notable exceptions (Kessler *et al.*, 2004; Mitchell-Olds *et al.*, 2008; Zheng & Dicke, 2008; Schuman & Baldwin, Chapter 15; Dicke *et al.*, Chapter 16), understanding the functional genes or gene regions responsible for community and ecosystem phenotypes in natural systems remains relatively unexplored (Whitham *et al.*, 2008; Zheng & Dicke, 2008). One approach to this is to use naturally occurring hybrid systems, since gene introgression following hybridisation can be a strong evolutionary force (Anderson, 1949; Stebbins, 1959; Arnold, 1992, 1997; Rieseberg & Willis, 2007). Introgression can result in genotypes that are more fit than their parents and are able to colonise new habitats (Lewontin & Birch, 1966; Burke & Arnold, 2001; Martinsen *et al.*,

2001; Schweitzer *et al.*, 2002; Rieseberg *et al.*, 2003, Arnold, 2004) as well as genotypes that have increased susceptibility or resistance to pests (Whitham, 1989; Fritz *et al.*, 2003; McIntyre & Whitham, 2003; Hochwender & Fritz, 2004; Whitney *et al.*, 2006). Because evolution of plants can result in related changes to communities and ecosystem function, and this may be associated with changes in the genetic interactions among species (Thompson, 2005; Shuster *et al.*, 2006), gene introgression may also be an important mechanism of rapid evolution in plants affecting communities and ecosystems.

Using a model *Populus* hybridising system in the wild (Figure 14.3a), Martinsen *et al.* (2001) showed that 7 of 33 nuclear markers had exceptional patterns of introgression, with *P. fremontii* alleles found in morphologically pure *P. angustifolia* populations 40–120 km from the morphological hybrid zone. Of these widely introgressed molecular markers, we found that a cryptically introgressing gene region associated with a single *P. fremontii* molecular marker (RFLP-755) occurred in 20% of *P. angustifolia* at the highest elevation sites (Figure 14.3b). For this gene region, we applied Long's (1991) test for marker heterogeneity along the cline (accounting for variance owing to genetic drift), which supported Martinsen *et al.* (2001) and suggested that the genes associated with the RFLP marker 755 are introgressing in *P. angustifolia* along the Weber River owing to natural selection. Therefore, one may expect to see phenotypic differences in quantitative traits in trees from the same randomly mating population that do and do not have the *P. fremontii* allele for this molecular marker (Figure 14.3).

From associated common garden experiments with replicated tree genotypes of known pedigree randomly collected from multiple sites across the elevational gradient, there was support for the hypothesis that this introgressed molecular marker was associated with sequence variation that had phenotypic effects associated with reduced concentration of foliar salicortin, a phenolic defence (Palo, 1984) and a greater degree of secondary branching (Figures 14.3e and f). There was a negative phenotypic correlation between the degree of asexual regeneration and foliar salicortin ($r^2 = 0.09$, $P = 0.017$) suggesting that the *BRC2* gene in which the molecular marker occurs (or other linked genes) was associated with a trade-off between growth and defence. Introgression of this region was also associated with extended effects at the community level, affecting both species richness and abundance of leaf-modifying arthropods (Figures 14.3c and d, respectively). Leaf-modifying arthropods (i.e. gall-forming herbivores) were twice as abundant, and overall there were 23% more species on introgressed trees carrying the *P. fremontii* allele for RFLP-755 compared with non-introgressed trees (Figures 14.3c and d). However, the effect of the molecular marker on residual richness, after

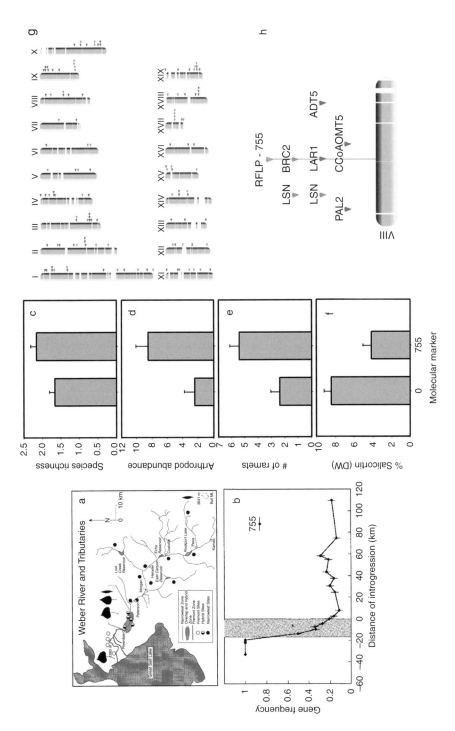

Figure 14.3 Cryptic introgression of a *Populus fremontii* molecular marker into *P. angustifolia* results in changes to biodiversity in a *Populus* hybridising system. In a natural *Populus* hybrid zone along the Weber River, UT, genes from low-elevation *P. fremontii* introgress through the hybrid zone into the high-elevation *P. angustifolia* populations (a). Of the 33 molecular markers used by Martinsen *et al.* (2001) to study introgression in this

system, molecular marker 'RFLP-75' occurred in 20% of the *P. angustifolia* trees sampled at the highest elevation sites, where all of the trees appear to be pure *P. angustifolia* (b). In a 15-year-old common garden setting (Schweitzer *et al.*, 2002), we examined trees sampled from throughout the pure *P. angustifolia* zone with 'RFLP-755' and compared them to trees without 'RFLP-755' (17 genotypes without 'RFLP-755' replicated 1–5 times compared to 10 genotypes with 'RFLP-755' replicated 1–7 times) to determine if the introgression of this region had quantitative phenotypic consequences. Trees with the 'RFLP-755' marker had (c) greater species richness of gall-forming arthropods (assessed by visual survey; GLM with exponential transformation: genotype (marker): $\chi^2 = 71.09$, $P < 0.0001$; marker: $\chi^2 = 32.36$, $P < 0.0001$); (d) greater abundance of gall-forming arthropods (visual survey; GLM; genotype (marker): $\chi^2 = 35.38$, $P < 0.08$; marker: $\chi^2 = 9.05$, $P < 0.0026$); (e) more asexual regeneration via secondary branching, with 56% more ramets (generalised linear model (exponential transformation); genotype (marker): $\chi^2 = 239.82$, $P < 0.0001$; marker: $\chi^2 = 85.56$, $P < 0.0001$) and (f) a 51% reduction in foliar salicortin (generalised linear model; genotype (marker): $\chi^2 = 84.98$, $P < 0.0001$, marker: $\chi^2 = 48.19$, $P < 0.0001$), measured using HPLC (Rehill *et al.*, 2006) and expressed as % of dry weight (DW). Bars represent marker means (±1 SE). A search of the *Populus* genome sequence (g) showed that the marker '755' is located on linkage group VIII within a BRC2-like transcription factor that regulates secondary shoot formation in Arabidopsis. (h) In addition, a LAR1 gene from the phenylpropanoid pathway is located within about 49 kb of RFLP-755. Other more distal genes on this linkage group are also shown. See plate section for colour version.

accounting for abundance, was not significant, indicating that the richness effect was driven by abundance. Multiple regression showed that arthropod abundance was positively correlated with the number of ramets ($r^2 = 0.13$, $P = 0.019$), but not salicortin, suggesting that the increase in arthropod abundance was due to shoot branching (Price, 1991) rather than shifts in phytochemistry. Importantly, it is still unclear to which morphological phenotype the arthropods are responding, e.g. increased density of shoots, longer shoots, bigger leaves. Such cascading community-level consequences of cryptic introgression and rapid evolution are consistent with the hypothesis that the genetic mechanisms driving community phenotypes may be less complex for some species than previously appreciated, and that evolutionary change through introgression may be a significant driver at the community level.

From among-population differentiation to cryptically introgressing candidate genes, genetic variation may be expressed and evolutionary processes may act in a variety of ways to affect patterns of community structure and ecosystem function. These results confirm that evolutionary dynamics can operate on ecological timescales and may have profound implications for how we understand the ecological consequences of historic evolutionary dynamics (Hendry & Kinnison, 1999; Kennison & Hendry, 2001; Yoshida *et al.*, 2003; Hairston *et al.*, 2005).

14.5   Eco-evolutionary feedbacks

Elucidating the links between ecological interactions and evolutionary processes (eco-evolutionary dynamics) is important for understanding how genetic variation in one species affects communities and ecosystems, and how communities and ecosystems may feed back to shape genetic variation in subsequent generations. Feedbacks are fundamental to the co-evolutionary process (Thompson, 2005), local adaption (Clausen *et al.*, 1940; Johnson *et al.*, 2010) and the maintenance of biodiversity (Duffy & Forde, 2009; Laine, 2009), but little is understood about how they can stabilise (negative feedbacks) or destabilise (positive feedbacks) eco-evolutionary patterns. Because genetic variation in one species can influence soil microbial communities and nutrient availability, terrestrial arthropod communities, and associated diversity and productivity of other neighbouring plants, one may expect feedbacks to occur from a variety of sources, e.g. plant–herbivore, plant–soil, plant–plant feedbacks. However, the specific 'home versus away' tests to examine feedbacks are not common and are easier in some fields than in others. An emerging body of work focusing on feedbacks between plants and soils is promising and supports the notion that feedbacks are fundamental to the co-evolutionary process as well as local adaptation (Kulmatiski *et al.*, 2008; Johnson *et al.*, 2010; Pregitzer *et al.*, 2010). For example, using *Andropogon gerardii* ecotypes collected from phosphorus (P)-limited and nitrogen

(N)-limited grasslands and grown with all possible 'home and away' combinations of soils and arbuscular mycorrhizal fungal communities, Johnson *et al.* (2010) showed that soil fertility was a key driver of local adaptation of the symbioses such that mycorrhizal exchange of the most limiting soil nutrient resource was maximised. Positive feedbacks, such as these, are commonly associated with strong patterns of co-evolution and local adaptation (Laine, 2009). While this study and others in plant–soil feedbacks have indicated that soils can act as selective agents affecting plant evolution (Johnson *et al.*, 2010; Pregitzer *et al.*, 2010), understanding the role of genetic variation in the plant–soil feedback as well as the extended consequences of the feedback on subsequent generations has only recently begun.

Dominant plant species can have legacy effects on soils that can feed back to affect the plant's own fitness (Binkley & Giardina, 1998; Bever *et al.*, 2009) but the role of intra-specific variation on plant–soil feedback is less understood (Murren *et al.*, 2006; Sambatti & Rice, 2006). A growing number of studies indicate that plant genetic variation can affect soil microbial communities and even soil nutrient availability and cycling (Schweitzer *et al.*, 2004, 2008a, b; Madritch *et al.*, 2009). For example, Schweitzer *et al.* (2004, 2008a, b) showed that condensed tannins, which varied among pure and hybrid cottonwoods, had a strong impact on soil microbial communities and net N mineralisation. Specifically, *P. angustifolia* genotypes had high foliar condensed tannins and low net N mineralisation rates, whereas *P. fremontii* genotypes had low foliar condensed tannins and high net N mineralisation rates, with hybrids intermediate in both.

In a follow-up study examining how seedlings of *P. angustifolia* may respond to soil conditions established by mature *P. angustifolia* relative to soils from *P. fremontii* and hybrids, Pregitzer *et al.* (2010) found a 'home-field advantage' or positive feedback for *P. angustifolia* seedling genetic families when grown in soil that had been previously conditioned by *P. angustifolia* trees. Seedling families grown in their native soil had higher survival and higher performance than seedlings grown in soils conditioned by other *Populus* species and hybrids, even though the native soils were the most N-limited (i.e. highest C:N ratio and highest microbial N pools; Schweitzer *et al.*, 2004, 2008b). *Populus angustifolia* seedlings grown in *P. angustifolia* soils were twice as likely to survive, grew 24% taller, and had 27% more leaves and a 29% greater total aboveground biomass. Consistent with increased survival and performance, seedlings grown in their home-soil resulted in greater additive genetic variance for performance traits compared with seedlings grown in away-soils. Because soil conditions to which the seedlings were adapted were a consequence of inputs by leaves, roots and root exudates by the parent species which alter soil microbial communities and nutrient cycling (Schweitzer *et al.*, 2004, 2008a, b; Whitham *et al.*, 2006), these results provide evidence that soils can act (1) as agents of selection affecting the performance and fitness of seedlings; (2) to affect the amount of genetic variation that is expressed in a population which affects a population's evolutionary responses to future selective events; and (3) to generate local adaptation that can be a consequence of plant–soil feedbacks even in closely related species (Pregitzer *et al.*, 2010). Importantly, because genetic variation in one species can have community and ecosystem phenotypes that affect genetic variation in subsequent generations, it becomes difficult to determine whether the relative fitness of an individual depends on feedbacks from multiple levels of organisation (e.g. population, community or ecosystem; Goodnight, 1990a, b; Wade, 2007; Genung *et al.*, 2010). This suggests that placing community and ecosystem genetics in a broad environmental context may be an important frontier.

14.6 Conclusions and future directions

There is growing evidence and support for the extended effects of genetic variation in plant and animal species which can influence associated biodiversity and ecosystem function. In plants, it has now been shown that variation in plant communities, arthropod communities, microbial communities, floral visitors and soil nutrient availability can all be affected by plant genetic variation (i.e. there is genetically based variation in the extended phenotype of plants). Moreover, variability in phytochemistry has been shown to be an important link between plant genetics and extended phenotypes in multiple systems, as well as the feedback between the extended phenotype and plant genetic variation. However, much work remains in this area to understand how mechanistic the role of phytochemistry is in the ecological and evolutionary dynamic. Results to date also indicate that across plant and animal systems, multiple modes of evolution in one species can alter community structure, ecosystem function and evolution of associated species. Importantly, extended phenotypes can result in feedbacks on the individual that created them, much like beavers building lodges, prairie dogs building towns, caddisflies building cases or birds building nests, but these feedbacks can also affect the fitness and performance of the next generation. Evidence of a dynamic feedback altering additive genetic variation in the population provides evidence for the manner by which evolutionary processes can maintain or reduce the effects of phenotypes on extended phenotypes. Such results suggest that genetic variation should be fundamental to the strength of feedbacks, and the strength of feedbacks should change with evolutionary processes.

Growing evidence that evolution can alter community structure and ecosystem function, that there are specific functional genes that underlie community and ecosystem phenotypes, and feedbacks can emerge from extended phenotypes, indicates that biological systems are much more tightly linked than previously appreciated.

Acknowledgements

Thanks to Randy Bangert for conversation and ideas over the past couple of years that helped develop our thoughts about this project.

References

Agrawal, A. F., Brodie, E. D. III and Wade, M. J. (2001) On indirect genetic effects in structured populations. *American Naturalist*, **158**, 308–324.

Aguilar-Martinez, J. A., Poza-Carrión, C. and Cubas, P. (2007) *Arabidopsis BRANCHED1* acts as an integrator of branching signals within axillary buds. *Plant Cell*, **19**, 458–472.

Anderson, E. (1949) *Introgressive Hybridization*. New York: Wiley.

Arnold, M. L. (1992) Natural hybridization as an evolutionary process. *Annual Review of Ecology and Systematics*, **23**, 237–261.

Arnold, M. L. (1997) *Natural Hybridization and Evolution*. New York: Oxford University Press.

Arnold, M. L. (2004) Natural hybridization and the evolution of domesticated, pest, and disease organisms. *Molecular Ecology*, **13**, 997–1007.

Bailey, J. K., Hendry, A., Kennison, M. *et al.* (2009a) From genes to ecosystems: an emerging synthesis of eco-evolutionary dynamics. *New Phytologist*, **184**, 746–749.

Bailey, J. K., Schweitzer, J. A., Koricheva, J. *et al.* (2009b) From genes to ecosystems: synthesizing the effects of plant genetic factors across systems. *Philosophical Transactions of the Royal Society B*, **364**, 1607–1616.

Barbour, R. C., O'Reilly-Wapstra, J. M., De Little, D. W. *et al.* (2009a) A geographic mosaic of genetic variation within a foundation tree species and its community-level consequences. *Ecology*, **90**, 1762–1772.

Barbour, R. C., Forster, L. G., Baker, S. C. and Potts, B. M. (2009b). Biodiversity consequences of genetic variation in bark characteristics within a foundation tree species. *Conservation Biology*, **23**, 1146–1155.

Barbour, R. C., Baker, S. C., O'Reilly-Wapstra, J. M. *et al.* (2009c) A footprint of tree-genetics on the biota of the forest floor. *Oikos*, **118**, 1917–1923.

Bever, J. D., Richardson, S., Lawrence, B. M., Holmes, J. and Watson M. (2009) Preferential allocation to beneficial symbiont with spatial structure maintains mycorrhizal mutualism. *Ecology Letters*, **12**, 13–21.

Binkley, D. and Giardina, C. (1998) Why do tree species affect soils? The warp and woof of tree–soil interactions. *Biogeochemistry*, **42**, 89–106.

Binkley, D. and Hart, S. (1989) The components of nitrogen availability assessments in forest soils. *Advances in Soil Science*, **10**, 57–116.

Bradley, K. L. and Pregitzer, K. S. (2007) Ecosystem assembly and terrestrial carbon balance under elevated CO_2. *Trends in Ecology and Evolution*, **22**, 538–547.

Burke, J. M. and Arnold, M. L. (2001) Genetics and the fitness of hybrids. *Annual Review of Genetics*, **35**, 31–52.

Cavender-Bares, J., Kozak, K., Fine, P. and Kembel, S. (2009) The merging of community ecology and phylogenetic biology. *Ecology Letters*, **12**, 693–715.

Clausen, J., Keck, D. D. and Heisey, W. M. (1940) Experimental studies on the nature of species. I. Effect of varied environments on western North American plants. *Carnegie Institution of Washington Publications*, **520**, 1–452.

Duffy, M. A. and Forde, S. E. (2009) Ecological feedbacks and the evolution of resistance. *Journal of Animal Ecology*, **78**, 1106–1112.

Endler, J. A. (1977) *Geographic Variation, Speciation and Clines*. Princeton, NJ: Princeton University Press.

Evans, L. M., Allan, G. J., Shuster, S. M., Woolbright, S. A. and Whitham, T. G. (2008) Tree hybridization and genotypic variation drive cryptic speciation of a specialized mite herbivore. *Evolution*, **62**, 3027–3040.

Foster, S. A., McKinnon, G. E., Steane, D. A., Potts, B. M. and Vaillancourt, R. E. (2007) Parallel evolution of dwarf ecotypes in the forest tree *Eucalyptus globulus*. *New Phytologist*, **175**, 370–380.

Fritz, R. S., Hochwender, C. G., Brunsfeld, S. J. and Roche, B. M. (2003) Genetic architecture of susceptibility to herbivores in hybrid willows. *Journal of Evolutionary Biology*, **16**, 1115–1126.

Genung, M. A., Lessard, J. P., Brown, C. B. *et al.* (2010) Non-additive effects of genotypic diversity increase pollinator visitation. PLoS ONE, **5**, e8711.

Genung, M. A., Schweitzer, J. A., Úbeda, F. *et al.* (2011) Genetic variation and community change – selection, evolution, and feedbacks. *Functional Ecology*, **25**, 408–419.

Goodnight, C. J. (1990a) Experimental studies of community evolution II. The ecological basis of the response to community selection. *Evolution*, **44**, 1625–1636.

Goodnight, C. J. (1990b) Experimental studies of community evolution I: The response to selection at the community level. *Evolution*, **44**, 1614–1624.

Hairston, N. G. Jr, Ellner, S. P., Geber, M. A., Yoshida, T. and Fox, J. A. (2005) Rapid evolution and the convergence of ecological and evolutionary time. *Ecology Letters*, **8**, 1114–1127.

Harmon, L. J., Matthews, B., DesRoches, S. *et al.* (2009) Evolutionary diversification in stickleback affects ecosystem functioning. *Nature*, **458**, 1167–1170.

Hendry, A. P. and Kinnison, M. T. (1999) The pace of modern life: measuring rates of contemporary microevolution. *Evolution*, **53**, 1637–1653.

Hochwender, C. G. and Fritz, R. S. (2004) Plant genetic differences influence herbivore community structure: evidence from a hybrid willow system. *Oecologia*, **138**, 547–557.

Hodkinson, I. D. (2005) Terrestrial insects along elevation gradients: species and community responses to altitude. *Biological Reviews*, **80**, 489–513.

Hughes, A. R., Inouye, B. D., Johnson, M. T. J., Underwood, N. and Vellend, M. (2008) Ecological consequences of genetic diversity. *Ecology Letters*, **11**, 609–623.

Johnson, M. T. J. (2008) Bottom-up effects of plant genotype on aphids, ants and predators. *Ecology*, **89**, 145–154.

Johnson, M. T. J. and Stinchcombe, J. R. (2007) An emerging synthesis between community ecology and evolutionary biology. *Trends in Ecology and Evolution*, **22**, 250–257.

Johnson, M. T. J., Vellend, M. and Stinchcombe, J. R. (2009) Evolution in plant populations as drivers of ecological change in arthropod communities. *Philosophical Transactions of the Royal Society B*, **364**, 1593–1605.

Johnson, N. C., Wilson, G. W. T., Bowker, M. A., Wilson, J. and Miller R. M. (2010) Resource limitation is a driver of local adaptation in mycorrhizal symbioses. *Proceedings of the National Academy of Sciences USA*, **107**, 2093–2098.

Kessler, A., Halitschke, R. and Baldwin, I. T. (2004) Silencing the jasmonate cascade: induced plant defenses and insect populations. *Science*, **305**, 665–668.

Kinnison, M. T. and Hendry, A. P. (2001) The pace of modern life II: from rates of contemporary microevolution to pattern and process. *Genetica*, **112–113**, 145–164.

Kulmatiski, A., Beard, K. H., Stevens, J. and Cobbold, S. M. (2008) Plant–soil feedbacks: a meta-analytical review. *Ecology Letters*, **11**, 980–992.

Laine, A. (2009) Role of coevolution in generating biological diversity: spatially divergent selection trajectories. *Journal of Experimental Biology*, **60**, 2957–2970.

Lewontin, R. C. and Birch, L. C. (1966) Hybridization as a source of variation for adaptation to new environments. *Evolution*, **20**, 315–336.

Lojewski, N. R., Fischer, D. G., Bailey, J. K. *et al.* (2009) Genetics of aboveground productivity in two riparian tree species and their hybrids. *Tree Physiology*, doi:10.1093/treephys/tpp046.

Long, J. C. (1991) The genetic structure of admixed populations. *Genetics*, **127**, 417–428.

Madritch, M. D., Greene, S. L. and Lindroth, R. L. (2009) Genetic mosaics of ecosystem functioning across aspen-dominated landscapes. *Oecologia*, **160**, 119–127.

Martinsen, G. D., Whitham, T. G., Turek, R. J. and Keim, P. (2001) Hybrid populations selectively filter gene introgression between species. *Evolution*, **55**, 1325–1335.

McIntyre, P. J. and Whitham, T. G. (2003) Plant genotype affects long-term herbivore population dynamics and extinction: conservation implications. *Ecology*, **84**, 311–322.

Mitchell-Olds, T., Feder, M. and Wray, G. (2008) Evolutionary and ecological functional genomics, *Heredity*, **100**, 101–102.

Moore, A. J., Brodie, E. D. III and Wolf, J. B. (1997) Interacting phenotypes and the evolutionary process. I. Direct and indirect effects of social interactions. *Evolution*, **51**, 1352–1362.

Murren, C. J., Douglass, L., Gibson, A. and Dudash, M. R. (2006) Individual and combined effects of Ca/Mg ratio and water on trait expression in *Mimulus guttatus*. *Ecology*, **87**, 2591–2602.

Palkovacs, E. P., Marshall, M. C., Lamphere, B. A. *et al.* (2009) Experimental evaluation of evolution and coevolution as agents of ecosystem change in Trinidadian streams. *Philosophical Transactions of the Royal Society B*, **364**, 1617–1628.

Palo, R. T. (1984) Distribution of birch Betula spp., willow Salix spp., and poplar Populus spp. secondary metabolites and their potential role as chemical defense against herbivores. *Journal of Chemical Ecology*, **10**, 499–520.

Parker, J. D., Burkpile, D. E. and Hay, M. E. (2006) Opposing effects of native and exotic herbivores on plant invasions. *Science*, **311**, 1459–1461.

Pennings, S. C. and Silliman, B. R. (2005) Linking biogeography and community ecology: latitudinal variation in plant–herbivore interaction strength. *Ecology*, **86**, 2310–2319.

Post, D. M. and Palkovacs, E. P. (2009) Eco-evolutionary feedbacks in community and ecosystem ecology: interactions between the ecological theatre and the evolutionary play. *Philosophical Transactions of the Royal Society B*, **364**, 1629–1640.

Post, D. M., Palkovacs, E. P., Schielke, E. G. and Dodson, S. I. (2008) Intraspecific phenotypic variation in a predator affects zooplankton community structure and cascading trophic interactions. *Ecology*, **89**, 2019–2032.

Pregitzer, C. C., Bailey, J. K., Hart, S. C. and Schweitzer, J. A. (2010) Soils as agents of selection: feedbacks between plants and soils alter seedling survival and performance. *Evolutionary Ecology*, **24**, 1045–1059.

Price, P. W. (1991) The plant vigor hypothesis and herbivore attack. *Oikos*, **62**, 244–251.

Rehill, B., Whitham, T. G., Martinsen, G. D. *et al.* (2006) Developmental trajectories in cottonwood phytochemistry. *Journal of Chemical Ecology*, **32**, 2269–2285.

Rieseberg, L. H. and Willis, J. H. (2007) Plant speciation. *Science*, **317**, 910–914.

Rieseberg, L. H., Raymond, O., Rosenthal, D. M. *et al.* (2003) Major ecological transitions in annual sunflowers facilitated by hybridization. *Science*, **301**, 1211–1216.

Ryan, M. E., Johnson, J. R. and Fitzpatrick, B. M. (2009) Invasive hybrid tiger salamander genotypes impact native amphibians. *Proceedings of the National Academy of Sciences USA*, **106**, 11166–11171.

Sambatti, J. B. M. and Rice, K. J. (2006) Local adaptation, patterns of selection, and gene flow in the Californian serpentine sunflower *Helianthus exilis*. *Evolution*, **60**, 696–710.

Schluter, D. (1996) Ecological speciation in postglacial fishes. *Philosophical Transactions of the Royal Society B*, **351**, 807–814.

Schluter, D. and Nagel, L. M. (1995) Parallel speciation by natural selection. *American Naturalist*, **146**, 292–301.

Schweitzer, J. A., Martinsen, G. D. and Whitham, T. G. (2002) Cottonwood hybrids gain fitness traits of both parents: a mechanism for their long-term persistence. *American Journal of Botany*, **89**, 981–990.

Schweitzer, J. A., Bailey, J. K., Rehill, B. J. *et al.* (2004) Genetically based trait in a dominant tree affects ecosystem processes. *Ecology Letters*, **7**, 127–134.

Schweitzer, J. A., Bailey, J. K., Fischer, D. G. *et al.* (2008a) Soil microorganism–plant interactions: a heritable relationship between plant genotype and associated soil microorganisms. *Ecology*, **89**, 773–781.

Schweitzer, J. A., Madritch, M. D., Bailey, J. K. *et al.* (2008b) Review – ecological impacts of foliar condensed tannins: a genes-to-ecosystem approach. *Ecosystems*, **11**, 1005–1020.

Shuster, S. M., Lonsdorf, E. V., Wimp, G. M., Bailey, J. K. and Whitham, T. G. (2006) Community heritability measures the evolutionary consequences of indirect genetic effects on community structure. *Evolution*, **60**, 146–158.

Stebbins, G. L. (1959) The role of hybridization in evolution. *Proceedings of the American Philosophical Society*, **103**, 231–251.

Stireman, J. O. III, Nason, J. D. and Heard, S. (2005) Host-associated genetic differentiation in phytophagous insects: general phenomenon or isolated exceptions? Evidence from a goldenrod-insect community. *Evolution*, **59**, 2573–2587.

Stireman, J. O. III, Nason, J. D., Heard, S. *et al.* (2006) Cascading host-associated genetic differentiation in parasitoids of phytophagous insects. *Proceedings of the Royal Society of London B*, **273**, 523–530.

Storfer, A. (1999) Gene flow and local adaptation in a sunfish–salamander system. *Behavioral Ecology and Sociobiology*, **46**, 273–279.

Tetard-Jones, C., Kertesz, M. A., Gallois, P. and Preziosi, R. F. (2007) Genotype-by-genotype interactions modified by a third species in a plant–insect system. *American Naturalist*, **170**, 492–499.

Thompson, J. N. (2005) *The Geographic Mosaic Theory of Coevolution*. Chicago, IL: University of Chicago Press.

Wade, M. J. (2007) The co-evolutionary genetics of ecological communities. *Nature Reviews Genetics*, **8**, 185–195.

Whitham, T. G. (1989) Plant hybrid zones as sinks for pests. *Science*, **244**, 1490–1493.

Whitham, T. G., Young, W. P., Martinsen, G. D. *et al.* (2003) Community and ecosystem genetics: a consequence of the extended phenotype. *Ecology*, **84**, 559–573.

Whitham, T. G., Bailey, J. K., Schweitzer, J. A. *et al.* (2006) A framework for community and ecosystem genetics: from genes to ecosystems. *Nature Reviews Genetics*, **7**, 510–523.

Whitham, T. G., DiFazio, S. P., Schweitzer, J. A. *et al.* (2008) Extending genomics to natural communities and ecosystems. *Science*, **320**, 492–495.

Whitney, K. D., Randell, R. A. and Rieseberg, L. H. (2006) Adaptive introgression of herbivore resistance traits in the weedy sunflower, *Helianthus annuus*. *American Naturalist*, **167**, 794–807.

Yoshida, T., Jones, L. E., Ellner, S. P. and Hairston, N. G. Jr (2003) Rapid evolution drives ecological dynamics in a predator–prey system. *Nature*, **424**, 303–306.

Zheng, S. J. and Dicke, M. (2008) Ecological genomics of plant–insect interactions: from gene to community. *Plant Physiology*, **146**, 812–817.

CHAPTER FIFTEEN

Asking the ecosystem if herbivory-inducible plant volatiles (HIPVs) have defensive functions

MEREDITH C. SCHUMAN and IAN T. BALDWIN

Max Planck Institute for Chemical Ecology

15.1 Introduction

Plant volatiles (PVs) comprise cues exchanged among plants and members of their ecological communities, including other plants, microorganisms and insects. Moreover, some PVs may protect plants against oxidative and thermal damage. Volatiles that are specifically herbivory-inducible (HIPVs) can betray the location of feeding herbivores to their natural enemies, and some HIPVs may defend plants by repelling herbivores or attracting natural enemies. However, the fitness benefits of HIPVs have not been clearly demonstrated in any plant system, so it remains unclear whether they function as indirect defences (Allison & Hare, 2009; Dicke & Baldwin, 2010). Indeed, HIPVs can be detrimental to plants, causing them to be more apparent to and attract herbivores as well as non-beneficial natural enemies that may interfere with other mutualistic interactors, such as pollinators (Halitschke *et al.*, 2008; Kessler & Halitschke, 2009). And it is not clear whether the carnivores found in native plant populations can cope with variability in HIPV emissions. Within single populations of a species, there can be significant variation in the production of PVs among individuals (e.g. Skoula *et al.*, 2000; Delphia *et al.*, 2009), and also after herbivore attack (Schuman *et al.*, 2009), raising the question of whether HIPVs are reliable indicators of herbivory. Do natural enemies learn which compounds are relevant in each population, or are they innately programmed to respond to certain HIPVs? Do plants that emit different or greater amounts of HIPVs than their neighbours risk making themselves more apparent to herbivores and other detrimental visitors, or benefit from greater apparency to beneficial natural enemies? The best way to answer these questions is to ask the ecosystem in which the plant evolved; however, PV research has a history of anthropomorphic metaphors and utilitarian motivations which we suggest may prevent researchers from placing their experiments in the proper ecological context. This chapter will describe an approach which attempts to 'phytopomorphise'

The Ecology of Plant Secondary Metabolites: From Genes to Global Processes, eds. Glenn R. Iason, Marcel Dicke and Susan E. Hartley. Published by Cambridge University Press. © British Ecological Society 2012.

the researcher by using field experiments with wild-type (WT) and appropriate transformed lines of the wild tobacco *Nicotiana attenuata*, in its native ecosystem.

15.2   Challenges in elucidating the functions of PVs
15.2.1   The 1001 functions of PVs?

Most PVs are derived from fatty acids, amino acids (primarily L-phenylalanine) or the terpenoid precursor isopentenyl diphosphate. Figure 15.1 depicts biosynthesis of fatty acid-derived and terpenoid PVs in *N. attenuata*, which reflects what is known about these biosynthetic pathways in other plants. Biosynthesis of

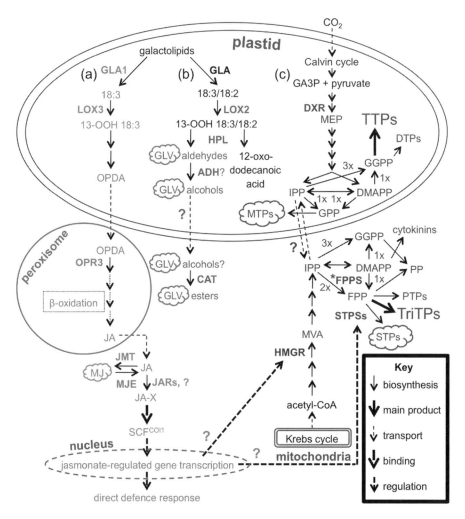

Figure 15.1 The biosynthetic pathways in *Nicotiana attenuata* for jasmonates (orange), GLVs (green) and terpenoids (purple), and putative mechanisms for the influence of jasmonate signalling on sesquiterpene emission. Enzymes and products in black are not specific for a particular pathway. Plant volatiles (PVs) are marked by clouds.

(a) Jasmonate biosynthesis: 18:3 polyunsaturated fatty acids are released from the plastid membrane by wound- or herbivory-induced lipase activity. Free fatty acids are peroxidated and converted to 12-oxy-phytodienoic acid (OPDA) which is transported to the peroxisome (probably as a CoA conjugate), where it is processed to yield the 12-carbon jasmonic acid (JA). JA moves to the cytosol and is modified, for example by methylation to produce methyl jasmonate (MJ), which can be demethylated back to JA; or conjugation to amino acids. Jasmonate conjugates (JA-X) bind to the SCF^{COI1} ubiquitination complex which marks jasmonate ZIM-domain (JAZ) repressors of the jasmonate response for degradation. Only JA-Ile has been shown to actively bind to SCF^{COI1}, but herbivore-induced terpene emission is probably elicited by the interaction of a different conjugate with SCF^{COI1}. Degradation of JAZ repressors permits the transcription of jasmonate-regulated genes and the activation of defence responses. GLA1, glycolipase 1; LOX3, lipoxygenase 3; OPR3, OPDA reductase 3; JMT, jasmonate methyl transferase; MJE, methyl jasmonate esterase; JAR, JA-amino acid conjugate synthase.

(b) GLV biosynthesis: a different GLA cleaves 18:3 and 18:2 polyunsaturated fatty acids from the plastid membrane and these are peroxidated by LOX2. The resulting 18C 13-peroxides are cleaved into *cis*-3-hexenal (18:3) or 1-hexanal (18:2) and 12-oxo-dodecanoic acid, the precursor of traumatin. The *cis*-3-hexenal can be isomerised to *trans*-2-hexenal. Aldehydes are converted in the plastid or in the cytosol to the corresponding alcohols; alcohols are esterified in the cytosol. HPL, hydroperoxide lyase; ADH, alcohol dehydrogenase; CAT, CoA-acetyltransferase.

(c) Terpenoid biosynthesis: there are two terpenoid biosynthetic pathways in plants. The first committed product of the plastidial pathway is methylerythritol phosphate (MEP) which is converted to the 5C terpenoid building blocks isopentenyl diphosphate (IPP) and dimethylallyl diphosphate (DMAPP); these can be interconverted by an isomerase. Different numbers of IPP molecules and one molecule of DMAPP are combined by prenyl transferases to create terpenoid precursors. There is probably exchange of IPP between the cytosolic and plastidial pathways, but the mechanism and regulation are not well understood. GA3P, glyceraldehyde-3-phosphate; DXP, deoxyxylulose 5-phosphate; DXR, DXP reductoisomerase (rate-limiting enzyme); GPPS, geranyl diphosphate (GPP) synthase (prenyl transferase); MTPs, monoterpenes (10C); GGPPS, geranylgeranyl diphosphate (GGPP) synthase (prenyl transferase); DTPs, diterpenes (20C), some of which have sufficient vapour pressure to be volatile, derivatives include gibberellins and plastoquinone; TTPs, tetraterpenes (40C), derivatives include phylloquinones, chlorophylls, tocopherols, carotenoids, ABA and volatile carotenoid derivatives.

The first committed product of the cytosolic pathway is mevalonic acid (MVA) which is converted in three steps to IPP, which can be isomerised to DMAPP. All or part of the cytosolic pathway may take place in vesicles or peroxisomes: HMGR in *Arabidopsis thaliana* is anchored in ER-derived vesicles with the active site facing the cytosol. *Farnesyl diphosphate (FPP) is also synthesised in the mitochondria from cytosolic substrate; some sesquiterpenes (STPs, 15C), homoterpenes (odd number of C, e.g. nerolidol, DMNT) and ubiquinone are produced there. The jasmonate signalling pathway may up-regulate STP emission after herbivore attack either by increasing HMGR activity, sesquiterpene synthase (STPS) activity or both. Cytosolic GGPP and also FPP are substrates for protein prenylation (PP). DMAPP is also used for cytokinin biosynthesis. HMG-CoA, hydroxymethylglutyryl coenzyme A; HMGR. HMG-CoA reductase (rate-limiting enzyme); FPPS, FPP synthase (prenyl transferase); PTPs, polyterpenes (>40C), e.g. dolichol; TriTPs, triterpenes (30C) including sterols, the most abundant products of the cytosolic pathway. See plate section for colour version.

amino-acid-derived PVs is more complex and less well characterised; see Dudareva *et al.* (2006) for a review. Compounds with lower vapour pressure whose primary function is not thought to be mediated by volatilisation, such as the alkaloid nicotine (Figure 15.2), also contribute to plant volatile bouquets at sufficiently high temperatures. Perhaps the most abundant HIPV, methanol released from pectin by pectin methylesterase (von Dahl *et al.*, 2006; Körner *et al.*, 2009), does not fit into the three main PV biosynthetic categories. PV biosynthesis and storage are reviewed in Dudareva *et al.* (2006) and Baldwin (2010).

PVs are emitted from vegetative, reproductive and root tissues, as well as from specialised structures such as glandular trichomes or resin ducts which produce so-called 'essential oils' or resins (Mahmoud & Croteau, 2002). Root emissions have so far been poorly characterised (Wenke *et al.*, 2010). PV emission may be controlled developmentally and diurnally as well as via induced signalling following wounding, herbivory, abiotic stress and other elicitation events (Karban & Baldwin, 1997, Table 4.4; Arimura *et al.*, 2005; Holopainen & Gershenzon, 2010; Loreto & Schnitzler, 2010; Niinemets, 2010; Peñuelas & Staudt, 2010; Koricheva & Barton, Chapter 3). As the ubiquity of their emission suggests, PVs are thought to function in nearly every aspect of plants' biotic and abiotic interactions, including:

- alerting pollinators, herbivores, and predators and parasitoids of herbivores to the plant's presence (attraction or apparency) (Raguso, 2008; Unsicker *et al.*, 2009);
- alerting plants to a neighbouring plant's presence or condition (plant–plant communication or the 'talking plants' phenomenon) (Baldwin *et al.*, 2006; Heil & Karban, 2010);
- alerting predators and parasitoids to the presence of herbivores, and/or deterring herbivores from colonising a plant (indirect defence or 'crying for help') (Dicke & van Loon, 2000; Turlings & Wäckers, 2004; Heil, 2008);
- alerting herbivores to the presence of each other (Dicke, 1986; Dicke & van Loon, 2000; Kessler & Baldwin, 2001; de Moraes *et al.*, 2001);
- alerting neighbouring plants or remote tissues of the same plant to the presence of herbivores (systemic signalling) (Arimura *et al.*, 2000; Kost & Heil, 2006; Frost *et al.*, 2008; Heil & Ton, 2008);
- increasing or decreasing the competitive ability of neighbouring plants (allelopathy) (Mizutani, 1999; Singh *et al.*, 2003);
- increasing or decreasing the palatability of plant tissue to herbivores (feeding stimulant or anti-feedant) (Halitschke *et al.*, 2004; Laothawornkitkul *et al.*, 2008);
- reducing growth of or colonisation by microbes (antimicrobial) (Deng *et al.*, 1993; Cowan, 1999; Khosla & Keasling, 2003; Shiojiri *et al.*, 2006); and
- alleviating oxidative stress due to drought, high ultraviolet exposure, high ozone levels, heat or other stresses that impair metabolism in photosynthetic tissues (antioxidant) (Vickers *et al.*, 2009; Loreto & Schnitzler, 2010; Bagnoli *et al.*, Chapter 6).

Figure 15.2 Interactions mediated by PVs in the *N. attenuata* system. PVs: 1, *trans*-α-bergamotene (TAB); 2, *trans*-2-hexenal; 3, *cis*-3-hexenol; 4, ethylene; 5, nicotine; 6, benzylacetone; 7, methyl jasmonate (MJ).

Feeding by either (a) larvae of the specialist *Manduca* spp. (Lepidoptera, Sphingidae; shown is *M. quinquemaculata*) or (b) generalist mirids (species *Tupiocoris notatus*) induces herbivory-specific PVs (HIPVs) comprising particular mixtures of (1) TAB and other terpenes, and GLVs of which the most abundant are (2) *trans*-2-hexenal and (3) *cis*-3-hexenol. *Geocoris* spp. (Hemiptera, Lygaeidae) predate herbivores on *N. attenuata* and respond to HIPVs (1–3); shown is (c) a *G. pseudopallens* adult on a leaf with a *Manduca* egg and neonate larva. The gaseous hormone (4) ethylene, which is also elicited by some herbivores, plays a role in regulating both (5) nicotine production and plant–plant competition. (5) Nicotine, found both in vegetative and reproductive tissue, repels (d) nectar-robbing carpenter bees (*Xylocopa* spp.) and defends against generalist herbivores including (e) the budworm *Heliothis virescens*. (f) Hummingbirds (here *Archilochus alexandri*) are daytime pollinators of *N. attenuata* and dislike (5) nicotine, which reduces the amount of nectar they remove per flower visited, and forces them to visit more flowers. (g) *Manduca* adults (here *M. sexta*) are also pollinators and are attracted to *N. attenuata* flowers by (6) benzylacetone. The hormone (7) MJ is released in large amounts by (h) damaged sagebrush (*Artemisia tridentata*) and has been shown to increase defences in *N. attenuata* growing nearby; (2, 3) GLVs released from damaged *N. attenuata* influence gene transcription in (i) conspecifics. To date, little is known about belowground interactions among *N. attenuata*, conspecifics, other plant species, herbivores and microbes, but root volatiles are expected to play a role. *Nicotiana attenuata* plants © C. von Dahl (big), R. Halitschke (small); animals and *A. tridentata* © D. Kessler. See plate section for colour version.

This chapter will focus on the possible functions of HIPVs, but first we will review what is known about the functions of PVs in general. For other topics, the reader is directed to the above references and to Gershenzon *et al.* (Chapter 4), Bagnoli *et al.* (Chapter 6), van Dam (Chapter 10) and Dicke *et al.* (Chapter 16).

In order for a metabolite to have an evolutionary function, by definition, it must increase plant fitness when employed by a plant in its natural setting (Karban & Baldwin, 1997). Beneficial or detrimental effects to the community interacting with the plant are only relevant to the evolved function of PVs insofar as they influence plant fitness. Yet there is little research showing any effects of HIPV emissions on plants' evolutionary fitness (Allison & Hare, 2009; Dicke & Baldwin, 2010) despite repeated calls to the community to demonstrate exactly this, beginning with the first discussion of tri-trophic plant–herbivore–carnivore/parasitoid interactions by Price *et al.* (1980).

All areas of scientific research move forward by developing tools which provide advanced resolution and control over the systems studied. PV research is no exception. The field has amassed a significant body of knowledge about mechanisms of biosynthesis and emission, and possible roles of PVs in mediating signalling and ecological interactions, but our ability to demonstrate the proposed functions of PVs is limited by our current toolbox. We need a set of tools which allows us to *discover* the functions of PVs in native plants, in the context of their evolved ecologies.

Some of the first field studies that set out to demonstrate defensive functions of HIPV emissions used agricultural plants in agricultural fields, rather than native plants in a natural setting (e.g. de Moraes *et al.*, 1998; Thaler, 1999; Rasmann *et al.*, 2005) and, thus, are difficult to interpret as showing evolved functions. The first field studies on wild plants (e.g. Rhoades, 1983) lacked the required controls to attribute functions conclusively to individual compounds or groups of compounds. Genetic and chemical tools to manipulate HIPV emission have allowed us to move beyond foundational studies correlating plant damage or HIPV profiles to observed effects. Studies with transgenic plants have demonstrated effects of specific HIPVs and groups of HIPVs on herbivores (e.g. Aharoni *et al.*, 2003; Halitschke *et al.*, 2004), and their predators and parasitoids (e.g. Kappers *et al.*, 2005; Schnee *et al.*, 2006; Degenhardt *et al.*, 2009), but most do so in an agricultural or laboratory setting and thus, again, the interactions demonstrated cannot be clearly interpreted as evolved functions of HIPVs. Many of these studies also use transgenic plants engineered to constitutively emit HIPVs through ectopic expression of HIPV biosynthetic enzymes. Although ectopic constitutive expression is a useful instrument to manipulate PV emission, it removes herbivore inducibility from the information usually conveyed by HIPVs; controlling the exposure of animals to constitutively emitting plants can perhaps mimic induction.

Finally, and most importantly, very few studies attempt to quantify fitness effects of HIPV emission. Van Loon *et al.* (2000) and Hoballah and Turlings (2001) are frequently miscited as demonstrating fitness effects of HIPVs, but both studies only examine the effect of parasitised versus unparasitised larvae on plant fitness: HIPVs played no role in either study. To our knowledge, only one study has quantified the effect of specific endogenous PVs on the fitness of a plant in its native environment (Kessler *et al.*, 2008).

15.2.2   What do we know about PV function?

Now that tools have been developed to manipulate *in planta* PV emissions in increasingly realistic ways, tool development must focus on manipulating the ecological context of HIPV-mediated interactions. Tests of HIPV function require simplified models, but we must be aware of the information we are omitting from our models, and the impact these omissions have on our ability to draw conclusions about natural interactions. HIPV-mediated interactions demonstrated in simplified settings consisting of one plant, one herbivore and one parasitoid or predator cannot be assumed to occur in nature, nor to significantly affect plant fitness in a natural setting where additional interactions within and among trophic levels may either amplify or negate the effect of a particular tri-trophic interaction (Dicke & Baldwin, 2010). Constitutive patterns of PV emission, engineered by ectopic expression of genetic elements, are not an adequate model of the inducible, tissue- and/or temporally specific emissions seen in native plants, in which the timing of emission both conveys additional information and limits its potential recipients. Ideally, simplification in model-building is driven by an intimate understanding of which details are most important in a natural setting, and this should guide the development of the next generation of tools to advance the field of PV functional research.

We lack molecular and chemical tools necessary to attribute functions to PVs in field studies, and thus to discover strategies employed by plants and their interactors in nature which may not be obvious to us. These necessary tools include genetic mutants of native plants, deficient or enhanced in the production, release or perception of specific compounds, which can be compared with otherwise identical wild-type (WT) or empty vector control (EVC) plants in their native ecologies; combined with transgenic or chemical manipulation of the same compounds in order to recreate a WT phenotype in mutants, or a mutant phenotype in WT plants. To interpret roles of different PVs correctly, we must also conduct unbiased, comprehensive analyses of PV profiles from WT and genetically modified plants. For a discussion of unbiased approaches to PV analysis see van Dam and Poppy (2008).

The current large body of knowledge regarding PV biosynthesis and regulation provides hints to the functions of PVs in nature and thus to the design of the tools for our new toolbox. What has been well demonstrated in previous studies? It includes the following:

- that the production of PVs may be controlled via changes to enzyme abundance and activity or substrate flux (Arimura *et al.*, 2000; Dudareva & Negre, 2005; Arimura *et al.*, 2008);
- that the emission of many terpenoid PVs (Figure 15.1) is strongly diurnal and may be closely connected to diurnal patterns of primary metabolism (Halitschke *et al.*, 2000; Rodríguez-Concepción, 2006; Arimura *et al.*, 2008);
- that some PVs are emitted exclusively, or in much greater amounts after herbivore damage (Pare & Tumlinson, 1999) or oviposition (Hilker & Meiners, 2006) and that the changes in emission are elicited by chemicals associated with these events (Alborn *et al.*, 1997, 2003; Gaquerel *et al.*, 2009);
- that feeding or oviposition (Hilker *et al.*, 2002; Fatouros *et al.*, 2009) of a herbivore on plant tissue can attract or arrest predators or parasitoids;
- that many specialist parasitoids innately respond to HIPVs induced by feeding or oviposition of their prey, although this is not always the case (Allison & Hare, 2009), and generalists can quickly learn to respond to prey-associated HIPVs (Lewis & Tumlinson, 1988; Dukas & Duan, 2000; de Boer & Dicke, 2005);
- that different herbivores may either cause similar (Kessler & Baldwin, 2004) or different (de Moraes *et al.*, 1998) HIPV emission profiles, which has consequences for herbivore, predator and parasitoid visitation;
- that competition among predators and parasitoids may interfere with the effects of HIPVs on herbivore parasitisation or predation (Rosenheim, 1998; Finke & Denno, 2004);
- that HIPVs can also attract other herbivores (Bolter *et al.*, 1997; Carroll *et al.*, 2006; Halitschke *et al.*, 2008) or mediate competition among herbivores (Denno *et al.*, 1995; Kaplan & Denno, 2007);
- that there is considerable variation among individual plants (e.g. Skoula *et al.*, 2000; Gouinguene *et al.*, 2001; Hare, 2007; Delphia *et al.*, 2009; Schuman *et al.*, 2009) or populations in PV emission, and that these differences may be associated with differences in biosynthesis (Tholl *et al.*, 2005), signalling (Wu *et al.*, 2008) or environmental conditions (Takabayashi *et al.*, 1994);
- that plants engineered to constitutively produce PVs which are usually induced by herbivory can attract previously trained predators or parasitoids in situations of forced interaction (e.g. Kappers *et al.*, 2005; Schnee *et al.*, 2006); and
- that herbivory can also induce changes to floral phenotypes (Kessler & Halitschke, 2009) and PV profiles, for example, causing a change from herbivorous to non-herbivorous pollinator attraction (Kessler *et al.*, 2010).

However, there are few studies demonstrating the natural functions of PVs, in particular:

- whether plants derive an overall fitness benefit from producing particular floral and vegetative PVs under natural conditions, or whether there are other reasons for the emission of some PVs in wild and in cultivated plants (Allison & Hare, 2009; Dicke & Baldwin, 2010);
- whether the production patterns of PVs (if emission is induced, when and by what) and their coordination with the production of other defences enhance plant fitness (Euler & Baldwin, 1996; Halitschke *et al.*, 2000; Kahl *et al.*, 2000; Gierl & Frey, 2001);
- how predators, parasitoids, floral visitors and herbivores develop responses to particular PVs (Allison & Hare, 2009);
- the extent to which floral visitors are affected by vegetative PVs, and vegetative visitors are affected by floral PVs (Heil, 2008);
- consequences of diverse PV profiles within single populations and between populations or ecotypes (Steppuhn *et al.* 2008; Schuman *et al.*, 2009);
- whether plants benefit or suffer from emitting different amounts of PVs than their neighbours: for example, a plant that emits much greater levels of HIPVs benefits from attracting more predators or attracting predators away from its neighbours, or its neighbours benefit from the attraction of too many predators for the actual prey reward (Shiojiri *et al.*, 2010); and
- whether plants that constitutively emit an HIPV can 'fool' predators into 'thinking' they have been induced by herbivores. If so, what is the evolutionarily stable balance of these 'cry wolf' plants with 'honest' inducible emitters (Shiojiri *et al.*, 2010)?

To address these open questions, we need a toolbox which allows us to manipulate PV emissions precisely and rigorously in a natural setting, and experimental designs which take plants' ecological communities into account.

15.3   The ask-the-ecosystem approach

15.3.1   *Nicotiana attenuata* and its community as an ecological model system

Our group has developed methods that allow us to ask plants and their interactors about functions of PV emission without making prior assumptions about what the answers might be, by using transgenic and chemical techniques to manipulate PV emission and then comparing WT with manipulated plants in the native ecosystem. This approach allows us to learn about both expected and unexpected evolved interactions. The knowledge gained is prerequisite to understanding how PV-mediated interactions function in the

real world, and that understanding is important for employing these mechanisms successfully in human-made systems (Degenhardt *et al.*, 2003).

We began by selecting an ecological model plant, the wild coyote tobacco
Nicotiana attenuata (Solanaceae) (Figure 15.2). *Nicotiana attenuata* is a fire-chaser
which forms monocultures in nitrogen-rich post-fire soils, similar to agricultural plants in artificially fertilised fields. Its relationship to cultivated solanaceous species (*N. tabacum, Solanum tuberosum*) and the wild tobacco *N. sylvestris*
allows us to benefit from the ecological, chemical and molecular genetic tools
developed in these systems. By closely studying the plant, its ecosystem and
its natural diversity, we can learn what questions can best be addressed in this
system, and which traits seem the most important for plants' evolutionary
success. Figure 15.2 presents a summary of *N. attenuata*'s HIPV-mediated interactions. Interactions with floral visitors are included because these can be
affected by herbivory: changes to floral PV profiles are also part of herbivory-
inducible changes to PVs (Kessler *et al.*, 2010).

The jasmonate phytohormones are particularly important in moderating
defence responses against herbivory, which is likely to be the environmental
stress which exerts the greatest selective pressure on *N. attenuata*. Research
in our system showed that *N. attenuata* plants employ and benefit from
jasmonate-mediated defences when attacked by herbivores, both in nature
and in the laboratory (Baldwin, 1998; van Dam *et al.*, 2000; Stork *et al.*, 2009),
and yet that plants demonstrate significant genetic and defence-related phenotypic variation within and between populations (Halitschke *et al.*, 2000;
Glawe *et al.*, 2003; Bahulikar *et al.*, 2004; Schuman *et al.*, 2009). Jasmonates
control the elicitation of both direct and indirect defences in *N. attenuata*,
and in combination with salicylic acid and the volatile phytohormone ethylene (see Figure 15.2), orchestrate specific responses to attacking specialist
and generalist herbivores (Kahl *et al.*, 2000; Diezel *et al.*, 2009). In particular,
upon attack by the nicotine-resistant solanaceous specialist *Manduca sexta*,
N. attenuata down-regulates its production of nicotine and up-regulates the
emission of putative indirect defence compounds such as the sesquiterpene
trans-α-bergamotene (TAB). Unlike nicotine accumulation, TAB emission is
not down-regulated by ethylene or inhibitors of jasmonate biosynthesis,
although emission is jasmonate-dependent (Halitschke *et al.*, 2000; 2008;
Kahl *et al.*, 2000).

In order to dissect variation in and regulation of direct and indirect defences, we have generated plants altered in the biosynthesis or perception of
individual signalling- and defence-related compounds (Krügel *et al.*, 2002, and
references above). Comparing altered plants to the corresponding WT or EVC
plants (Schwachtje *et al.*, 2008) allows us to determine the function of the
manipulated trait(s) in a single genetic background. Altering the same trait in
multiple genetic backgrounds allows us to study the influence of genetic

background on trait function, and thus explore the consequences of natural variation for plant defence (Steppuhn *et al.*, 2008). We have additionally built up a toolkit of techniques and chemical elicitors including standardised procedures for wounding and for application of herbivore oral secretions (OS), supplementation of methyl jasmonate (MJ) or perfuming with specific PVs, which allow us to manipulate plants' behaviour reproducibly and 'recover' some of the phenotypes 'lost' from transformants, thus robustly demonstrating that the manipulated genes and their products are responsible for the observed phenotypic effects.

15.3.2 HIPV-mediated interactions of *N. attenuata*

We discovered a great diversity of volatile emission following treatment with wounding plus *M. sexta* OS, or supplementation with MJ in populations of *N. attenuata* (Schuman *et al.*, 2009). Nevertheless, several studies by our group have indicated the importance of particular HIPVs in defence and apparency to pollinators, florivores, nectar robbers, herbivores and predators (Figure 15.2; Kessler & Baldwin, 2001; Halitschke *et al.*, 2008; Kessler *et al.*, 2008). We have evidence that generalist *Geocoris* predatory insects in *N. attenuata*'s habitat (Figure 15.2c) must learn to respond to HIPVs (S. Allmann & I.T. Baldwin, unpublished data), yet we see that *Geocoris* consistently respond to the same groups of compounds: GLVs, which are characteristic of leaf damage in all plants and can carry a herbivory-specific signature (Allmann & Baldwin, 2010), and induced sesquiterpenes, which are emitted most strongly beginning 1 day after herbivore attack (Halitschke *et al.*, 2008; Skibbe *et al.*, 2008). TAB is thought to be the most important of *N. attenuata*'s sesquiterpenes for *Geocoris* (Halitschke *et al.*, 2000). However, a naturally occurring genotype of *N. attenuata* deficient in TAB was able to attract predators just as well as a TAB-producing genotype in an area where TAB production should be common (Steppuhn *et al.*, 2008). Thus, native *Geocoris* predators seem to be telling us they learn to respond to HIPVs when they are associated with prey. Many questions remain open in our understanding of *N. attenuata*'s HIPV-mediated interactions, in particular those introduced at the end of Section 15.2.2 regarding the influence of particular HIPVs on plant fitness and the community effects of HIPV emissions for plants in populations.

15.4 Discovering HIPV functions in a natural system

As an example of the ask-the-ecosystem approach applied to HIPVs, we present an experimental set-up designed to separate the effects of TAB emission from the emission of GLVs and total HIPVs, and from the induction of direct defences. For this experiment, we have generated lines of *N. attenuata* manipulated in three biosynthetic genes: lipoxygenase 3 (*lox3*) which provides substrate for jasmonates (Figure 15.1a), *lox2* which provides substrate for GLVs

(Figure 15.1b) and *TPS10*, a sesquiterpene synthase (STPS, Figure 15.1c) from *Zea mays* which produces TAB and its linear precursor, *trans*-β-farnesene (TBF) (Figure 15.3; Köllner *et al.*, 2009). Silencing *lox3* produces plants deficient in jasmonate signalling which cannot induce direct defences and jasmonate-mediated HIPVs, including TAB. GLV emission during herbivore feeding is not regulated by jasmonate signalling, so in order to reduce total HIPVs and provide a 'passive' background for engineered *TPS10* terpene emission, we have additionally silenced *lox2* (IR*lox2/3* plants have been characterised together with S. Allmann.) To manipulate TAB emission independently from endogenous signalling, in collaboration with J. Gershenzon, we have chosen to ectopically over-express either *TPS10* or a point mutant version of the enzyme (*TPS10M*) which cannot cyclise TBF to TAB (Figure 15.3; Köllner *et al.*, 2009), under the control of a 35S promotor. We predict that this will allow us to manipulate TAB without altering *N. attenuata*'s native VOC emissions, because the ectopically expressed enzyme will not be part of the regulatory framework that governs the expression and activity of native enzymes controlling HIPV emissions. Comparing plants expressing *TPS10* with those expressing *TPS10M* allows us to separate the fitness effects of emitting TAB, an endogenous HIPV, from those of emitting TBF, a structurally related PV emitted at very low or undetectable levels from WT *N. attenuata*. Preliminary data from these transformed lines (Figure 15.3) confirm that 35S::*TPS10* and

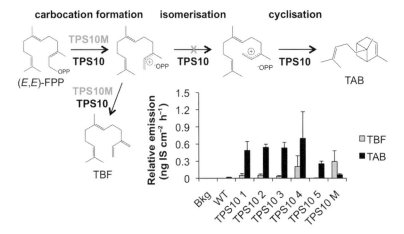

Figure 15.3 Mechanism of TBF and TAB biosynthesis by the sesquiterpene synthase ZmTPS10 and the point mutant ZmTPS10M, and constitutive TBF and TAB emission from wild-type (WT) and multiple lines transformed with 35S::*TPS10* or 35S::*TPS10M* constructs. Bkg, background control (ambient air). Scale is relative to nanograms of tetralin internal standard (IS). Biosynthetic pathway simplified from Figure 4 of Köllner *et al.* (2009).

35S::*TPS10M* lines produce the product profile that we predicted, with little or no effect on the emission of other PVs (data not shown).

We use this set of transformed lines to ask *N. attenuata* and its ecological community about the effects of altering TAB emission, whether the effect of TAB emission depends on the action of jasmonate-mediated defences and on the total PV background, and to what extent the quality and effects of HIPV emission depend on the presence and phenotypes of neighbour plants (see below). Including plants impaired in inducible direct defences and HIPVs grants us an additional level of sensitivity: TAB emission may have a stronger effect in the absence of other defences, and other HIPVs; and impaired plants are likely to experience larger neighbour-dependent effects on their levels of herbivore damage, and perhaps on their fitness, than are WT plants.

Neighbour-dependent effects are the effects of the presence and identity of neighbour plants on growth, HIPV emission and interactions with herbivores and predators. Neighbour-dependent effects may be mediated by competition, plant–plant communication, and the herbivore- and predator-attractive or repellent properties of neighbours. To discover these effects, we compare populations (both mixed- and monocultures) versus individual plants. Individuals and populations are randomly distributed across a field plot in *N. attenuata*'s native habitat. The random order is modified to ensure that no two populations or individuals of the same type are directly adjacent, so that an animal finding itself at any point in the experimental plot may always choose among different types.

Our set of transformed lines (Table 15.1) gives us all of the comparisons we need in order to dissect the effects of TAB and TBF emission in the context of a WT HIPV and defence background, or in 'passive' backgrounds lacking WT HIPV and defence profiles. Taking these lines to *N. attenuata*'s native environment, we can ask the plant and its community what the evolved functions of TAB, GLVs and other HIPVs might be. Once we have observed effects in the field, we can confirm them under more controlled conditions in the laboratory and glasshouse. Although this is an open and exploratory experiment, we can propose several hypotheses based on ideas about HIPV function. These hypotheses (Table 15.2) guide our choice of phenotypes and planting design.

15.5   Conclusions and outlook

Since the first reports in the 1980s of 'talking trees' and plants 'crying for help', continuing through to today's growing concern with the interaction of PVs and climate change (Peñuelas, 2008), PVs have been a hot topic which intrigues scientists and non-scientists alike. Humans love to anthropomorphise, and what better way to anthropomorphise plants than through stories in which they talk, listen, sniff each other out and cry for help? While the delightful metaphors which have been created by the field of PV research

Table 15.1 Transformants used to investigate effects of enhancing terpenoid emission in the presence or absence of other HIPVs and defences. Note that we introduce metaphors in the Phenotype column; these are meant to help the reader develop an intuitive understanding of our experimental phenotypes, but our experiment is designed to allow us to test the accuracy of these metaphors! A distinction is made between constitutive (C) and herbivore-inducible (I) metabolite levels. 'Odoriferous' 35S::TPS plants constitutively emit similar amounts of terpenoids as herbivore-induced WT plants, and may be able to induce even greater emission upon induction (see Figure 15.1). 'Unresponsive' IRlox2/3 plants cannot induce direct or indirect defences upon herbivore attack and have lower constitutive defence levels; everything but GLV emission can be 'restored' to a WT phenotype by MJ application. IRlox2/3 × 35S::TPS 'odoriferous unresponsive' plants constitutively emit TAB and TBF in an IRlox2/3 phenotypic background.

Genotype	Phenotype	(HI)PVs								Signalling jasmonates		Direct defence	
		GLVs		Terpenoids									
				Total		TAB		TBF					
		C	I	C	I	C	I	C	I	C	I	C	I
Constitutive/inducible													
WT	Wild-type	nd	+	o	+	o	+	o	+	o	+	o	—
35S::*TPS10*	'Odoriferous' plants: ectopic TAB and TBF emission	nd	+	+	+++	++	+++	+	++	o	+	o	+
35S::*TPS10M*		nd	+	+	+++	+	++	++	+++	o	+	o	+
IRlox2/3	'Unresponsive' plants: lacking induced defence including HIPVs	nd	nd	—	—(+)	—	—(+)	—	—(+)	—	—(+)	—	—(+)
IRlox2/3 × 35S::*TPS10*	'Odoriferous unresponsive' plants: ectopic TAB and TBF emission in the absence of induced defence and HIPVs	nd	nd	++	++(+++)	++	++(+++)	+	+(++)	—	—(+)	—	—(+)
IRlox2/3 × 35S::*TPS10M*		nd	nd	++	++(+++)	+	+(++)	++	++(+++)	—	—(+)	—	—(+)

Key
o Untreated WT levels
+ Greater than untreated WT levels, ++ > + > o
− Less than corresponding WT levels, − <−
nd Not detectable

Table 15.2 *Hypotheses and predicted outcomes for the example field experiment.*

Hypothesis	Prediction
Plant–plant interactions	
Growth in competition with other plants affects VOC emissions.	The same genotype will have different PV profiles when growing in a population versus alone.
GLV, TAB and TBF emission enhance/decrease competitive ability.	35S::*TPS10* will grow best/worst and be fittest/least fit in competition with other genotypes; IR*lox2/3* will grow worst/best and be least fit/fittest in competition with other genotypes.
Plant–herbivore interactions	
Herbivores are repelled by indirect defences.	Plant phenotypes which attract the most *Geocoris* will attract the fewest herbivores and have greater fitness than phenotypes with the same direct defence background.
Herbivores are most attracted to plants producing the greatest amount of PVs.	Number of herbivores: 35S::*TPS* > IR*lox2/3* × 35S::*TPS* > WT > IR*lox2/3*.
High terpene emission deters herbivores from feeding.	Amount of damage per herbivore attack: IR*lox2/3* > WT > IR*lox2/3* × 35S::*TPS* > 35S::*TPS*.
Low GLV emission deters herbivores from feeding.	IR*lox2/3* and IR*lox2/3* × 35S::*TPS* treated with MJ will suffer the least damage per herbivore attack. If GLV emission is more important than direct defence, then plants with an IR*lox2/3* background will suffer less damage even without MJ treatment.
Herbivores cause most damage on plants with the least direct defence.	IR*lox2/3* and IR*lox2/3* × 35S::*TPS* will suffer the most damage per herbivore attack.
Herbivores cause most damage on plants with the least indirect defence.	Plant phenotypes which attract the fewest *Geocoris* will accumulate the most herbivore damage: IR*lox2/3* > IR*lox2/3* × 35S::*TPS* > IR*lox2/3* × 35S::*TPS10* > 35S::*TPS10M* > WT or 35S::*TPS10*. This will be reflected in realised plant fitness, and will not be affected by MJ treatment.
Plant–predator interactions	
Geocoris predators learn to respond to HIPVs.	If a change in emission is sufficient for a learned association, then *Geocoris* will be most attracted to plants producing high levels of TAB and WT levels of GLVs, which will be the most common herbivore-inducible VOCs in the experiment: herbivore-induced 35S::*TPS10* plants.
	If presence/absence of an HIPV is necessary for associative learning, *Geocoris* will respond to GLVs and GLV-emitting genotypes, but may not respond to TAB; they will be most attracted to induced WT plants.
Geocoris shows an innate response to GLVs and TAB.	*Geocoris* will be most attracted to herbivore-induced 35S::*TPS10* plants.

have inspired effective PR campaigns – which is both good and important for research science – perhaps they have negatively influenced the focus of the research. This is because in order to understand what plants are doing and why, and in order to create proper simplified models for experiments, plant researchers should not be anthropomorphising their subjects, but rather 'phytopomorphising' themselves.

Table 15.1 introduces a set of different HIPV emission/defence phenotypes used as ecological molecular tools in an experiment aimed at discovering natural functions of HIPVs. The potential to move from an anthropomorphic towards a 'phytopomorphic' understanding lies in choosing a set of plant phenotypes that lets us test the accuracy of our metaphors. As we gather data on the functions of the traits we have manipulated to create each phenotype, the metaphors which grant us 'intuitive access' to the phenotypes are likely to change.

It is also important to keep in mind that a large portion of *N. attenuata*'s HIPV emission is not manipulated in our example experiment. This portion includes methanol, ethylene, the floral compounds benzylacetone and nicotine (Figure 15.2) (nicotine is also volatilised from leaf tissue), and several mono- and sesquiterpenes which are constitutively emitted and not regulated by *N. attenuata* after herbivory. We are not assuming that the compounds we have manipulated, namely GLVs, TAB and total HIPVs, are the most important PVs for *N. attenuata*'s ecological interactions, but we are using our toolbox to ask *how* important they are.

References

Aharoni, A., Giri, A. P., Deuerlein, S. *et al.* (2003) Terpenoid metabolism in wild-type and transgenic *Arabidopsis* plants. *Plant Cell*, **15**, 2866–2884.

Alborn, H. T., Turlings, T. C. J., Jones, T. H. *et al.* (1997) An elicitor of plant volatiles from beet armyworm oral secretion. *Science*, **276**, 945–949.

Alborn, H. T., Brennan, M. M. and Tumlinson, J. H. (2003) Differential activity and degradation of plant volatile elicitors in regurgitant of tobacco hornworm (*Manduca sexta*) larvae. *Journal of Chemical Ecology*, **29**, 1357–1372.

Allison, J. D. and Hare, J. D. (2009) Learned and naive natural enemy responses and the interpretation of volatile organic compounds as cues or signals. *New Phytologist*, **184**, 768–782.

Allmann, S. and Baldwin, I. T. (2010) Insects betray themselves in nature to predators by rapid isomerization of green leaf volatiles. *Science*, **329**, 1075–1078.

Arimura, G., Ozawa, R., Shimoda, T. *et al.* (2000) Herbivory-induced volatiles elicit defence genes in lima bean leaves. *Nature*, **406**, 512–515.

Arimura, G., Kost, C. and Boland, W. (2005) Herbivore-induced, indirect plant defences. *Biochimica et Biophysica Acta: Molecular and Cell Biology of Lipids*, **1734**, 91–111.

Arimura, G., Kopke, S., Kunert, M. *et al.* (2008) Effects of feeding *Spodoptera littoralis* on lima bean leaves: IV. Diurnal and nocturnal damage differentially initiate plant volatile emission, *Plant Physiology*, **146**, 965–973.

Bahulikar, R. A., Stanculescu, D., Preston, C. A. and Baldwin, I. T. (2004) ISSR and AFLP analysis of the temporal and spatial population structure of the post-fire annual, *Nicotiana attenuata*, in SW Utah. *BMC Ecology*, **4**, 12.

Baldwin, I. T. (1998) Jasmonate-induced responses are costly but benefit plants under attack in native populations. *Proceedings of the National Academy of Sciences USA*, **95**, 8113–8118.

Baldwin, I. T. (2010) Plant volatiles. *Current Biology*, **20**, R392–R397.

Baldwin, I. T., Halitschke, R., Paschold, A., von Dahl, C. C. and Preston, C. A. (2006) Volatile signaling in plant–plant interactions: 'talking trees' in the genomics era. *Science*, **311**, 812–815.

Bolter, C. J., Dicke, M., van Loon, J. J. A., Visser, J. H. and Posthumus, M. A. (1997) Attraction of Colorado potato beetle to herbivore-damaged plants during herbivory and after its termination. *Journal of Chemical Ecology*, **23**, 1003–1023.

Carroll, M. J., Schmelz, E. A., Meagher, R. L. and Teal, P. E. A. (2006) Attraction of *Spodoptera frugiperda* larvae to volatiles from herbivore-damaged maize seedlings. *Journal of Chemical Ecology*, **32**, 1911–1924.

Cowan, M. M. (1999) Plant products as antimicrobial agents. *Clinical Microbiology Reviews*, **12**, 564–582.

de Boer, J. G. and Dicke, M. (2005) Information use by the predatory mite *Phytoseiulus persimilis* (Acari, Phytoseiidae), a specialised natural enemy of herbivorous spider mites. *Applied Entomology and Zoology*, **40**, 1–12.

de Moraes, C. M., Lewis, W. J., Pare, P. W., Alborn, H. T. and Tumlinson, J. H. (1998) Herbivore-infested plants selectively attract parasitoids. *Nature*, **393**, 570–573.

de Moraes, C. M., Mescher, M. C. and Tumlinson, J. H. (2001) Caterpillar-induced nocturnal plant volatiles repel conspecific females. *Nature*, **410**, 577–580.

Degenhardt, J., Gershenzon, J., Baldwin, I. T. and Kessler, A. (2003) Attracting friends to feast on foes: engineering terpene emission to make crop plants more attractive to herbivore enemies. *Current Opinion in Biotechnology*, **14**, 169–176.

Degenhardt, J., Hiltpold, I., Köllner, T. G. *et al.* (2009) Restoring a maize root signal that attracts insect-killing nematodes to control a major pest. *Proceedings of the National Academy of Sciences USA*, **106**, 13213–13218.

Delphia, C. M., Rohr, J. R., Stephenson, A. G., de Moraes, C. M. and Mescher, M. C. (2009) Effects of genetic variation and inbreeding on volatile production in a field population of horsenettle. *International Journal of Plant Sciences*, **170**, 12–20.

Deng, W., Hamilton-Kemp, T. R., Nielson, M. T. *et al.* (1993) Effects of six-carbon aldehydes and alcohols on bacterial proliferation. *Journal of Agricultural and Food Chemistry*, **41**, 506–510.

Denno, R. F., McClure, M. S. and Ott, J. R. (1995) Interspecific interaction in phytophagous insects: competition reexamined and resurrected. *Annual Review of Entomology*, **40**, 297–331.

Dicke, M. (1986) Volatile spider-mite pheromone and host-plant kairomone, involved in spaced-out gregariousness in the spider mite *Tetranychus urticae*. *Physiological Entomology*, **11**, 251–262.

Dicke, M. and Baldwin, I. T. (2010) The evolutionary context for herbivore-induced plant volatiles, beyond the 'cry for help'. *Trends in Plant Science*, **15**, 167–175.

Dicke, M. and van Loon, J. J. A. (2000) Multitrophic effects of herbivore-induced plant volatiles in an evolutionary context. *Entomologia Experimentalis et Applicata*, **97**, 237–249.

Diezel, C., von Dahl, C. C., Gaquerel, E. and Baldwin, I. T. (2009) Different lepidopteran elicitors account for cross-talk in herbivory-induced phytohormone signalling. *Plant Physiology*, **150**, 1576–1586.

Dudareva, N. and Negre, F. (2005) Practical applications of research into the regulation of plant volatile emission. *Current Opinion in Plant Biology*, **8**, 113–118.

Dudareva, N., Negre, F., Nagegowda, D. A. and Orlova, I. (2006) Plant volatiles: recent advances and future perspectives. *Critical Reviews in Plant Sciences*, **25**, 417–440.

Dukas, R. and Duan, J. J. (2000) Potential fitness consequences of associative learning in a parasitoid wasp. *Behavioral Ecology*, **11**, 536–543.

Euler, M. and Baldwin, I. T. (1996) The chemistry of defense and apparency in the corollas of *Nicotiana attenuata*. *Oecologia*, **107**, 102–112.

Fatouros, N., Pashalidou, F., Cordero, W. A. *et al.* (2009) Anti-aphrodisiac compounds of male butterflies increase the risk of egg parasitoid attack by inducing plant synomone production. *Journal of Chemical Ecology*, **35**, 1373–1381.

Finke, D. L. and Denno, R. F. (2004) Predator diversity dampens trophic cascades. *Nature*, **429**, 407–410.

Frost, C. J., Mescher, M. C., Dervinis, C. *et al.* (2008) Priming defense genes and metabolites in hybrid poplar by the green leaf volatile *cis*-3-hexenyl acetate. *New Phytologist*, **180**, 722–734.

Gaquerel, E., Weinhold, A. and Baldwin, I. T. (2009) Molecular interactions between the specialist herbivore *Manduca sexta* (Lepidoptera, Sphigidae) and its natural host *Nicotiana attenuata*. VIII. An unbiased GCxGC-ToFMS analysis of the plant's elicited volatile emissions. *Plant Physiology*, **149**, 1408–1423.

Gierl, A. and Frey, M. (2001) Evolution of benzoxazinone biosynthesis and indole production in maize. *Planta*, **213**, 493–498.

Glawe, G. A., Zavala, J. A., Kessler, A., van Dam, N. M. and Baldwin, I. T. (2003) Ecological costs and benefits correlated with trypsin protease inhibitor production in *Nicotiana attenuata*. *Ecology*, **84**, 79–90.

Gouinguene, S., Degen, T. and Turlings, T. C. J. (2001) Variability in herbivore-induced odour emissions among maize cultivars and their wild ancestors (teosinte). *Chemoecology*, **11**, 9–16.

Halitschke, R., Kessler, A., Kahl, J., Lorenz, A. and Baldwin, I. T. (2000) Ecophysiological comparison of direct and indirect defenses in *Nicotiana attenuata*. *Oecologia*, **124**, 408–417.

Halitschke, R., Ziegler, J., Keinänen, M. and Baldwin, I. T. (2004) Silencing of hydroperoxide lyase and allene oxide synthase reveals substrate and defense signaling crosstalk in *Nicotiana attenuata*. *The Plant Journal*, **40**, 35–46.

Halitschke, R., Stenberg, J. A., Kessler, D., Kessler, A. and Baldwin, I. T. (2008) Shared signals – 'alarm calls' from plants increase apparency to herbivores and their enemies in nature. *Ecology Letters*, **11**, 24–34.

Hare, J. D. (2007) Variation in herbivore and methyl jasmonate-induced volatiles among genetic lines of *Datura wrightii*. *Journal of Chemical Ecology*, **33**, 2028–2043.

Heil, M. (2008) Indirect defence via tritrophic interactions. *New Phytologist*, **178**, 41–61.

Heil, M. and Karban, R. (2010) Explaining evolution of plant communication by airborne signals. *Trends in Ecology and Evolution*, **25**, 137–144.

Heil, M. and Ton, J. (2008) Long-distance signalling in plant defence. *Trends in Plant Science*, **13**, 264–272.

Hilker, M. and Meiners, T. (2006) Early herbivore alert: insect eggs induce plant defense. *Journal of Chemical Ecology*, **32**, 1379–1397.

Hilker, M., Kobs, C., Varama, M. and Schrank, K. (2002) Insect egg deposition induces *Pinus sylvestris* to attract egg parasitoids. *Journal of Experimental Biology*, **205**, 455–461.

Hoballah, M. E. F. and Turlings, T. C. J. (2001) Experimental evidence that plants under caterpillar attack may benefit from attracting parasitoids. *Evolutionary Ecology Research*, **3**, 553–565.

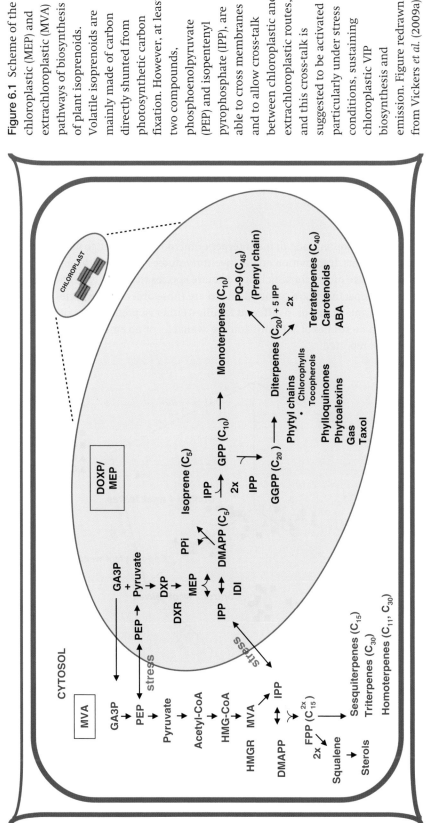

Figure 6.1 Scheme of the chloroplastic (MEP) and extrachloroplastic (MVA) pathways of biosynthesis of plant isoprenoids. Volatile isoprenoids are mainly made of carbon directly shunted from photosynthetic carbon fixation. However, at least two compounds, phosphoenolpyruvate (PEP) and isopentenyl pyrophosphate (IPP), are able to cross membranes and to allow cross-talk between chloroplastic and extrachloroplastic routes, and this cross-talk is suggested to be activated particularly under stress conditions, sustaining chloroplastic VIP biosynthesis and emission. Figure redrawn from Vickers *et al.* (2009a).

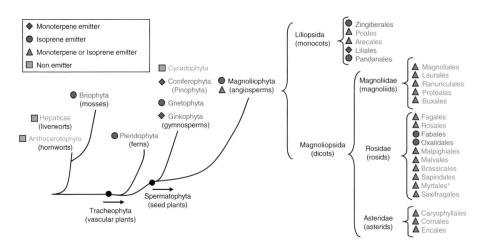

Figure 6.2 The trait isoprene and/or monoterpene emission is assumed to have evolved several times in plants. Information on VIP inventory studies is given at the order level. In green and blue are orders for which one or more species that emit isoprene or monoterpenes, respectively, are reported. In red are those orders with species that emit (within the same family or in different families) either isoprene or monoterpenes. Orders for which no studies of VIP emission are available, or no emission has been detected, are not reported.

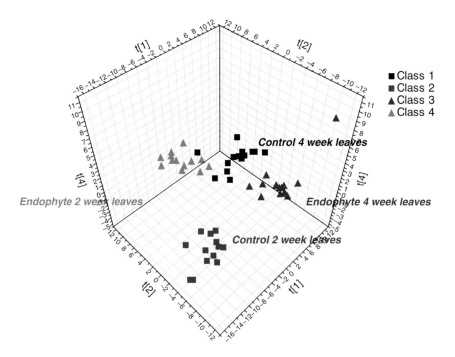

Figure 11.4 Partial least squares-discriminant analyses (PLS-DA) of changes in the metabolome of *Cirsium avenae* leaves infected with *Chaetomium cochlioides*. Biochemicals in the leaf extracts were profiled using UPLC-TOFMS (-ESI mode) and the datasets analysed by PLS-DA. Each symbol represents data from one plant. The axes represent the first three latent variables, t_1, t_2, t_3, of the PLS-DA model, and these account for 26.8%, 28.7% and 28.4 % of the Y variation.

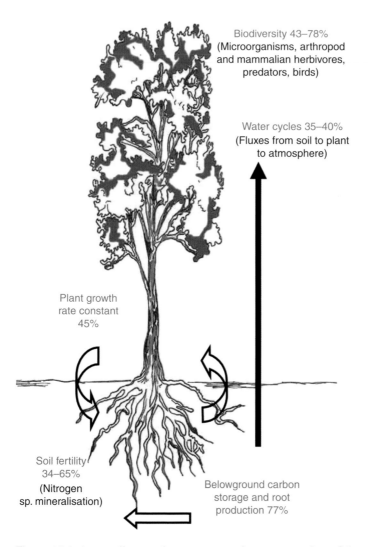

Biodiversity 43–78%
(Microorganisms, arthropod
and mammalian herbivores,
predators, birds)

Water cycles 35–40%
(Fluxes from soil to plant
to atmosphere)

Plant growth
rate constant
45%

Soil fertility
34–65%
(Nitrogen
sp. mineralisation)

Belowground carbon
storage and root
production 77%

Figure 14.1 Across diverse plant systems, a large proportion of the variation in biodiversity and ecosystem function in common environment trials can be explained by plant genetic factors. For example, in *Populus angustifolia*, plant genetic factors can affect soil nutrient availability and belowground carbon storage as well as species interactions with arthropods, avian predators, mammalian herbivores like beaver and even soil microbes.

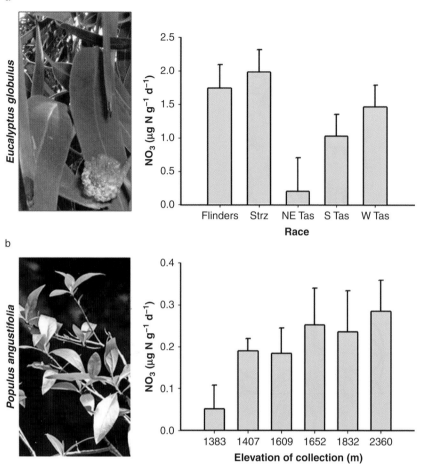

Figure 14.2 Evolution in host plants can alter biodiversity and ecosystem function. Common garden experiments indicate that genetic differentiation along elevational and latitudinal gradients can alter soil nutrients. See page 275 for full details.

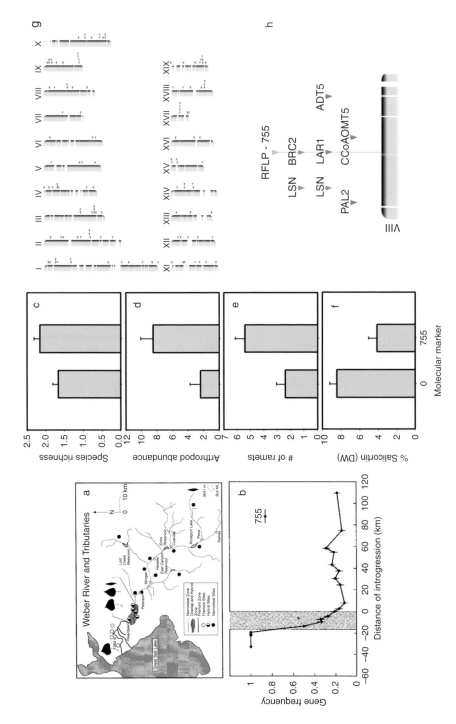

Figure 14.3 Cryptic introgression of a *Populus fremontii* molecular marker into *P. angustifolia* results in changes to biodiversity in a *Populus* hybridising system. See page 278 for full details.

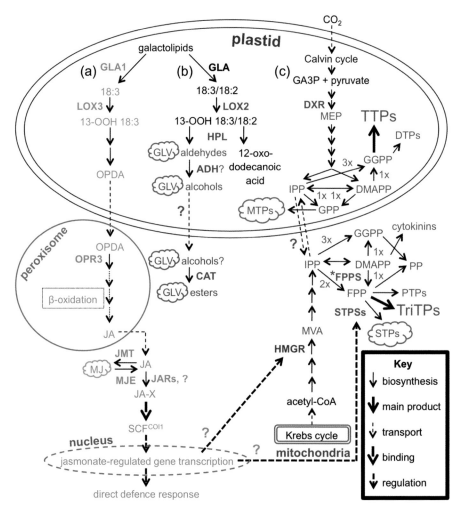

Figure 15.1 The biosynthetic pathways in *Nicotiana attenuata* for jasmonates (orange), GLVs (green) and terpenoids (purple), and putative mechanisms for the influence of jasmonate signalling on sesquiterpene emission. Enzymes and products in black are not specific for a particular pathway. Plant volatiles (PVs) are marked by clouds. See page 288 for full details.

Figure 15.2 Interactions mediated by PVs in the *N. attenuata* system. See page 290 for full details.

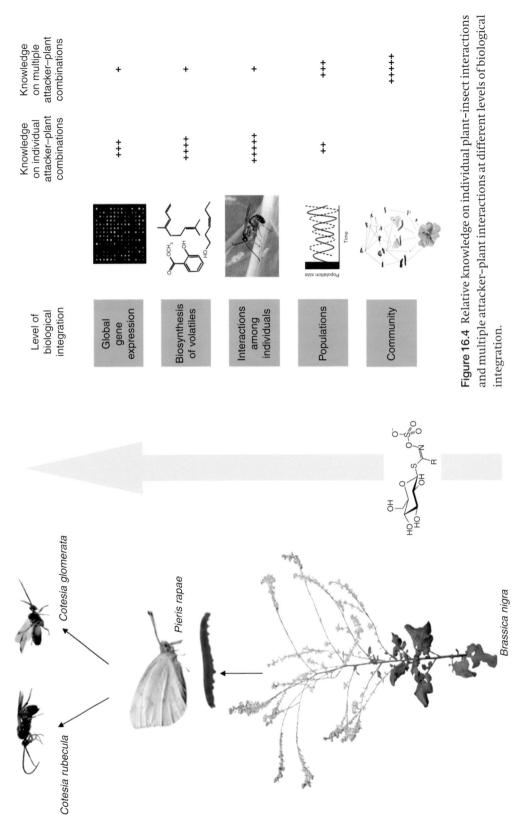

Figure 16.4 Relative knowledge on individual plant–insect interactions and multiple attacker–plant interactions at different levels of biological integration.

Level of biological integration		Knowledge on individual attacker–plant combinations	Knowledge on multiple attacker–plant combinations
Global gene expression		+++	+
Biosynthesis of volatiles		++++	+
Interactions among individuals		+++++	+
Populations		++	+++
Community			++++

Figure 16.3 Glucosinolates can influence different organisms in the food chain.

Cotesia glomerata

Cotesia rubecula

Pieris rapae

Brassica nigra

Holopainen, J. K. and Gershenzon, J. (2010) Multiple stress factors and the emission of plant VOCs. *Trends in Plant Science*, **15**, 176–184.

Kahl, J., Siemens, D. H., Aerts, R. J. *et al.* (2000) Herbivore-induced ethylene suppresses a direct defense but not a putative indirect defense against an adapted herbivore. *Planta*, **210**, 336–342.

Kaplan, I. and Denno, R. F. (2007) Interspecific interactions in phytophagous insects revisited: a quantitative assessment of competition theory. *Ecology Letters*, **10**, 977–994.

Kappers, I. F., Aharoni, A., van Herpen, T. W. J. M. *et al.* (2005) Genetic engineering of terpenoid metabolism attracts bodyguards to *Arabidopsis. Science*, **309**, 2070–2072.

Karban, R. and Baldwin, I. T. (1997) *Induced Responses to Herbivory*. Chicago, IL: University of Chicago Press.

Kessler, A. and Baldwin, I. T. (2001) Defensive function of herbivore-induced plant volatile emissions in nature. *Science*, **291**, 2141–2144.

Kessler, A. and Baldwin, I. T. (2004) Herbivore-induced plant vaccination. Part I. The orchestration of plant defenses in nature and their fitness consequences in the wild tobacco *Nicotiana attenuata. The Plant Journal*, **38**, 639–649.

Kessler, A. and Halitschke, R. (2009) Testing the potential for conflicting selection on floral chemical traits by pollinators and herbivores, predictions and case study. *Functional Ecology*, **23**, 901–912.

Kessler, D., Gase, K. and Baldwin, I. T. (2008) Field experiments with transformed plants reveal the sense of floral scents. *Science*, **321**, 1200–1202.

Kessler, D., Diezel, C. and Baldwin, I. T. (2010) Changing pollinators as a means of escaping herbivores. *Current Biology*, **20**, 237–242.

Khosla, C. and Keasling, J. D. (2003) Metabolic engineering for drug discovery and development. *Nature Reviews Drug Discovery*, **2**, 1019–1025.

Köllner, T. G., Gershenzon, J. and Degenhardt, J. (2009) Molecular and biochemical evolution of maize terpene synthase 10, an enzyme of indirect defense. *Phytochemistry*, **70**, 1139–1145.

Körner, E., von Dahl, C., Bonaventure, G. and Baldwin, I. T. (2009) Pectin methylesterase NaPME1 contributes to the emission of methanol during insect herbivory and to the elicitation of defence responses in *Nicotiana attenuata. Journal of Experimental Botany*, **60**, 2631–2640.

Kost, C. and Heil, M. (2006) Herbivore-induced plant volatiles induce an indirect defence in neighbouring plants. *Journal of Ecology*, **94**, 619–628.

Krügel, T., Lim, M., Gase, K., Halitschke, R. and Baldwin, I. T. (2002) *Agrobacterium*-mediated transformation of *Nicotiana attenuata*, a model ecological expression system. *Chemoecology*, **12**, 177–183.

Laothawornkitkul, J., Paul, N. D., Vickers, C. E. *et al.* (2008) The role of isoprene in insect herbivory. *Plant Signaling and Behavior*, **3**, 1141–1142.

Lewis, W. J. and Tumlinson, J. H. (1988) Host detection by chemically mediated associative learning in a parasitic wasp. *Nature*, **331**, 257–259.

Loreto, F. and Schnitzler, J.-P. (2010) Abiotic stresses and induced BVOCs. *Trends in Plant Science*, **15**, 154–166.

Mahmoud, S. S. and Croteau, R. B. (2002) Strategies for transgenic manipulation of monoterpene biosynthesis in plants. *Trends in Plant Science*, **7**, 366–373.

Mizutani, J. (1999) Selected allelochemicals. *Critical Reviews in Plant Sciences*, **18**, 653–671.

Niinemets, U. (2010) Mild versus severe stress and BVOCs: thresholds, priming and consequences. *Trends in Plant Science*, **15**, 145–153.

Pare, P. W. and Tumlinson, J. H. (1999) Plant volatiles as a defense against insect herbivores. *Plant Physiology*, **121**, 325–332.

Peñuelas, J. (2008) An increasingly scented world. *New Phytologist*, **180**, 735–738.

Peñuelas, J. and Staudt, M. (2010) BVOCs and global change. *Trends in Plant Science*, **15**, 133–144.

Price, P. W., Bouton, C. E., Gross, P. *et al.* (1980) Interactions among three trophic levels: influence of plants on interactions between insect herbivores and natural enemies. *Annual Review of Ecology and Systematics*, **11**, 41–65.

Raguso, R. A. (2008) Wake up and smell the roses: the ecology and evolution of floral scent. *Annual Review of Ecology, Evolution and Systematics*, **39**, 549–569.

Rasmann, S., Kollner, T. G., Degenhardt, J. *et al.* (2005) Recruitment of entomopathogenic nematodes by insect-damaged maize roots. *Nature*, **434**, 732–737.

Rhoades, D. F. (1983) Responses of alder and willow to attack by tent caterpillars and webworms: evidence for pheromonal sensitivity of willows. In P. A. Hedin (ed.) *Plant Resistance to Insects* Volume 208. Washington, DC: American Chemical Society, 55–68.

Rodríguez-Concepción, M. (2006) Early steps in isoprenoid biosynthesis: multilevel regulation of the supply of common precursors in plant cells. *Phytochemistry Reviews*, **5**, 1–15.

Rosenheim, J. A. (1998) Higher-order predators and the regulation of insect herbivore populations. *Annual Review of Entomology*, **43**, 421–447.

Schnee, C., Köllner, T. G., Held, M. *et al.* (2006) The products of a single maize sesquiterpene synthase form a volatile defense signal that attracts natural enemies of maize herbivores. *Proceedings of the National Academy of Sciences USA*, **103**, 1129–1134.

Schuman, M. C., Heinzel, N., Gaquerel, E., Svatoš, A. and Baldwin, I. T. (2009) Polymorphism in jasmonate signaling partially accounts for the variety of volatiles produced by *Nicotiana attenuata* plants in a native population. *New Phytologist*, **183**, 1134–1148.

Schwachtje, J., Kutschbach, S. and Baldwin, I. T. (2008) Reverse genetics in ecological research. *PLoS ONE*, **3**, e1543.

Shiojiri, K., Kishimoto, K., Ozawa, R. *et al.* (2006) Changing green leaf volatile biosynthesis in plants: an approach for improving plant resistance against both herbivores and pathogens. *Proceedings of the National Academy of Sciences USA*, **103**, 16672–16676.

Shiojiri, K., Ozawa, R., Kugimiya, S. *et al.* (2010) Herbivore-specific, density-dependent induction of plant volatiles: honest or 'cry wolf' signals? *PLoS ONE*, **5**, e12161.

Singh, H. P., Batish, D. and Kohli, R. K. (2003) Allelopathic interactions and allelochemicals: new possibilities for sustainable weed management. *Critical Reviews in Plant Sciences*, **22**, 239–311.

Skibbe, M., Qu, N., Gális, I. and Baldwin, I. T. (2008) Induced plant defenses in the natural environment: *Nicotiana attenuata* WRKY3 and WRKY6 coordinate responses to herbivory. *Plant Cell*, **20**, 1984–2000.

Skoula, M., Abbes, J. E. and Johnson, C. B. (2000) Genetic variation of volatiles and rosmarinic acid in populations of *Salvia fruticosa* mill growing in Crete. *Biochemical Systematics and Ecology*, **28**, 551–561.

Steppuhn, A., Schuman, M. C. and Baldwin, I. T. (2008) Silencing jasmonate signalling and jasmonate-mediated defences reveals different survival strategies between two *Nicotiana attenuata* accessions. *Molecular Ecology*, **17**, 3717–3732.

Stork, W., Diezel, C., Halitschke, R., Gális, I. and Baldwin, I. T. (2009) An ecological analysis of the herbivory-elicited JA burst and its metabolism: plant memory processes and predictions of the moving target model. *PLoS ONE*, **4**, e4697.

Takabayashi, J., Dicke, M. and Posthumus, M. A. (1994) Volatile herbivore-induced terpenoids in plant–mite interactions: variation caused by biotic and abiotic factors. *Journal of Chemical Ecology*, **20**, 1329–1354.

Thaler, J. S. (1999) Jasmonate-inducible plant defences cause increased parasitism of herbivores. *Nature*, **399**, 686–688.

Tholl, D., Chen, F., Petri, J., Gershenzon, J. and Pichersky, E. (2005) Two sesquiterpene synthases are responsible for the complex mixture of sesquiterpenes emitted from *Arabidopsis* flowers. *The Plant Journal*, **42**, 757–771.

Turlings, T. C. J. and Wäckers, F. (2004) Recruitment of predators and parasitoids by herbivore-injured plants. In R. T. Cardé and J. G. Millar (eds.) *Advances in Insect Chemical Ecology*. Cambridge: Cambridge University Press, 21–75.

Unsicker, S. B., Kunert, G. and Gershenzon, J. (2009) Protective perfumes: the role of vegetative volatiles in plant defense against herbivores. *Current Opinion in Plant Biology*, **12**, 479–485.

van Dam, N. M. and Poppy, G. M. (2008) Why plant volatile analysis needs bioinformatics: detecting signal from noise in increasingly complex profiles. *Plant Biology*, **10**, 29–37.

van Dam, N. M., Hadwich, K. and Baldwin, I. T. (2000) Induced responses in *Nicotiana attenuata* affect behavior and growth of the specialist herbivore *Manduca sexta*. *Oecologia*, **122**, 371–379.

van Loon, J. J. A., de Boer, J. G. and Dicke, M. (2000) Parasitoid–plant mutualism: parasitoid attack of herbivore increases plant reproduction. *Entomologia Experimentalis et Applicata*, **97**, 219–227.

Vickers, C. E., Gershenzon, J., Lerdau, M. T. and Loreto, F. (2009) A unified mechanism of action for volatile isoprenoids in plant abiotic stress. *Nature Chemical Biology*, **5**, 283–291.

von Dahl, C. C., Hävecker, M., Schlögl, R. and Baldwin, I. T. (2006) Caterpillar-elicited methanol emission, a new signal in plant–herbivore interactions? *The Plant Journal*, **46**, 948–960.

Wenke, K., Kai, M. and Piechulla, B. (2010) Belowground volatiles facilitate interactions between plant roots and soil organisms. *Planta*, **231**, 499–506.

Wu, J., Hettenhausen, C., Schuman, M. C. and Baldwin, I. T. (2008) A comparison of two *Nicotiana attenuata* accessions reveals large differences in signaling induced by oral secretions of the specialist herbivore *Manduca sexta*. *Plant Physiology*, **146**, 927–939.

CHAPTER SIXTEEN

Dynamics of plant secondary metabolites and consequences for food chains and community dynamics

MARCEL DICKE, RIETA GOLS and ERIK H. POELMAN
Laboratory of Entomology, Wageningen University

16.1 Introduction

A central issue in ecology is to identify the mechanisms driving and maintaining community diversity. Studies of plant–insect associations have played an important role in understanding ecological and evolutionary processes that underlie community dynamics (Whitham *et al.*, 2006; Poelman *et al.*, 2008b). Plants are autotrophic organisms that produce organic matter from carbon dioxide, water and sunlight, and as such they are at the base of most food webs. There is a wealth of animals that feed on plants, and insects are by far the most speciose of these. There are an estimated 6 million insect species, of which half are herbivorous (Schoonhoven *et al.*, 2005). Terrestrial plant–animal communities, therefore, represent a large proportion of the communities on Earth.

Plants produce a multitude of organic compounds ranging from simple molecules such as ethylene and methanol to complex terpenoids and nitrogen-containing alkaloids. More than 100 000 chemical products are known to be produced by plants (Schoonhoven *et al.*, 2005) with estimates ranging up to 200 000 (Pichersky & Gang, 2000), and these compounds affect many interactions with community members. Recent studies have provided ample information on the molecular basis and ecology of plant defences against insects (Kessler & Baldwin, 2002; Arimura *et al.*, 2005; Dicke *et al.*, 2009). Constitutive defences of plants differentially affect various insect herbivores. These defences can affect the behaviour of herbivores during host-plant selection and their performance after plant tissues have been ingested (Kessler *et al.*, 2004; Schoonhoven *et al.*, 2005).

Plants also display phenotypic plasticity, which includes plasticity in their chemical phenotype. Upon oviposition or damage by herbivores, plants respond to their attacker with a range of phenotypic changes (Hilker & Meiners, 2006; Schaller, 2008). Induced defence is observed as alterations in a set of traits that lead to a reduction in the negative effects of the attacker on plant fitness (Schoonhoven *et al.*, 2005). Many of these induced plant defences

The Ecology of Plant Secondary Metabolites: From Genes to Global Processes, eds. Glenn R. Iason, Marcel Dicke and Susan E. Hartley. Published by Cambridge University Press. © British Ecological Society 2012.

have been studied in detail at both the molecular level and at the level of physiological consequences for insect performance, which revealed that plants can fine-tune their responses to specific attackers (Heidel & Baldwin, 2004; de Vos *et al.*, 2005; Zarate *et al.*, 2007; Textor & Gershenzon, 2009; Walling, 2009; Broekgaarden *et al.*, 2010). Induced chemical defences include induced direct defence that affects the performance of the attacker directly, e.g. by interference with its physiology (Kessler & Baldwin, 2002; Schaller, 2008). In contrast, induced indirect defences involve the promotion of the effectiveness of natural enemies of the attacker, e.g. by producing volatile organic compounds that attract carnivorous enemies of the attacking herbivore to the herbivore-infested plant (Heil, 2008; Dicke & Baldwin, 2010). Each of these induced defences can affect a range of plant–insect interactions and, thus, can have extensive effects on community dynamics (Whitham *et al.*, 2006; Poelman *et al.*, 2008b; Dicke & Baldwin, 2010; Goheen & Palmer, 2010).

Community ecologists increasingly address intra-specific genetic variation in plant species and its consequences for biodiversity at higher trophic levels (Hochwender & Fritz, 2004; Wimp *et al.*, 2005). It appears that genotypic effects can be important in shaping insect herbivore communities (Johnson & Agrawal, 2005; Bangert *et al.*, 2006a) and genotypic effects have been shown to extend to the third trophic level and beyond (Bailey *et al.*, 2006; Bukovinszky *et al.*, 2008). Molecular ecologists have made extensive progress in unravelling mechanisms underlying individual plant–attacker interactions (Kessler & Baldwin, 2002; Pieterse & Dicke, 2007; Zheng & Dicke, 2008; Textor & Gershenzon, 2009) as well as multiple plant–attacker interactions (Kessler & Baldwin, 2004; de Vos *et al.*, 2006; Voelckel & Baldwin, 2004; Zhang *et al.*, 2009). As a result of the recent expansion of their interests, the research fields of community ecology and molecular biology have transgressed each other's boundaries, and it has been established that variation in molecular mechanisms underlying plant defences can structure biodiversity at higher trophic levels (Kessler *et al.*, 2004; Whitham *et al.*, 2008; Zheng & Dicke, 2008).

In this review we will address the effects of plant secondary metabolites (PSMs) on the ecology of plant–insect communities. We will first address the occurrence of plant-mediated indirect interactions among community members and subsequently concentrate on a specific system, *Brassica*–insect communities. The effects of PSMs on food chains and community dynamics will form the main emphasis of this review. Finally, we will discuss future directions of this research field.

16.2   Plant-mediated indirect interactions

Apart from direct interactions, plant–insect communities are characterised by an extensive network of indirect interactions. For instance, plants affect the phenotype of their attackers when the herbivores sequester secondary plant metabolites which consequently influence the interactions between the herbivores and their carnivorous enemies (Nishida, 2002; Gols & Harvey, 2009). Moreover, plants may also alter their own phenotype in response to attack by herbivores leading to altered herbivore–carnivore interactions, e.g. by the emission of herbivore-induced plant volatiles that attract the carnivorous enemies of the herbivores (Dicke & Baldwin, 2010). Consequently, plants and animals extensively interact directly and indirectly, and these interactions are quite dynamic because of the underlying plasticity that characterises them (Agrawal, 2001). In the past 15 years it has become increasingly clear that plant-mediated interactions among community members are common, and that they underlie many important ecological phenomena such as competition among herbivores, predation, parasitism and apparent competition (Denno *et al.*, 1995; van der Putten *et al.*, 2001; Kaplan & Denno, 2007; Karban, 2008; Dicke *et al.*, 2009; Dicke & Baldwin, 2010; Utsumi *et al.*, 2010). For example, root-feeding insects, such as larvae of the cabbage root fly *Delia radicum* which infest *Brassica nigra* plants, influence oviposition choices of various aboveground leaf-feeding herbivores and thus may influence aboveground community dynamics (Soler *et al.*, 2009). Moreover, infestation of bean plants by the whitefly *Bemisia tabaci* interferes with the spider mite-induced attraction of predatory mites by the bean plants (Zhang *et al.*, 2009). Whitefly feeding interferes with the induction of the phytohormone jasmonic acid (JA) and consequently attenuates the induction of JA-dependent genes involved in signal-transduction and in the biosynthesis of the plant volatile (*E*)-β-ocimene. The reduced emission of (*E*)-β-ocimene is responsible for a reduced attraction of predatory mites that feed on spider mites (Zhang *et al.*, 2009). Therefore, whitefly-infested plants represent a relatively enemy-free space to spider mites, which may explain why spider mites prefer whitefly-infested over uninfested plants as food source.

Aside from trophic interactions, PSMs can influence insect community structure and diversity. Genotypic variation in the quantitative and qualitative composition of secondary metabolites directly affects the abundance and species diversity of insects associated with the plant (Bangert *et al.*, 2006b; Poelman *et al.*, 2009b; Bailey *et al.*, Chapter 14). Because herbivores are in close reciprocal interaction with plants, their abundance and species composition on plants are affected by plant characteristics such as morphology and plant secondary chemistry. Changes in herbivore density and quality, as influenced through the food plant, may as a result alter the abundance and diversity of members of the third trophic level. Intra-specific variation in secondary plant chemistry may even cascade through the food chain and affect species interactions at the fourth trophic level in insect communities (Bukovinszky *et al.*, 2008) and extend to vertebrate predators (Bailey *et al.*, 2006; Gruner & Taylor,

2006). Extensive knowledge on the link between plant genotype and insect community comes from large-scale field experiments on cottonwood (*Populus* sp.) species (Whitham *et al.*, 1999; Bangert *et al.*, 2006b). Cottonwoods were hybridised and backcrossed with their parents to obtain a genetic gradient in similarity among cottonwoods. When genetic distance between genotypes increased, the phytochemistry and arthropod community composition changed accordingly, revealing a strong link between plant genotype, phytochemistry and the associated community (Whitham *et al.*, 1999; Bangert *et al.*, 2006b). The composition and concentration of secondary metabolites in plants is, however, subject to changes in environmental conditions and often alters with plant age (Koricheva & Barton, Chapter 3). Seasonal phenotypic variation in expressed plant chemistry largely affects the dynamics of insect communities. Phenolic glucoside concentration in cottonwood (*Populus* sp.) declines by 25% over the course of the season, making its impact primarily important in interactions with early-season herbivores (Wimp *et al.*, 2007). Other plant traits, such as leaf shape, biomass and flowering characteristics (Johnson & Agrawal, 2005), may become more important in determining arthropod communities when the concentration of plant chemicals decreases.

16.3 *Brassica*–insect interactions

The cosmopolitan Brassicaceae family comprises around 3700 species. The ecology of interactions between insects and brassicaceous plants has received ample investigations over the past decades (Tahvanainen & Root, 1972; Root, 1973; Fox & Eisenbach, 1992; Agrawal, 2000; Gols *et al.*, 2008; Newton *et al.*, 2009a, 2009b; Soler *et al.*, 2009; Poelman *et al.*, 2010). Furthermore, mechanistic aspects of the effects of PSMs, such as the family-specific glucosinolates (Figure 16.1), on plant–insect interactions have been extensively investigated for *Brassica* species (Hopkins *et al.*, 2009). More recently, studies on the ecology of *Brassica*–insect interactions have also been extended to underlying mechanisms at the molecular genetic level in *Brassica* plants (Broekgaarden *et al.*, 2007; Poelman *et al.*, 2008a) and insect herbivores feeding on *Brassica* plants, such as *Pieris* spp. and aphids (Wittstock *et al.*, 2004; Ramsey *et al.*, 2007; Wheat *et al.*, 2007).

16.3.1   From *Arabidopsis* to *Brassica*

To fully understand how PSMs shape community dynamics, profound knowledge of the underlying mechanisms is essential, and rapid progress is being made in this area (Kessler & Baldwin, 2002; van Zandt & Agrawal, 2004; Poelman *et al.*, 2008b; Whitham *et al.*, 2008; Gershenzon *et al.*, Chapter 4; Schuman & Baldwin, Chapter 15). Yet the absence of extensive, well-characterised mutants hampers a directed experimental approach. The only

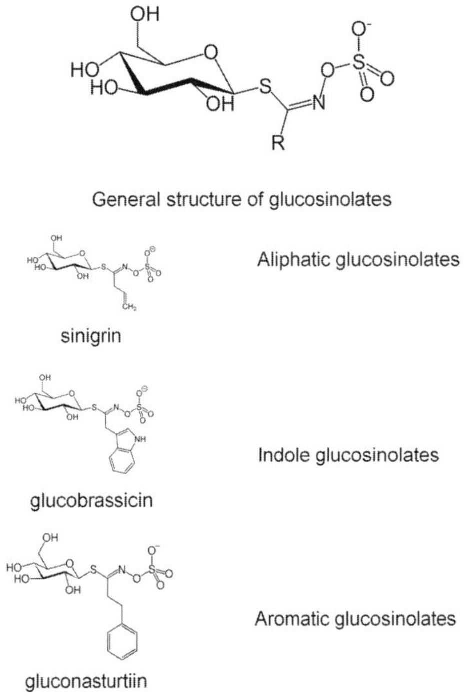

General structure of glucosinolates

sinigrin

Aliphatic glucosinolates

glucobrassicin

Indole glucosinolates

gluconasturtiin

Aromatic glucosinolates

Figure 16.1 Glucosinolates are the characteristic group of secondary plant metabolites of plants in the Brassicaceae family. The three main groups are aliphatic glucosinolates (derived from the amino acid methionine), indole glucosinolates (derived from the amino acid tryptophane) and aromatic glucosinolates (derived from the amino acids phenylalanine or tyrosine).

plant for which ample mutants are available is *Arabidopsis thaliana*. Although this plant is less interesting from an ecological point of view because of its early phenology, it provides a valuable resource for mechanistic studies. Its response to insect herbivory has many similarities with responses of other plants (Mitchell-Olds, 2001; van Poecke & Dicke, 2004). By using *Arabidopsis* it has become clear that different herbivorous insects and microbial pathogens induce very different transcriptomic responses, but that the transcriptomic response is mainly regulated by the three phytohormones jasmonic acid, salicylic acid and ethylene (de Vos *et al.*, 2005). Yet it is the combination of these (and other?) phytohormones that is likely to affect the transcriptional outcome as a result of, for example, cross-talk between signal-transduction pathways (de Vos *et al.*, 2005; Pieterse *et al.*, 2009). Research exploiting specific mutants has demonstrated the roles of certain glucosinolates (de Vos *et al.*, 2008; Kim *et al.*, 2008), nitriles (Mumm *et al.*, 2008) and plant volatiles (Snoeren *et al.*, 2010) in the interactions of *Arabidopsis* plants with insects. Moreover, tools that have been developed for *Arabidopsis*, such as full-genome microarrays, can also be used to investigate mechanisms in other brassicaceous plants, such as *Brassica oleracea* or *Brassica nigra* (Lee *et al.*, 2004; Broekgaarden *et al.*, 2007, 2010; Fatouros *et al.*, 2008). These exciting developments, in combination with the extensive knowledge on the ecology of insect–plant interactions in *Brassica* plants (Root, 1973; Gols *et al.*, 2008; Newton *et al.*, 2009b; Soler *et al.*, 2009), make brassicaceous plants highly suited to study the ecology of insect–plant interactions from genes to the community.

16.3.2   Volatile PSMs in *Brassica*-insect interactions

Volatiles constitute an important component of a plant's chemical phenotype. This phenotype is highly plastic as herbivory, pathogen infestation and abiotic stress can all modify the emission of plant volatiles (Eigenbrode *et al.*, 2002; Dicke & Baldwin, 2010; Holopainen & Gershenzon, 2010). Such changes are known to deter or attract herbivores by making the plant status more apparent to herbivores (Eigenbrode *et al.*, 2002; Dicke & Baldwin, 2010; Kessler & Halitschke, 2007; Holopainen & Gershenzon, 2010; Poelman *et al.*, 2010). Moreover, these induced volatile changes will interact directly with higher trophic-level organisms that are attracted to volatiles from damaged plants and in particular to plants damaged by their herbivorous hosts or prey (Heil, 2008; Dicke & Baldwin, 2010). The induction of plant volatiles may lead to more diverse and more abundant higher trophic-level organisms.

Brassica plants emit herbivore-induced volatiles in response to feeding by, for example, caterpillars (Mattiacci *et al.*, 1995; Bruinsma *et al.*, 2009; Gols *et al.*, 2009) or aphids (Blande *et al.*, 2007; Girling & Hassall, 2008). This blend can include fatty acid derived green-leaf volatiles, terpenoids, isothiocyanates, nitriles, di- and trisulfides and aromatic phenylpropanoids such as methyl salicylate (Figure 16.2). These volatiles are induced by mechanical damage and by herbivore elicitors from herbivore oral secretions. A β-glucosidase from the oral secretion of *Pieris brassicae* caterpillars induces a similar volatile blend as feeding damage by the caterpillars does (Mattiacci *et al.*, 1995). The parasitisation of *Pieris* caterpillars can benefit plants in terms of damage reduction and increase in fitness (Harvey, 2000; van Loon *et al.*, 2000; Smallegange *et al.*, 2008). Several other community members may interfere with this plant–parasitoid attraction. For instance, larvae of the cabbage root fly that feed on *B. nigra* roots result in a reduced attraction of *Cotesia glomerata* parasitoids to plants infested by their host, the folivorous caterpillar *P. brassicae* (Soler *et al.*, 2007b). Moreover, parasitisation of *Pieris* caterpillars also alters the plant volatile-mediated interaction between *Pieris* caterpillars and their parasitoids. *Brassica* plants respond differently to feeding by parasitised than by unparasitised caterpillars, which results in a different composition of the volatile blend that is emitted. *Cotesia* parasitoids differentiate between the volatile blends: they are more attracted to plants infested with unparasitised

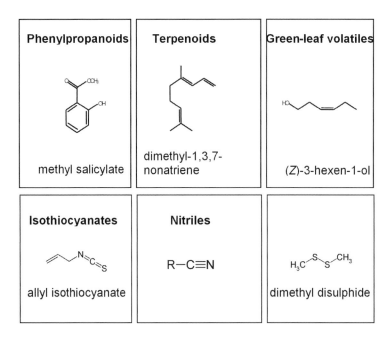

Figure 16.2 Classes of herbivore-induced *Brassica* volatiles with characteristic examples.

host caterpillars than to plants infested with parasitised caterpillars (Fatouros *et al.*, 2005). In this way the parasitoids can exploit volatile PSMs to avoid competition with other parasitoids. There is ample evidence that non-volatile host-related cues are used for this host discrimination. The work on *Brassica–Pieris–Cotesia* interactions is the first example of the involvement of volatile PSMs in host discrimination by parasitoids (Fatouros *et al.*, 2005).

The attraction of parasitoids towards herbivore-infested *Brassica* plants varies with plant population and plant accession (Gols *et al.*, 2009; Poelman *et al.*, 2009b). Some genotypes or populations are more attractive, when the plants are damaged by herbivory, than others. Such variation in the emission of plant volatiles is likely to play a role in differential community dynamics (Poelman *et al.*, 2008a, 2009a, 2010).

16.3.3   Non-volatile PSMs in *Brassica* tri-trophic interactions

PSMs play an important role in providing plants with protection against a range of attackers. However, adaptation of the attackers to these compounds has resulted in specialisation in many attacker species, including insects. Specialised insects often use secondary plant compounds specific for their food plants as oviposition and feeding stimulants (Schoonhoven *et al.*, 2005). Moreover, some insects sequester plant chemicals from their food plants and employ them as defence against their own attackers (Nishida, 2002). The

Brassicaceae contain important crop species such as cabbages and various types of oil-containing seed varieties. Because a number of insect species specialised on brassicaceous plant species have become economically serious pests, their interactions with these plants and their natural enemies are well studied (Gols & Harvey, 2009; Hopkins *et al.*, 2009). Plant species in this family characteristically biosynthesise secondary plant compounds known as gluco-sinolates. The defensive properties of these plant chemicals emerge after tissue damage when the glucosinolates come in contact with myrosinase enzymes, which hydrolyse the glucosinolates into more toxic metabolites. Counter-adaptations in insect herbivores specialised on glucosinolate-containing plant species rely on the production of special enzymes that prevent or convert hydrolysis by myrosinases (Ratzka *et al.*, 2002; Wittstock *et al.*, 2004). Herbivore species known for glucosinolate sequestration are the aphids *Brevicoryne brassicae* and *Lipaphis erysimi* (Bridges *et al.*, 2002), the harle-quin bug *Murganthia histrionica* (Aliabadi *et al.*, 2002) and larvae of the sawfly *Athalia rosae* (Muller, 2009).

Predators and parasitoids can be exposed to plant toxins that have been ingested or sequestered by the herbivore, with potential negative consequences for their own development (Harvey, 2005; Ode, 2006) (Figure 16.3). Alternatively, a reduced performance of the herbivore on toxic plants may reduce the quality as hosts (parasitoids) or prey (predators). On the other hand, the 'immune system' of the insect herbivore, which reduces the success of parasitism, may be compromised when feeding on a poor-quality host plant. For instance, previous damage by *Pieris rapae* caterpillars to *Brassica* plants resulted in an increased production of glucosinolates and a reduction in growth of subsequent caterpillars, and reduced the encapsulation response of the caterpillars to parasitism by *C. glomerata* parasitoids (Bukovinszky *et al.*, 2009).

Several studies have shown that growth and development of insects at different trophic levels that are associated with brassicaceous plant species are affected by within-species and among-species variation in food-plant quality (Francis *et al.*, 2001; Harvey *et al.*, 2003; Soler *et al.*, 2005, 2007a; Gols *et al.*, 2008a, 2009). However, in these studies the investigated insects were all specialist herbivores that efficiently detoxify or excrete glucosinolates in their faeces. Therefore, a direct causal correlation between specific glucosi-nolate compounds and insect performance seems unlikely. Surprisingly, Gols *et al.* (2008a) found a negative correlation between the performance of a specialist herbivore, *P. rapae*, and its larval endoparasitoid *Cotesia rubecula* with concentrations of the indole glucosinolate 1-methoxy-3-indolylmethyl glucosinolate. Moreover, in the same study, the performance of a generalist herbivore, *Mamestra brassicae*, and its parasitoid *Microplitis mediator* negatively correlated with concentrations of aliphatic glucosinolates. In a field study with natural *B. oleracea* populations, Newton *et al.* (2009a) found that colony

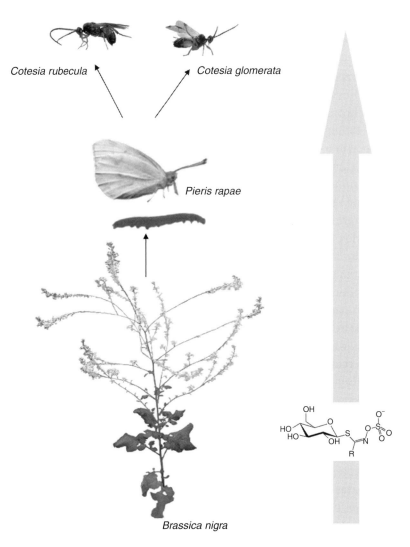

Figure 16.3 Glucosinolates can influence different organisms in the food chain. See plate section for colour version.

sizes of the aphid *B. brassicae*, which sequesters glucosinolates from its food plant, were smaller on plants without sinigrin. However, the presence or absence of sinigrin in the plant did not affect the abundance of the aphid's natural enemy *Diaeretiella rapae* on these plants. These results indicate that if glucosinolate compounds affect the performance of insect herbivores and their natural enemies, then their effects depend on the system studied. Studies incorporating different glucosinolates in artificial diets (Agrawal & Kurashige, 2003) or using mutant plants that are silenced with respect to the biosynthesis of specific glucosinolates may reveal whether glucosinolates or variation in other plant chemicals are responsible for variation in insect performance. Some brassicaceous plants have been shown to contain secondary metabolites other than glucosinolates that render these species unsuitable for growth of Brassicaceae specialists (Renwick, 2002). Recently, Harvey *et al.* (2010) demonstrated that larval mortality of the specialists *P. brassicae* and *P. rapae* is close to 100% when developing on the invasive brassicaceous *Bunias orientalis*, whereas larvae of the generalist *M. brassicae* develop normally. These results indicate that the role of specific secondary plant chemicals as defensive compounds against insect herbivores is not straightforward. Moreover, plants have to deal with a range of attackers, and defence strategies aimed at a single or a limited number of enemies could be disadvantageous. It becomes more and more evident that there are strengths of studying ecological and evolutionary consequences of plant–animal interactions in a community context that may not be revealed in studies investigating pairwise interactions (Strauss, 1991; Poelman *et al.*, 2008b; Dicke & Baldwin, 2010).

16.3.4   Plant chemicals in community dynamics

An array of different secondary compounds can often be found in individual plants. In brassicaceous plants, individual compounds may differently affect various herbivores. The family-specific secondary metabolites, i.e. glucosinolates and their breakdown products, deter generalist herbivores whereas specialists exploit these compounds in host-plant selection (Hopkins *et al.*, 2009). Furthermore, glucosinolates are a diverse group of compounds leading to various chemical profiles in which different dominant compounds prevail. Each of these compounds in a chemical profile may have different effects on herbivores (Bidart-Bouzat & Kliebenstein, 2008). Short side-chain glucosinolates may, for example, defend plants against a large range of herbivores compared with compounds with a longer side chain. *Brassica oleracea* plants with high concentrations of the short side-chain alkenyl glucosinolate, glucoiberin, harboured a low herbivore diversity. In contrast, plants with a glucosinolate profile containing high concentrations of elongated side-chain compounds, which are biosynthetically linked to glucoiberin, had higher diversity of herbivores (Poelman *et al.*, 2009b). Plants with such longer side-chain compounds might deter particular herbivores, while they have neutral effects on or even attract other herbivores. Typically, differential effects of individual compounds do not coincide with categories of host-plant specialisation of herbivores: they do not necessarily result in opposite directional effects on specialist and generalist herbivores (Newton *et al.*, 2009b). Because individual secondary metabolites can have contrasting effects on different herbivores, spatial and temporal variation in herbivore communities may provide a variable selection pressure on the possession of different compounds in plants (O'Reilly-Wapstra *et al.*, Chapter 2).

Herbivore communities may not only select for diversity of PSMs within individual plants, but may also maintain different genotypes with absolute differences in presence of particular compounds within plant populations, i.e. so-called plant chemotypes (Nielsen, 1997; van Leur *et al.*, 2006). These chemotypes may differ in a small number of genes or alleles of those genes that convert secondary metabolites during biosynthesis, and consequently will have a different associated insect community (Nielsen, 1997; van Leur *et al.*, 2006). Especially, the genetically determined composition of the secondary metabolite profile of a plant provides a phenotype that strongly links to the arthropod community that colonises a plant. However, when plants are attacked by a first herbivore early in the season, they usually respond to this herbivore by enhancing their defences. Those responses include relocation and increase in concentration of secondary metabolites and, importantly, the responses vary with herbivore species. Each early-season herbivore may induce different phenotypic changes in plants (de Vos *et al.*, 2005; Zarate *et al.*, 2007). The induced phenotypes include altered secondary metabolite composition and will, thereby, imply that plants interact differently with their associated community. Such so-called indirect plant-mediated interactions (Ohgushi, 2008) between herbivores have been recorded in the field and have been identified between herbivores of the same feeding guild as well as across feeding guilds (Denno *et al.*, 2000; Kessler & Baldwin, 2004). In the field, plants that are induced by early-season herbivory inflicted by leaf-chewing herbivores were found to become more attractive to specialist herbivores and were colonised by fewer generalist herbivores than undamaged plants (van Zandt & Agrawal, 2004; Poelman *et al.*, 2008b, 2010). Herbivore species-specific changes in secondary plant chemistry in response to herbivory may be maintained when other subsequent herbivores feed on the plants (Viswanathan *et al.*, 2007) and those changes may result in stronger effects on the associated community composition than effects of variation in plant genotypes (Poelman *et al.*, 2010). Owing to relocation and systemic responses of plants to attackers, plant-associated communities above and below ground are also tightly linked. Induced changes inflicted by aboveground organisms affect root quality and thus root-associated organisms (and vice versa) (Kaplan *et al.*, 2008). Not only do induced changes in secondary metabolites link above- and belowground communities, induced changes will extend to all community members including both plant attackers and beneficial organisms such as predators, parasitoids and pollinators (Kessler & Halitschke, 2007; Lucas-Barbosa *et al.*, 2011). Thus, both genotypic and phenotypic variation in PSMs play a major role in driving plant-associated community composition and influence ecosystems biodiversity (Poelman *et al.*, 2008a; Schweitzer *et al.*, 2008).

16.4   PSMs in insect–plant communities: conclusions and future perspectives

Plants produce large numbers of organic chemicals. The chemical composition of a plant is highly dynamic, being influenced by ontogeny, abiotic and biotic stresses. In response to stress, plants can produce volatile as well as non-volatile secondary metabolites. This dynamic chemical phenotype of plants has far-reaching effects on the animal community that is associated with plants. The interactions of plants and insects from different trophic levels are extensively influenced by PSMs. Herbivore-induced plant volatiles affect the behaviour of insects, while induced non-volatile secondary metabolites affect the performance of herbivorous insects as well as insects at higher levels in the food chain. *Brassica* volatiles include fatty-acid-derived green-leaf volatiles, glucosinolate-derived isothiocyanates and nitriles and terpenoids. The best-studied non-volatile secondary metabolites are the family-specific glucosinolates, found in the brassicas, and it is their interaction with the associated insect communities which is the focus of our research group. To investigate how induced changes in plant secondary metabolites influence insect community dynamics, ecologists exploit multivariate statistics and complement this with tools derived from molecular biological approaches, such as well-characterised mutants, and plants that are transiently silenced in specific genes of interest.

There have been ample discussions on the *raison d'être* of PSMs since the thought-provoking paper by Fraenkel (1959). For a long time, the effects of PSMs have been investigated for interactions between one plant and one attacker. One of the most important discoveries in the past 50 years has been that a plant's composition of secondary metabolites is a plastic characteristic that is dependent on, for example, plant ontogeny, age, biotic and abiotic stress (Barton & Koricheva, 2010; Dicke & Baldwin, 2010; Holopainen & Gershenzon, 2010; Loreto & Schnitzler, 2010). This underpins the dynamic nature of the plant's chemical phenotype. There is a rapidly increasing body of information on the mechanisms of these chemical changes for individual plant–insect interactions (Kessler & Baldwin, 2002), while the effects of multiple infestations have only recently begun to be investigated (Dicke *et al.*, 2009) (Figure 16.4).

With the growing body of evidence on the differential effects of individual PSMs on distinct community members, it is becoming clear that any change in PSMs has extensive effects on the behaviour and performance of the plant-associated animal community. Moreover, the effects of individual compounds depend on the chemical background within which the individual compound

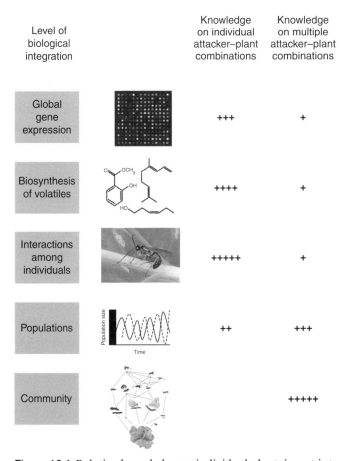

Figure 16.4 Relative knowledge on individual plant–insect interactions and multiple attacker–plant interactions at different levels of biological integration. See plate section for colour version.

is present (Mumm & Hilker, 2005; Schroeder & Hilker, 2008). Consequently, the chemical complexity and its dynamics are also likely to play an important role in the various effects of environmentally and stress-induced changes in secondary plant chemistry. A plant's chemical phenotype is highly dynamic, because previous herbivore attack may alter the chemical phenotype with concomitant effects on subsequent herbivores (Poelman *et al.*, 2008a, 2010). Subsequently, the secondary (and later) herbivores may modify the plant responses to the first herbivore and consequently the interactions with other herbivores and their natural enemies (Kessler & Baldwin, 2004; Dicke *et al.*, 2009; Zhang *et al.*, 2009).

Thus, plant secondary chemistry as well as plant–animal communities is highly dynamic, and it may seem that this hampers a systematic study of the role of PSMs in the community ecology of plant–animal interactions. However, it is exactly these dynamics that make the study of the role of PSMs in plant–animal community ecology so exciting. After all, understanding these dynamics provides an important step to understanding the driving forces in community diversity. The effects of specific PSMs on plant–insect interactions are effectively studied by ecologists in comprehensive, directed experiments at different levels of biological organisation and with the aid of modern multivariate statistics (Mercke *et al.*, 2004; Gaquerel *et al.*, 2009; Jansen *et al.*, 2009; Poelman *et al.*, 2009b; Kabouw *et al.*, 2010) (Figure 16.4). These methods provide valuable insights into the forces shaping the dynamics of plant–animal communities. Yet it is important to validate the hypotheses that these correlative data have provided through further directed experimentation. A major breakthrough that has been made in the past decade is that ecologists have embraced the developments in molecular biology and have employed these to investigate the mechanisms underlying individual interactions between a plant and a herbivore (Kessler & Baldwin, 2002; Wittstock *et al.*, 2004; Pieterse & Dicke, 2007). This has been done especially for *A. thaliana*, *Nicotiana attenuata, Populus* spp. and *Solanum esculentum* (Snoeren *et al.*, 2007; Chen *et al.*, 2009; Galis *et al.*, 2009). A valuable tool for the investigation of mechanisms is the availability of well-characterised mutants, or plants in which the effects of specific genes are eliminated by anti-sense silencing or virus-induced gene silencing (Roda & Baldwin, 2003; Kessler *et al.*, 2004; Saedler & Baldwin, 2004; Mumm *et al.*, 2008; Snoeren *et al.*, 2009; Zheng *et al.*, 2010; Schuman & Baldwin, Chapter 15). Transient gene silencing techniques will become available for many more plant species and, thus, will expand the possibilities for taking a directed approach to understand how environmentally induced changes in gene transcription relate to a plant's chemical phenotype and what the consequences are for community dynamics. These new molecular techniques are expected to yield tools to manipulatively generate sets of genotypes that differ quantitatively in the expression of a gene of interest. That will provide ecologists with much more interesting tools than the 'digital' tools represented by knock-out or over-expressor genotypes where the functional expression of one gene is either eliminated or added. It will provide further complementary approaches for ecologists and molecular geneticists to address the effects of PSMs on plant–insect interactions and community dynamics.

Acknowledgements

We thank two anonymous reviewers and Glenn Iason for comments on a previous version of the manuscript that helped to improve the text. Our research benefits from funds from the Council for Earth and Life Sciences of the Netherlands Organisation for Scientific Research.

References

Agrawal, A. A. (2000) Benefits and costs of induced plant defense for *Lepidium virginicum* (Brassicaceae). *Ecology*, **81**, 1804–1813.

Agrawal, A. A. (2001) Phenotypic plasticity in the interactions and evolution of species. *Science*, **294**, 321–326.

Agrawal, A. A. and Kurashige, N. S. (2003) A role for isothiocyanates in plant resistance against the specialist herbivore *Pieris rapae*. *Journal of Chemical Ecology*, **29**, 1403–1415.

Aliabadi, A., Renwick, J. A. A. and Whitman, D. W. (2002) Sequestration of glucosinolates by harlequin bug *Murgantia histrionica*. *Journal of Chemical Ecology*, **28**, 1749–1762.

Arimura, G., Kost, C. and Boland, W. (2005) Herbivore-induced, indirect plant defences. *Biochimica et Biophysica Acta*, **1734**, 91–111.

Bailey, J. K., Wooley, S. C., Lindroth, R. L. and Whitham, T. G. (2006) Importance of species interactions to community heritability: a genetic basis to trophic-level interactions. *Ecology Letters*, **9**, 78–85.

Bangert, R. K., Allan, G. J., Turek, R. J. *et al.* (2006a) From genes to geography: a genetic similarity rule for arthropod community structure at multiple geographic scales. *Molecular Ecology*, **15**, 4215–4228.

Bangert, R. K., Turek, R. J., Rehill, B. *et al.* (2006b) A genetic similarity rule determines arthropod community structure. *Molecular Ecology*, **15**, 1379–1391.

Barton, K. E. and Koricheva, J. (2010) The ontogeny of plant defense and herbivory: characterizing general patterns using meta-analysis. *American Naturalist*, **175**, 481–493.

Bidart-Bouzat, M. G. and Kliebenstein, D. J. (2008) Differential levels of insect herbivory in the field associated with genotypic variation in glucosinolates in *Arabidopsis thaliana*. *Journal of Chemical Ecology*, **34**, 1026–1037.

Blande, J. D., Pickett, J. A. and Poppy, G. M. (2007) A comparison of semiochemically mediated interactions involving specialist and generalist *Brassica*-feeding aphids and the braconid parasitoid *Diaeretiella rapae*. *Journal of Chemical Ecology*, **33**, 767–779.

Bridges, M., Jones, A. M. E., Bones, A. M. *et al.* (2002) Spatial organization of the glucosinolate-myrosinase system in brassica specialist aphids is similar to that of the host plant. *Proceedings of the Royal Society B, Biological Sciences*, **269**, 187–191.

Broekgaarden, C., Poelman, E. H., Steenhuis, G. *et al.* (2007) Genotypic variation in genome-wide transcription profiles induced by insect feeding: *Brassica oleracea–Pieris rapae* interactions. *BMC Genomics*, **8**, 239.

Broekgaarden, C., Poelman, E. H., Voorrips, R. E., Dicke, M. and Vosman, B. (2010) Intraspecific variation in herbivore community composition and transcriptional profiles in field-grown *Brassica oleracea* cultivars. *Journal of Experimental Botany*, **61**, 807–819.

Bruinsma, M., Posthumus, M. A., Mumm, R. *et al.* (2009) Jasmonic acid-induced volatiles of *Brassica oleracea* attract parasitoids: effects of time and dose, and comparison with induction by herbivores. *Journal of Experimental Botany*, **60**, 2575–2587.

Bukovinszky, T., van Veen, F. J. F., Jongema, Y. and Dicke, M. (2008) Direct and indirect effects of resource quality on food web structure. *Science*, **319**, 804–807.

Bukovinszky, T., Poelman, E. H., Gols, R. *et al.* (2009) Consequences of constitutive and induced variation in plant nutritional quality for immune defence of a herbivore against parasitism. *Oecologia*, **160**, 299–308.

Chen, F., Liu, C. J., Tschaplinski, T. J. and Zhao, N. (2009) Genomics of secondary metabolism in Populus: interactions with biotic and abiotic environments. *Critical Reviews in Plant Sciences*, **28**, 375–392.

de Vos, M., van Oosten, V. R., van Poecke, R. M. P. *et al.* (2005) Signal signature and transcriptome changes of *Arabidopsis* during pathogen and insect attack. *Molecular Plant–Microbe Interactions*, **18**, 923–937.

de Vos, M., van Zaanen, W., Koornneef, A. *et al.* (2006) Herbivore-induced resistance against microbial pathogens in Arabidopsis. *Plant Physiology*, **142**, 352–363.

de Vos, M., Kriksunov, K. L. and Jander, G. (2008) Indole-3-acetonitrile production from indole glucosinolates deters oviposition by *Pieris rapae*. *Plant Physiology*, **146**, 916–926.

Denno, R. F., McClure, M. S. and Ott, J. R. (1995) Interspecific interaction in phytophagous insects: competition reexamined and resurrected. *Annual Review of Entomology*, **40**, 297–331.

Denno, R. F., Peterson, M. A., Gratton, C. *et al.* (2000) Feeding-induced changes in plant quality mediate interspecific competition between sap-feeding herbivores. *Ecology*, **81**, 1814–1827.

Dicke, M. and Baldwin, I. T. (2010) The evolutionary context for herbivore-induced plant volatiles: beyond the 'cry for help'. *Trends in Plant Science*, **15**, 167–175.

Dicke, M., van Loon, J. J. A. and Soler, R. (2009) Chemical complexity of volatiles from plants induced by multiple attack. *Nature Chemical Biology*, **5**, 317–324.

Eigenbrode, S. D., Ding, H. J., Shiel, P. and Berger, P. H. (2002) Volatiles from potato plants infected with potato leafroll virus attract and arrest the virus vector, *Myzus persicae* (Homoptera: Aphididae). *Proceedings of the Royal Society of London B: Biological Sciences*, **269**, 455–460.

Fatouros, N. E., van Loon, J. J. A., Hordijk, K. A., Smid, H. M. and Dicke, M. (2005) Herbivore-induced plant volatiles mediate in-flight host discrimination by parasitoids. *Journal of Chemical Ecology*, **31**, 2033–2047.

Fatouros, N. E., Broekgaarden, C., Bukovinszkine'Kiss, G. *et al.* (2008) Male-derived butterfly anti-aphrodisiac mediates induced indirect plant defense. *Proceedings of the National Academy of Sciences USA*, **105**, 10033–10038.

Fox, L. R. and Eisenbach, J. (1992) Contrary choices: possible exploitation of enemy-free space by herbivorous insects in cultivated vs. wild crucifers. *Oecologia*, **89**, 574–579.

Fraenkel, G. S. (1959) The *raison d'être* of secondary plant substances. *Science*, **129**, 1466–1470.

Francis, F., Lognay, G., Wathelet, J. P. and Haubruge, E. (2001) Effects of allelochemicals from first (Brassicaceae) and second (*Myzus persicae* and *Brevicoryne brassicae*) trophic levels on *Adalia bipunctata*. *Journal of Chemical Ecology*, **27**, 243–256.

Galis, I., Gaquerel, E., Pandey, S. P. and Baldwin, I. T. (2009) Molecular mechanisms underlying plant memory in JA-mediated defence responses. *Plant Cell and Environment*, **32**, 617–627.

Gaquerel, E., Weinhold, A. and Baldwin, I. T. (2009) Molecular Interactions between the specialist herbivore *Manduca sexta* (Lepidoptera, Sphingidae) and its natural host *Nicotiana attenuata*. VIII. An unbiased GCxGC-ToFMS analysis of the plant's elicited volatile emissions. *Plant Physiology*, **149**, 1408–1423.

Girling, R. D. and Hassall, M. (2008) Behavioural responses of the seven-spot ladybird *Coccinella septempunctata* to plant headspace chemicals collected from four crop *Brassicas* and *Arabidopsis thaliana*, infested with *Myzus persicae*. *Agricultural and Forest Entomology*, **10**, 297–306.

Goheen, J. R. and Palmer, T. M. (2010) Defensive plant-ants stabilize megaherbivore-driven landscape change in an African savanna. *Current Biology*, **20**, 1768–1772.

Gols, R. and Harvey, J. A. (2009a) Plant-mediated effects in the Brassicaceae on the performance and behaviour of parasitoids. *Phytochemistry Reviews*, **8**, 187–206.

Gols, R., Wagenaar, R., Bukovinszky, T. *et al.* (2008a) Genetic variation in defense chemistry in wild cabbages affects herbivores and their endoparasitoids. *Ecology*, **89**, 1616–1626.

Gols, R., Witjes, L. M. A., van Loon, J. J. A. *et al.* (2008b) The effect of direct and indirect defenses in two wild brassicaceous plant species on a specialist herbivore and its gregarious endoparasitoid. *Entomologia Experimentalis et Applicata*, **128**, 99–108.

Gols, R., van Dam, N. M., Raaijmakers, C. E., Dicke, M. and Harvey, J. A. (2009b) Are population differences in plant quality reflected in the preference and performance of two endoparasitoid wasps? *Oikos*, **118**, 733–743.

Gruner, D. S. and Taylor, A. D. (2006) Richness and species composition of arboreal arthropods affected by nutrients and predators: a press experiment. *Oecologia*, **147**, 714–724.

Harvey, J. A. (2000) Dynamic effects of parasitism by an endoparasitoid wasp on the development of two host species: implications for host quality and parasitoid fitness. *Ecological Entomology*, **25**, 267–278.

Harvey, J. A. (2005) Factors affecting the evolution of development strategies in parasitoid wasps: the importance of functional constraints and incorporating complexity. *Entomologia Experimentalis et Applicata*, **117**, 1–13.

Harvey, J. A., van Dam, N. M. and Gols, R. (2003) Interactions over four trophic levels: foodplant quality affects development of a hyperparasitoid as mediated through a herbivore and its primary parasitoid. *Journal of Animal Ecology*, **72**, 520–531.

Harvey, J. A., Biere, A., Fortuna, T. *et al.* (2010) Ecological fits, mis-fits and lotteries involving insect herbivores on the invasive plant, *Bunias orientalis. Biological Invasions*, **12**, 3045–3059.

Heidel, A. J. and Baldwin, I. T. (2004) Microarray analysis of salicylic acid- and jasmonic acid-signalling in responses of *Nicotiana attenuata* to attack by insects from multiple feeding guilds. *Plant Cell and Environment*, **27**, 1362–1373.

Heil, M. (2008) Indirect defence via tritrophic interactions. *New Phytologist*, **178**, 41–61.

Hilker, M. and Meiners, T. (2006) Early herbivore alert: insect eggs induce plant defense. *Journal of Chemical Ecology*, **32**, 1379–1397.

Hochwender, C. G. and Fritz, R. S. (2004) Plant genetic differences influence herbivore community structure: evidence from a hybrid willow system. *Oecologia*, **138**, 547–557.

Holopainen, J. K. and Gershenzon, J. (2010) Multiple stress factors and the emission of plant VOCs. *Trends in Plant Science*, **15**, 176–184.

Hopkins, R. J., van Dam, N. M. and van Loon, J. J. A. (2009) Role of glucosinolates in insect–plant relationships and multitrophic interactions. *Annual Review of Entomology*, **54**, 57–83.

Jansen, J. J., Allwood, J. W., Marsden-Edwards, E. *et al.* (2009) Metabolomic analysis of the interaction between plants and herbivores. *Metabolomics*, **5**, 150–161.

Johnson, M. T. J. and Agrawal, A. A. (2005) Plant genotype and environment interact to shape a diverse arthropod community on evening primrose (*Oenothera biennis*). *Ecology*, **86**, 874–885.

Kabouw, P., Biere, A., van der Putten, W. H. and van Dam, N. M. (2010) Intra-specific differences in root and shoot glucosinolate profiles among white cabbage (*Brassica oleracea* var. capitata) cultivars. *Journal of Agricultural and Food Chemistry*, **58**, 411–417.

Kaplan, I. and Denno, R. F. (2007) Interspecific interactions in phytophagous insects revisited: a quantitative assessment of competition theory. *Ecology Letters*, **10**, 977–994.

Kaplan, I., Halitschke, R., Kessler, A. *et al.* (2008) Physiological integration of roots and shoots in plant defense strategies links above- and belowground herbivory. *Ecology Letters*, **11**, 841–851.

Karban, R. (2008) Plant behaviour and communication. *Ecology Letters*, **11**, 727–739.

Kessler, A. and Baldwin, I. T. (2002) Plant responses to insect herbivory: the emerging molecular analysis. *Annual Review of Plant Biology*, **53**, 299–328.

Kessler, A. and Baldwin, I. T. (2004) Herbivore-induced plant vaccination. Part I. The orchestration of plant defenses in nature and their fitness consequences in the wild tobacco *Nicotiana attenuata*. *Plant Journal*, **38**, 639–649.

Kessler, A. and Halitschke, R. (2007) Specificity and complexity: the impact of herbivore-induced plant responses on arthropod community structure. *Current Opinion in Plant Biology*, **10**, 409–414.

Kessler, A., Halitschke, R. and Baldwin, I. T. (2004) Silencing the jasmonate cascade: induced plant defenses and insect populations. *Science*, **305**, 665–668.

Kim, J. H., Lee, B. W., Schroeder, F. C. and Jander, G. (2008) Identification of indole glucosinolate breakdown products with antifeedant effects on *Myzus persicae* (green peach aphid). *Plant Journal*, **54**, 1015–1026.

Lee, H. S., Wang, J. L., Tian, L. *et al.* (2004) Sensitivity of 70-mer oligonucleotides and cDNAs for microarray analysis of gene expression in Arabidopsis and its related species. *Plant Biotechnology Journal*, **2**, 45–57.

Loreto, F. and Schnitzler, J. P. (2010) Abiotic stresses and induced BVOCs. *Trends in Plant Science*, **15**, 154–166.

Lucas-Barbosa, D., van Loon, J. J. A. and Dicke, M. (2011) The effects of herbivore-induced plant volatiles on interactions between plants and visiting insects. *Phytochemistry*, **72**, 1647–1654.

Mattiacci, L., Dicke, M. and Posthumus, M. A. (1995) Beta-glucosidase: an elicitor of herbivore-induced plant odor that attracts host-searching parasitic wasps. *Proceedings of the National Academy of Sciences USA*, **92**, 2036–2040.

Mercke, P., Kappers, I. F., Verstappen, F. W. A. *et al.* (2004) Combined transcript and metabolite analysis reveals genes involved in spider mite induced volatile formation in cucumber plants. *Plant Physiology*, **135**, 2012–2024.

Mitchell-Olds, T. (2001) *Arabidopsis thaliana* and its wild relatives: a model system for ecology and evolution. *Trends in Ecology and Evolution*, **16**, 693–700.

Muller, C. (2009) Interactions between glucosinolate- and myrosinase-containing plants and the sawfly *Athalia rosae*. *Phytochemistry Reviews*, **8**, 121–134.

Mumm, R. and Hilker, M. (2005) The significance of background odour for an egg parasitoid to detect plants with host eggs. *Chemical Senses*, **30**, 337–343.

Mumm, R., Burow, M., Bukovinszkine'Kiss, G. *et al.* (2008) Formation of simple nitriles upon glucosinolate hydrolysis affects direct and indirect defense against the specialist herbivore, *Pieris rapae*. *Journal of Chemical Ecology*, **34**, 1311–1321.

Newton, E., Bullock, J. M. and Hodgson, D. (2009a) Bottom-up effects of glucosinolate variation on aphid colony dynamics in wild cabbage populations. *Ecological Entomology*, **34**, 614–623.

Newton, E. L., Bullock, J. M. and Hodgson, D. J. (2009b) Glucosinolate polymorphism in wild cabbage (*Brassica oleracea*) influences the structure of herbivore communities. *Oecologia*, **160**, 63–76.

Nielsen, J. K. (1997) Variation in defences of the plant *Barbarea vulgaris* and in counteradaptations by the flea beetle *Phyllotreta nemorum*. *Entomologia Experimentalis et Applicata*, **82**, 25–35.

Nishida, R. (2002) Sequestration of defensive substances from plants by Lepidoptera. *Annual Review of Entomology*, **47**, 57–92.

Ode, P. J. (2006) Plant chemistry and natural enemy fitness: effects on herbivore and natural enemy interactions. *Annual Review of Entomology*, **51**, 163–185.

Ohgushi, T. (2008) Herbivore-induced indirect interaction webs on terrestrial plants: the importance of non-trophic, indirect, and facilitative interactions. *Entomologia Experimentalis et Applicata*, **128**, 217–229.

Pichersky, E. and Gang, D. R. (2000) Genetics and biochemistry of secondary metabolites in plants: an evolutionary perspective. *Trends in Plant Science* **5**(10): 439–445.

Pieterse, C. M. J. and Dicke, M. (2007) Plant interactions with microbes and insects: from molecular mechanisms to ecology. *Trends in Plant Science*, **12**, 564–569.

Pieterse, C. M. J., Leon-Reyes, A., van der Ent, S. and van Wees, S. C. M. (2009) Networking by small-molecule hormones in plant immunity. *Nature Chemical Biology*, **5**, 308–316.

Poelman, E. H., Broekgaarden, C., van Loon, J. J. A. and Dicke, M. (2008a) Early season herbivore differentially affects plant defence responses to subsequently colonizing herbivores and their abundance in the field. *Molecular Ecology*, **17**, 3352–3365.

Poelman, E. H., van Loon, J. J. A. and Dicke, M. (2008b) Consequences of variation in plant defense for biodiversity at higher trophic levels. *Trends in Plant Science*, **13**, 534–541.

Poelman, E. H., Oduor, A. M. O., Broekgaarden, C. *et al.* (2009a) Field parasitism rates of caterpillars on *Brassica oleracea* plants are reliably predicted by differential attraction of *Cotesia* parasitoids. *Functional Ecology*, **23**, 951–962.

Poelman, E. H., van Dam, N. M., van Loon, J. J. A., Vet, L. E. M. and Dicke, M. (2009b) Chemical diversity in *Brassica oleracea* affects biodiversity of insect herbivores. *Ecology*, **90**, 1863–1877.

Poelman, E. H., van Loon, J. J. A., van Dam, N. M., Vet, L. E. M. and Dicke, M. (2010) Herbivore-induced plant responses in *Brassica oleracea* prevail over effects of constitutive resistance and result in enhanced herbivore attack. *Ecological Entomology*, **35**, 240–247.

Ramsey, J. S., Wilson, A. C. C., de Vos, M. *et al.* (2007) Genomic resources for *Myzus persicae*: EST sequencing, SNP identification, and microarray design. *Bmc Genomics*, **8**, 423.

Ratzka, A., Vogel, H., Kliebenstein, D. J., Mitchell-Olds, T. and Kroymann, J. (2002) Disarming the mustard oil bomb. *Proceedings of the National Academy of Sciences USA*, **99**, 11223–11228.

Renwick, J. A. A. (2002) The chemical world of crucivores: lures, treats and traps. *Entomologia Experimentalis et Applicata*, **104**, 35–42.

Roda, A. L. and Baldwin, I. T. (2003) Molecular technology reveals how the induced direct defenses of plants work. *Basic and Applied Ecology*, **4**, 15–26.

Root, R. B. (1973) Organization of a plant–arthropod association in simple and diverse habitats: the fauna of collards (*Brassica oleracea*). *Ecological Monographs*, **43**, 95–124.

Saedler, R. and Baldwin, I. T. (2004) Virus-induced gene silencing of jasmonate-induced direct defences, nicotine and trypsin proteinase-inhibitors in *Nicotiana attenuata*. *Journal of Experimental Botany*, **55**, 151–157.

Schaller, A. (ed.) (2008) *Induced Plant Resistance to Herbivory*. Berlin: Springer.

Schoonhoven, L. M., van Loon, J. J. A. and Dicke, M. (2005) *Insect–Plant Biology*. Oxford: Oxford University Press.

Schroeder, R. and Hilker, M. (2008) The relevance of background odor in resource location by insects: a behavioral approach. *Bioscience*, **58**, 308–316.

Schweitzer, J. A., Madritch, M. D., Bailey, J. K. *et al.* (2008) From genes to ecosystems: the genetic basis of condensed tannins and their role in nutrient regulation in a Populus model system. *Ecosystems*, **11**, 1005–1020.

Smallegange, R. C., van Loon, J. J. A., Blatt, S. E., Harvey, J. A. and Dicke, M. (2008) Parasitoid load affects plant fitness in a tritrophic system. *Entomologia Experimentalis et Applicata*, **128**, 172–183.

Snoeren, T. A. L., de Jong, P. W. and Dicke, M. (2007) Ecogenomic approach to the role of herbivore-induced plant volatiles in community ecology. *Journal of Ecology*, **95**, 17–26.

Snoeren, T. A. L., van Poecke, R. M. P. and Dicke, M. (2009) Multidisciplinary approach to unravelling the relative contribution of different oxylipins in indirect defense of *Arabidopsis thaliana*. *Journal of Chemical Ecology*, **35**, 1021–1031.

Snoeren, T. A. L., Mumm, R., Poelman, E. H. *et al.* (2010) The herbivore-induced plant volatile methyl salicylate negatively affects attraction of the parasitoid *Diadegma semiclausum*. *Journal of Chemical Ecology*, **36**, 479–489.

Soler, R., Bezemer, T. M., van der Putten, W. H., Vet, L. E. M. and Harvey, J. A. (2005) Root herbivore effects on above-ground herbivore, parasitoid and hyperparasitoid performance via changes in plant quality. *Journal of Animal Ecology*, **74**, 1121–1130.

Soler, R., Bezemer, T. M., Cortesero, A. M. *et al.* (2007a) Impact of foliar herbivory on the development of a root-feeding insect and its parasitoid. *Oecologia*, **152**, 257–264.

Soler, R., Harvey, J. A., Kamp, A. F. D. *et al.* (2007b) Root herbivores influence the behaviour of an aboveground parasitoid through changes in plant-volatile signals. *Oikos*, **116**, 367–376.

Soler, R., Schaper, S. V., Bezemer, T. M. *et al.* (2009) Influence of presence and spatial arrangement of belowground insects on host-plant selection of aboveground insects: a field study. *Ecological Entomology*, **34**, 339–345.

Strauss, S. Y. (1991) Direct, indirect, and cumulative effects of 3 native herbivores on a shared host plant. *Ecology*, **72**, 543–558.

Tahvanainen, J. O. and Root, R. B. (1972) The influence of vegetational diversity on the population ecology of a specialized herbivore, *Phyllotretra cruciferea* (Coleoptera: Chrysomelidae). *Oecologia*, **10**, 321–346.

Textor, S. and Gershenzon, J. (2009) Herbivore induction of the glucosinolate-myrosinase defense system: major trends, biochemical bases and ecological significance. *Phytochemistry Reviews*, **8**, 149–170.

Utsumi, S., Ando, Y. and Miki, T. (2010) Linkages among trait-mediated indirect effects: a new framework for the indirect interaction web. *Population Ecology*, **52**, 485–497.

van der Putten, W. H., Vet, L. E. M., Harvey, J. A. and Wäckers, F. L. (2001) Linking above- and belowground multitrophic interactions of plants, herbivores, pathogens, and their antagonists. *Trends in Ecology and Evolution*, **16**, 547–554.

van Leur, H., Raaijmakers, C. E. and van Dam, N. M. (2006) A heritable glucosinolate polymorphism within natural populations of *Barbarea vulgaris*. *Phytochemistry*, **67**, 1214–1223.

van Loon, J. J. A., De Boer, J. G. and Dicke, M. (2000) Parasitoid–plant mutualism: parasitoid attack of herbivore increases plant reproduction. *Entomologia Experimentalis et Applicata*, **97**, 219–227.

van Poecke, R. M. P. and Dicke, M. (2004) Indirect defence of plants against herbivores: using *Arabidopsis thaliana* as a model plant. *Plant Biology*, **6**, 387–401.

van Zandt, P. A. and Agrawal, A. A. (2004) Community-wide impacts of herbivore-induced plant responses in milkweed (*Asclepias syriaca*). *Ecology*, **85**, 2616–2629.

Viswanathan, D. V., Lifchits, O. A. and Thaler, J. S. (2007) Consequences of sequential attack for resistance to herbivores when plants have specific induced responses. *Oikos*, **116**, 1389–1399.

Voelckel, C. and Baldwin, I. T. (2004) Herbivore-induced plant vaccination. Part II. Array-studies reveal the transience of herbivore-specific transcriptional imprints and a distinct imprint from stress combinations. *Plant Journal*, **38**, 650–663.

Walling, L. L. (2009) Adaptive defense responses to pathogens and insects. *Plant Innate Immunity*, **51**, 551–612.

Wheat, C. W., Vogel, H., Wittstock, U. *et al.* (2007) The genetic basis of a plant–insect coevolutionary key innovation. *Proceedings of the National Academy of Sciences USA*, **104**, 20427–20431.

Whitham, T. G., Martinsen, G. D., Floate, K. D. *et al.* (1999) Plant hybrid zones affect biodiversity: tools for a genetic-based understanding of community structure. *Ecology*, **80**, 416–428.

Whitham, T. G., Bailey, J. K., Schweitzer, J. A. *et al.* (2006) A framework for community and ecosystem genetics: from genes to ecosystems. *Nature Reviews Genetics*, **7**, 510–523.

Whitham, T. G., DiFazio, S. P., Schweitzer, J. A. *et al.* (2008) Extending genomics to natural communities and ecosystems. *Science*, **320**, 492–495.

Wimp, G. M., Martinsen, G. D., Floate, K. D., Bangert, R. K. and Whitham, T. G. (2005) Plant genetic determinants of arthropod community structure and diversity. *Evolution*, **59**, 61–69.

Wimp, G. M., Wooley, S., Bangert, R. K. *et al.* (2007) Plant genetics predicts intra-annual variation in phytochemistry and arthropod community structure. *Molecular Ecology*, **16**, 5057–5069.

Wittstock, U., Agerbirk, N., Stauber, E. J. *et al.* (2004) Successful herbivore attack due to metabolic diversion of a plant chemical defense. *Proceedings of the National Academy of Sciences USA*, **101**, 4859–4864.

Zarate, S. I., Kempema, L. A. and Walling, L. L. (2007) Silverleaf whitefly induces salicylic acid defenses and suppresses effectual jasmonic acid defenses. *Plant Physiology*, **143**, 866–875.

Zhang, P. J., Zheng, S. J., van Loon, J. J. A. *et al.* (2009) Whiteflies interfere with indirect plant defense against spider mites in Lima bean. *Proceedings of the National Academy of Sciences USA*, **106**, 21202–21207.

Zheng, S. J. and Dicke, M. (2008) Ecological genomics of plant–insect interactions: from gene to community. *Plant Physiology*, **146**, 812–817.

Zheng, S. J., Snoeren, T. A. L., Hogewoning, S. W., van Loon, J. J. A. and Dicke, M. (2010) Disruption of plant carotenoid biosynthesis through virus-induced gene silencing affects oviposition behaviour of the butterfly *Pieris rapae. New Phytologist*, **186**, 733–745.

Index